Holography

Series in Optics and Optoelectronics

Handbook of Optoelectronic Device Modeling and Simulation
Lasers, Modulators, Photodetectors, Solar Cells, and Numerical Methods, Vol. 2
Joachim Piprek

Handbook of Optoelectronics, Second Edition
Concepts, Devices, and Techniques (Volume One)
John P. Dakin, Robert Brown

Handbook of GaN Semiconductor Materials and Devices
Wengang (Wayne) Bi, Haochung (Henry) Kuo, Peicheng Ku, Bo Shen

**Handbook of Optoelectronic Device Modeling and Simulation
(Two-Volume Set)**
Joachim Piprek

Handbook of Optoelectronics, Second Edition (Three-Volume Set)
John P. Dakin, Robert G. W. Brown

Optical MEMS, Nanophotonics, and Their Applications
Guangya Zhou, Chengkuo Lee

Thin-Film Optical Filters, Fifth Edition
H. Angus Macleod

Laser Spectroscopy and Laser Imaging
An Introduction
Helmut H. Telle, Ángel González Ureña

Fourier Optics in Image Processing
Neil Collings

Holography
Principles and Applications
Raymond K. Kostuk

For more information about this series, please visit:
https://www.crcpress.com/Series-in-Optics-and-Optoelectronics/book-series/TFOPTICSOPT

Holography

Principles and Applications

Raymond K. Kostuk

CRC Press
Taylor & Francis Group
Boca Raton London New York

CRC Press is an imprint of the
Taylor & Francis Group, an **informa** business

CRC Press
Taylor & Francis Group
6000 Broken Sound Parkway NW, Suite 300
Boca Raton, FL 33487-2742

First issued in paperback 2020

© 2019 by Taylor & Francis Group, LLC
CRC Press is an imprint of Taylor & Francis Group, an Informa business

No claim to original U.S. Government works

ISBN-13: 978-1-4398-5583-6 (hbk)
ISBN-13: 978-0-367-77957-3 (pbk)

Library of Congress Cataloging-in-Publication Data

Names: Kostuk, Raymond K., 1950- author.
Title: Holography : principles and applications / Raymond K. Kostuk.
Other titles: Series in optics and optoelectronics.
Description: Boca Raton, FL : CRC Press, Taylor & Francis Group, [2019] |
Series: Series in optics and optoelectronics
Identifiers: LCCN 2019007593 | ISBN 9781439855836 (hardback ; alk. paper) |
ISBN 1439855838 (hardback ; alk. paper) | ISBN 9780429185830 (ebook) |
ISBN 0429185839 (ebook)
Subjects: LCSH: Holography.
Classification: LCC QC449 .K65 2019 | DDC 621.36/75--dc23
LC record available at https://lccn.loc.gov/2019007593

Visit the Taylor & Francis Web site at
http://www.taylorandfrancis.com

and the CRC Press Web site at
http://www.crcpress.com

For

My Wonderful Wife Diane and Daughters Michelle and Marla

and

In Memory of Parents Walter and Eleanor

Contents

Preface

Holography is a very rich field that touches a broad range of disciplines ranging from chemistry to art. The thousands of papers written on the subject of holography since its invention by Denis Gabor in 1948 underscores this fact. Holography is also part of the public thinking thanks to popularizing the subject through appearance of the Princess Leia hologram in the first Star Wars movie, the "holodeck" in Star Trek, and different art exhibits that have been on display around the world. While holographic projections like those that appeared in Star Wars and Star Trek may not be completely possible, they have stimulated many researchers and engineers toward new types of holographic systems, some of which are discussed later in this book.

Technical interest in holography stems from the ability to encode more information about an optical field and optical functionality into a relatively thin recording material. This has led to holographic data storage systems, head-up displays, and point of sale scanners. In the past, one of the difficulties of implementing such applications has been the limitations of recording materials. However, the problem with materials seems to be changing with the advent of new photopolymers that do not require wet processing and do not change their properties during processing. Active recording materials such as photorefractive polymers and polymer-dispersed liquid crystals have also been developed and finding use in updatable displays and augmented reality eyewear. With the availability of these materials, I feel that many new applications of holography will be forthcoming.

In addition to the research and engineering applications, holography is a great tool for teaching optics because it encompasses a very broad range of optical phenomena. Wave theory, coherence, diffraction, interference, digital and analog optical recording devices and materials, lasers, the list can go on for quite some time. Furthermore, experiments can be performed with inexpensive equipment making lab experiments possible as part of a course. To facilitate this possibility, a number of experiments used in my class are included in Appendix E at the end of the book.

Over the years, I have been fortunate enough to work on many projects that apply holographic techniques and designs to different types of optical systems. These included optical interconnects, data storage, imaging, displays, and solar energy conversion. I have also taught a graduate course on holography at our University for over 30 years. In writing this book, I have tried to document the concepts that I found most useful for developing holographic designs and for teaching students. Certainly, most of the ideas were developed by others and hopefully I have given credit to their many contributions. However, I have attempted to explain the different concepts in a manner that has helped students (and me) to understand them in a clear and simple manner. In addition, I have provided background on many of the projects that our group has researched, especially on holographic optical elements and materials. To this end, Chapter 8 on materials and Chapter 11 on holographic optical elements are longer than the other chapters. In addition, five Appendices are provided to give more details on practical issues, experiments, and material preparation that I hope the reader will find useful in their studies and research of holography.

Preparing a book is a considerable undertaking and requires a great deal of motivation to bring to completion. For this, I would like to thank the help from many of my students at the University of Arizona over the past 30 plus years. Each one has contributed to the knowledge and understanding of different features that are essential to the design and implementation of a successful hologram. A special thanks to my former student Dr. Shanalyn Kemme for her computer-generated hologram example that appears in Chapter 6 and to Yuechen Wu for his algorithm on digital hologram reconstruction in Appendix E. I have also had the great fortune to be inspired and guided by several of the giants in the field of holography including my graduate research advisor Joseph W. Goodman, Stephen A. Benton, Richard Rallison, Nicholas J. Phillips, Hans I. Bjelkhagen, and Glenn T. Sincerbox and have tried to include their insight and passion for holography in this manuscript. In addition, I would like to thank many of my colleagues and friends at the University of Arizona including Mark Neifeld, Jennifer Barton, Pierre Blanche, Hyatt

Gibbs (deceased), Jack Gaskill, Nasser Peyghambarian, Masud Mansuripur, Ivan Djordjevic, Kathy Melde, Jerzy Rozenblit, and Bane Vasic. I apologize for others that I may have forgotten. In addition, I would like to thank Lou Han and Francesca McGowan at Taylor and Francis for their patience, and a special thanks to Joe Goodman again for inspiration especially throughout my career and during the last year of writing this book.

Finally, thanks to my wonderful wife Diane and my special daughters Michelle and Marla for their unwavering support and encouragement throughout the course of writing this book and in all matters important in life.

Raymond K. Kostuk
Tucson, Arizona

Author

Raymond K. Kostuk is a professor of the Electrical and Computer Engineering Department and the College of Optical Sciences at the University of Arizona. He received a Bachelor of Science Degree from the United States Coast Guard Academy and served as an officer in the United States Coast Guard for 10 years. During that time, he was sent to the University of Rochester where he received a Master of Science Degree in Optical Engineering from the Institute of Optics. After completing his military service, he went to Stanford University and obtained his Doctorate under the advisement of Professor Joseph W. Goodman working on applying multiplexed holograms to optical interconnect systems. After graduating, he spent a year at the IBM Almaden Research Center where he worked with Glenn Sincerbox on problems related to optical data storage. He then went to the University of Arizona and has continued to work on various types of optical systems with the main focus on holographic optical elements and holographic materials. He has published over 80 journal papers and several book chapters and patents primarily related to different aspects of holography. He is a fellow of the Optical Society of America and the Society of Photo Instrumentation Engineers (SPIE).

1

Introduction and Brief History of Holography

1.1 Introduction

The visual effects of holographic images can be quite striking. This results from their three-dimensional nature that gives them a lifelike character. Different parts of a scene can appear as the observer changes location similar to what occurs when viewing an object in real life. To illustrate this, consider the images in Figure 1.1 which shows two photographs of a hologram taken at two different angles. In the top image, only the woman's left eye can be seen while in the lower image both eyes are visible. The three-dimensional property of a holographic image gives rise to parallax effects and contrasts sharply with images formed by conventional refractive and reflective optical systems.

A refractive or reflective imaging system converts a single object point to a single image point following the conjugate imaging relations (Figure 1.2). In contrast, a hologram can convert a single source point into an entire image. This is a consequence of the ability of a hologram to record and reconstruct wave information about the object.

A hologram is formed in a photosensitive material much like that used for a conventional photograph. However, a photograph only records the intensity of a scene while a hologram records both the intensity and phase of an object wave coming from an object. Since the photosensitive material used for holography and photography are essentially the same in that it only responds to intensity variations, a method for recording both the intensity and phase of a wave had to be developed. This was the result of several important achievements in physics and optics during the nineteenth and early twentieth centuries culminating with the experiments demonstrating the first holographic images by Dennis Gabor in 1948 [1].

1.2 Historical Background

Dennis Gabor's invention of holography in 1948 [1] was a remarkable achievement (recognized with the Nobel Prize in 1971) that brought together a number of earlier ideas on image formation. The following discussion provides a brief overview of these early efforts. More comprehensive discussion on the history of holography can be found in reviews by Benton [2], Johnston [3,4], and Richardson and Wiltshire [5]. There is also an enlightening historical perspective in the classic book on holography by Collier, Burkhardt, and Lin [6] and in the book on silver halide materials by Bjelkhagen [7]. The following is a synopsis of these more comprehensive histories with some additional anecdotes by the author.

Perhaps the earliest inspiration for holography came from the work by Abbe (1873) who described the formation of an image with a two-step diffraction process [8]. In his analysis, Abbe considers a "grating like" object illuminated by a plane wave and then imaged by a lens. The lens forms a set of regularly spaced bright spots at the back focal plane of the lens. Each focused spot acts like a point source with spherical waves emanating from secondary sources. The image results from the diffraction pattern formed by the interfering waves from the secondary sources.

FIGURE 1.1 (See color insert.)   The figures above are photographs of a hologram taken from two different angles.

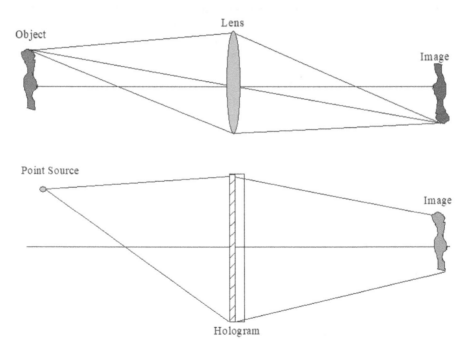

FIGURE 1.2   Contrast between refractive and holographic imaging. Upper figure shows a lens imaging an object, and the lower figure shows a hologram transforming light from a point source into an image.

During the same time that Abbe was performing his work, other developments in photography were taking place that also helped the development of holography. Most notably was the work of Lippmann who was investigating ways of making color photographs [7,9–11]. He came up with an ingenious method that used a highly reflective coating of mercury on the back surface of an emulsion and then formed a color image on the film with a camera lens. The different color areas set up standing wave patterns in the film. After development and viewing with white light, an image forms with colors similar to those in the original object.

In the early twentieth century, Sir Lawrence Bragg pursued the work of Abbe. Bragg was performing research on X-ray diffraction by atomic structures and developed a method of imaging the atomic structure by superimposing interference fringes from X-rays that were scattered by the regularly spaced atoms in crystals on a photographic plate [12]. In some of his methods, accurately placed holes were drilled in a screen at the locations of fringe "spots" formed by the X-ray interference. If the screen was then illuminated with coherent light and imaged through a lens, the optical far-field diffraction pattern on the back focal plane of the lens provided a magnified image of a unit cell of a crystal.

Gabor recognized Bragg for inspiring his work in his initial paper on holography [1]. At the time of his invention, Gabor was trying to correct for aberrations in the lenses used for electron microscopy. He eventually decided to try to use Abbe and Bragg's approach to eliminate the need for the lens altogether. In doing so, he came up with the unique contribution of using coherent light to partially illuminate the object and scatter light. The remaining non-scattered illumination serves as a reference beam that interferes with the object beam, and the interference pattern of the two waves is recorded on a sensitive film. He initially planned to do this recording with the electron beam and then to reconstruct it with an optical beam. However, to demonstrate the method, he performed both the recording and the reconstruction with optical beams and reported this in his seminal paper of 1948 [1]. This was the first demonstration of an optical hologram.

There were, however, some difficulties with Gabor's initial approach. Since light sources with long coherence length were not readily available, his method was restricted to in-line geometries. This also required the object to be highly transparent. Although work continued on the theory of holography during the 1950s most notably by H.M.A. El-Sum and A.W. Lohmann [13–15], experimental work did not significantly advance during this time. One important exception was the experiments by G.L. Rogers who also was the first to coin the word "holography" [16].

A true breakthrough in holography came from Emmett N. Leith and Juris Upatnieks who were working on high-resolution radar imaging at the University of Michigan's Willow Run Lab [17,18] during the early 1960s. Leith had studied Gabor's work and realized the significance of being able to record the phase of an object for imaging applications. He then set out to address the problem of separating the overlapping reconstructed beams that occur with Gabor's in-line hologram geometry. His work resulted in the off-axis hologram geometry that not only eliminated the overlapping reconstruction beams, but also allowed holograms to be made of three-dimensional opaque objects. It is interesting to note that holograms made by Leith and Upatnieks were accomplished without the use of a laser. Instead, they used a clever method to extend the coherence length of a mercury discharge lamp with a diffraction grating [6]. Their holograms captured the imagination of both scientists and the public when they displayed a hologram of a three-dimensional object at the Optical Society of America's annual meeting in 1964. Shortly afterward when lasers became readily available, off-axis and true three-dimensional holography was possible and generated a great deal of interest and experimentation [19].

Another pioneer of early holography was Yuri Denisyuk from the former Soviet Union. He focused his efforts on the formation of holograms using a reflection geometry [20,21]. Although he was not aware of Gabor's work, he was influenced by the dissertation by El-Sum and Lipmann photography. His efforts resulted in truly exceptional reflection holograms that can be viewed with white light due to the highly selective characteristics of reflection holograms. To achieve high-quality reflection holograms, he helped to develop fine grain silver halide emulsions capable of resolving up to 10,000 L/mm [22]. For his work on holography, Denisyuk received the Lenin Prize in 1970 which in the former Soviet Union was an award considered comparable to the Nobel Prize.

The 1960s also saw a great deal of other work that had a significant impact on the field of holography. In 1963, P.J. van Heerden wrote two papers that outlined the capability of volume holography for holographic data storage [23,24] and influenced an entirely new way of thinking of optical storage that continues to this day [25]. In 1965, K.A. Setson and R.L. Powell demonstrated the technique of interferometric holography for measuring wavelength scale defects and changes in physical surfaces [26,27]. This work has grown into perhaps one of the most widely used applications of holography. (Thomas Kreis has written an excellent text on this subject [28].) In 1967, two important developments in holography occurred. First was the demonstration of digital holography by Joseph W. Goodman and R.W. Lawrence that optically records an interference pattern with a digital camera [29]. The holographic image is then computationally reconstructed. This has given rise to the advent of digital or electronic holography which has become an area of intense research with the development and availability of high-performance digital cameras. The second contribution in 1967 was the demonstration of computer-generated holograms by A.W. Lohmann and D.P. Paris in which the hologram is mathematically described and printed in a medium [30]. A holographic image or wavefront is then optically reconstructed. The method allows the recording of objects that are not physically realizable and has been found very useful for testing aspheric optical surfaces [31,32]. In 1969, Steve Benton invented the "rainbow hologram" which allowed high fidelity viewing of images from transmission holograms with non-laser sources [33]. A key aspect of this design is recognizing that for viewing an object, horizontal parallax is more significant than vertical parallax. This allowed him to disperse light from a white light source vertically and view each color band horizontally without degrading image resolution. Finally, one other development during the 1960s was an important paper by Herwig Kogelnik on volume holography [34]. This paper integrated many concepts on diffraction by thick holograms and presented them in a clear and easy to understand manner. The concepts in this paper provide a good starting point for analyzing many types of holographic optical elements, and Kogelnik's paper is highly cited in subsequent papers on volume holography. Later work on rigorous coupled wave modeling by M.G. Moharam and T.K. Gaylord [35] provided an exact treatment of diffraction by gratings with a variety of physical characteristics. Their work serves as the most general framework for modeling and designing holographic systems and is the basis for many commercial simulation programs. Finally, during the past 10–15 years, the development of new photopolymer recording materials and advances in traditional films such as dichromated gelatin have led to a variety of new applications ranging from solar energy collection [36] to augmented reality displays and eyewear [37,38].

1.3 Philosophy and Content of the Book

The material for this book primarily comes from a graduate course on holography that has been taught for over 30 years at the College of Optical Sciences and the Electrical and Computer Engineering Department at the University of Arizona. Since the main experience of the author has been with volume holography and holographic optical elements, there is a certain bias in the book toward these areas. However, an effort is made to provide an overview of the important areas of holography such as digital and computer-generated hologram as well as holographic interferometry.

Chapter 2 reviews some of the basic concepts of optics that will be useful in understanding holography. Chapter 3 introduces some of the preliminary ideas of holography and associated terminology used throughout the rest of the book. Explaining the basic ideas at the outset allows more advanced material to be discussed later without the need to step back. The content of Chapter 3 can also be used as a framework for a short course or tutorial on holography.

Chapters 4 and 5 cover, respectively, two important analytical methods of holography: imaging and diffraction efficiency. The concepts of image formation by holograms are developed from the basic ideas of phase matching through advanced methods of raytracing, aberration analysis, and optical design. The analysis of hologram diffraction efficiency is considered from several perspectives including the Fourier analysis of thin gratings, approximate coupled wave methods, and rigorous coupled wave analysis. This provides a suitable framework for analyzing the diffraction efficiency of holograms at several levels.

Chapter 6 describes the basics of computer-generated hologram (CGH). It starts with a review of the sampling criteria necessary for retaining a desired spatial frequency content in the image. This is followed by a discussion of discrete Fourier transform methods that are required for modeling CGH properties. The spatial frequency recording requirements are evaluated for CGHs in both Fresnel and Fourier transform type geometries. The detour-phase hologram encoding method is used as an example to illustrate the process of recording both the phase and amplitude information for CGH. An example of a binary phase interferogram CGH is also provided. This work is followed by a discussion of Dammann gratings and dynamic holograms formed using spatial light modulators.

Chapter 7 covers the subject of digital holography and builds on the discussion of spatial frequency sampling requirements presented in Chapter 6. Several recording geometries for digital holography are examined with respect to the required spatial frequency content of the object and digital camera resolution capability. The Fresnel and convolution reconstruction methods are discussed as well as techniques to improve imaging quality such as zero-order suppression and phase shifting recording. The chapter concludes with examples of digital holography applications including microscopy and interferometry.

Chapter 8 is devoted to holographic materials and is meant to serve as a reference for someone interested either in the study of these materials or with specific holographic application needs. The materials covered include: emulsions such as silver halide and dichromated gelatin; photopolymers for volume hologram recording; photoresists for surface relief type gratings; inorganic and organic photorefractive materials; holographic polymer dispersed liquid crystals; thermoplastics; and sensitized glass fiber and photo-thermo-refractive glass.

Chapters 9 through 12 are devoted to holographic applications. Chapter 9 deals with holographic displays and includes: image plane and reflection holograms, various types of rainbow holograms, multicolor holograms, composite holograms such those developed by Zebra Imaging and stereograms, updateable holographic displays, heads-up displays, and holograms for augmented reality eyewear. Chapter 10 covers the basic concepts of holographic interferometry including: techniques such as double exposure and real-time holographic interferometry, the measurement of surface displacements, contours, and refractive index variation, phase shifting interferometric methods, and methods for analyzing holographic interferometry patterns. Chapter 11 deals with holographic optical elements. This chapter starts with a discussion of holographic lenses and provides a methodology for their design. It then describes the properties of holographic spectral filters, beam splitters, and polarization elements. This is followed by an overview of folded or substrate-mode holographic elements with applications in displays and communication systems. Next, volume holographic imaging and its use for subsurface tissue imaging are described as well as its use for the detection of cancer. Following this, the application of holographic optical elements in solar energy conversion is described including holographic solar concentrators, light trapping elements, and spectrum splitting to increase photovoltaic energy conversion efficiency and energy yield. The chapter concludes with an overview of holographic optical element applications in optical interconnects and communications systems such as fiber Bragg gratings and code division multiplexers. The final main chapter of the book, Chapter 12, deals with holographic data storage. It starts with a description of the basic method of holographic data storage and different recording configurations. This is followed by an overview of the most commonly used multiplexing methods and a discussion of their benefits and limitations. The recording material requirements and the basic metrics for evaluating data storage properties are also given. Techniques for obtaining high-quality holographic data storage including exposure scheduling and object beam conditioning with de-focus and phase mask methods are also reviewed. The chapter concludes with a discussion of representative holographic data storage systems.

In addition to the main chapters, five appendices are included to provide some practical background for analyzing, preparing materials, exposing, and processing holograms. These include: (A) commonly used Fourier transform algorithms; (B) practical considerations for holographic recording; (C) lasers used for holographic recording; (D) processing and development of commonly used recording materials; and (E) a set of holographic experiments suitable for teaching a class on holography or for personal experimentation.

Problems and exercises for strengthening concepts and a set of useful references follow each chapter.

REFERENCES

1. D. Gabor, "A new microscope principle," *Nature*, Vol. 161, 777 (1948).
2. S. A. Benton, "Holography reinvented," *SPIE Proc.*, Vol. 4737, 23–26 (2002).
3. S. F. Johnston, "From white elephant to Nobel Prize: Dennis Gabor's wavefront reconstruction," *Hist. Stud. Phys. Biol. Sci.*, Vol. 36, 35–70 (2005).
4. S. F. Johnston, "Reconstructing the history of holography," *Proc. SPIE*, Vol. 5005, 455–464 (2003).
5. M. J. Richardson and J. D. Wiltshire, *The Hologram: Principles and Techniques*, Ch. 10, Wiley Online Library (2017).

6. R. J. Collier, C. B. Burckhardt, and L. H. Lin, *Optical Holography*, Academic Press, New York (1971).
7. H. I. Bjelkhagen, *Silver-Halide Recording Materials for Holography and Their Processing*, Springer-Verlag, Berlin, Germany (1993).
8. E. Abbe, "Beitrage zur Theorie des Mikroskops und der Mikroskopischen wahrnehmung," *Archiv. Mikroskopische Anat.*, Vol. 9, 413–468 (1873).
9. M. G. Lippmann, "La photographie des couleurs," *Comptes Rendus Acad. Sci.*, Vol. 112, 274–275 (1891).
10. M. G. Lippmann, "Photographies colorees du spectre, sur albumin et sur gelatin bichromatees," *Comptes Rendus Acad. Sci.*, Vol. 115, 575 (1892).
11. M. G. Lippmann, "Sur la theorie de la photographie des couleurs simples et composes par la method interferentielle," *J. Physique*, Vol. 3, 97–107 (1894).
12. W. L. Bragg, "The x-ray microscope," *Nature*, Vol. 149, 470 (1942).
13. A. W. Lohmann, "Optical single-sideband transmission applied to the Gabor microscope," *Optica Acta*, Vol. 3, 97 (1956).
14. H. M. A. El-Sum, Reconstructed wavefront microscopy, PhD Thesis, Stanford University, Department of Physics 1952.
15. H. M. A. El-Sum and P. Kirkpatrick, "Microscopy by reconstructed wavefronts," *Phys. Rev.*, Vol. 85, 763 (1952).
16. G. L. Rogers, "Experiments in diffraction microscopy," *Proc. Roy. Soc.*, Vol. 63A, 193 (1952).
17. E. Leith and J. Upatnieks, "Reconstructed wavefronts and communication theory," *J. Opt. Soc. Amer.*, Vol. 52, 1123–1130 (1962).
18. E. Leith and J. Upatnieks, "Wavefront reconstruction with continuous-tone objects," *J. Opt. Soc. Amer.*, Vol. 53, 1377–1381 (1963).
19. R. P. Chambers and J. S. Courtney-Pratt, "Bibliography on holography," *J. SMPTE*, Vol. 75, 373–435 (1966).
20. Y. N. Denisyuk, "Photographic reconstruction of the optical properties of an object in its own scattered radiation field," *Sov. Phys. Dokl.*, Vol. 144, 1275–1278 (1962).
21. Y. N. Denisyuk, "Photographic reconstruction of the optical properties of an object by the wave field of its scattered radiation," *Opt. Spectrosc.*, Vol. 14, 279–284 (1963).
22. Y. N. Denisyuk and I. R. Protas, "Improved Lippmann photographic plates for recording stationary light waves," *Opt. Spectrosc.*, Vol. 14, 381–383 (1963).
23. P. J. van Heerden, "A new optical method for storing and retrieving information," *Appl. Opt.*, Vol. 2, 387–392 (1963).
24. P. J. van Heerden, "Theory of optical information storage in solids," *Appl. Opt.*, Vol. 2, 393–400 (1963).
25. K. Curtis, L. Dhar, A. Hill, W. Wilson, and M. Ayres, *Introduction to Holographic Data Recording*, John Wiley & Sons, Hoboken, NJ (2010).
26. R. L. Powell and K. A. Stetson, "Interferometric vibration analysis by wavefront reconstruction," *J. Opt. Soc. Am.*, Vol. 55, 1593–1598 (1965).
27. K. A. Stetson and R. L. Powell, "Interferometric hologram evaluation of real-time vibration analysis of diffuse objects," *J. Opt. Soc. Am.*, Vol. 55, 1694–1695 (1965).
28. T. Kreis, *Handbook of Holographic Interferometry-Optical and Digital Methods*, Wiley-VCH, Weinheim, Germany, 2005.
29. J. W. Goodman and R. W. Lawrence, "Digital image formation from electronically detected holograms," *Appl. Phys. Lett.*, Vol. 11, 77–79 (1967).
30. A. W. Lohmann and D. P. Paris, "Binary Fraunhofer holograms generated by computer," *Appl. Opt.*, Vol. 6, 1739–1748 (1967).
31. J. C. Wyant and V. P. Bennett, "Using computer generated holograms to test of aspheric wavefronts," *Appl. Opt.*, Vol. 11, 2833–2839 (1972).
32. J. C. Wyant and P. K. O'Neill, "Computer generated hologram: Null lens test of aspheric wavefronts," *Appl. Opt.*, Vol. 13, 2762–2765 (1974).
33. S. A. Benton, "Hologram reconstructions with extended incoherent sources," *J. Opt. Soc. Am.*, Vol. 59, 1545–1546 (1969).
34. H. Kogelnik, "Coupled wave theory for thick hologram gratings," *Bell Syst. Tech. J.*, Vol. 48, 2909–2947 (1969).

35. T. K. Gaylord and M. G. Moharam, "Analysis and applications of optical diffraction gratings," *Proc. IEEE*, Vol. 73, 894–937 (1985).
36. R. K. Kostuk et al., *Holographic Applications in Solar-Energy–Conversion Processes, SPIE Spotlight*, SPIE Press, Bellingham, WA (2016).
37. B. C. Kress and M. Shin, "Diffractive and holographic optics as optical combiners in head mounted displays," *Proceedings of the 2013 ACM Conference on Pervasive and Ubiquitous Computing Adjunct Publication*, 1479–1482, Zurich, Switzerland, September 8–13 (2013).
38. B. C. Kress and T. Starmer, "A review of head-mounted displays (HMD) technologies and applications for consumer electronics," *Proc. SPIE*, Vol. 8720A (2013).

2

Background of Physical and Geometrical Optics for Holography

2.1 Introduction

This chapter provides a review of the basic principles of physical and geometrical optics that are useful for an understanding of the field of holography. While not completely comprehensive, the material presented here provides the essentials of optics that are needed to understand some of the advanced topics on holography that follow later in the book. For a more comprehensive treatment of the principles of optics, the reader is encouraged to refer to the references at the end of this chapter [1–4].

2.2 Light as an Electromagnetic Wave

Light is a part of the electromagnetic spectrum, and the propagation of light can be described by Maxwell's equations. The general form for these equations are:

$$\nabla \times \vec{E} = -\frac{\partial \vec{B}}{\partial t};$$

$$\nabla \times \vec{H} = \frac{\partial \vec{D}}{\partial t} + \vec{J};$$

$$\nabla \cdot \vec{D} = \rho;$$

$$\nabla \cdot \vec{B} = 0;$$

(2.1)

where $\vec{E}$ is the electric field vector, $\vec{D}$ is the electric displacement vector, $\vec{B}$ is the magnetic induction vector, $\vec{H}$ is the magnetic field vector, $\vec{J}$ the electric current density, and ρ is the electric charge density. The symbol $\times$ is the vector cross product, $\cdot$ is the vector dot product, and $\nabla = \frac{\partial}{\partial x}\hat{x} + \frac{\partial}{\partial y}\hat{y} + \frac{\partial}{\partial z}\hat{z}$ is the vector gradient operator. The fields $\vec{E}, \vec{H}, \vec{B},$ and $\vec{D}$ are in general functions of both time and space. Several of the field values are related according to:

$$\vec{D} = \varepsilon \vec{E}$$

$$\vec{B} = \mu \vec{H}$$

$$\vec{J} = \sigma \vec{E},$$

(2.2)

where ε is the permittivity, μ is the permeability, and σ is the conductivity (Ohm's law) of the material in which the fields exist with:

$$\varepsilon = \varepsilon_r \varepsilon_o$$

$$\mu = \mu_r \mu_o.$$

(2.3)

where ε_r is the relative permittivity, and ε_o is the free-space or vacuum permittivity. Similarly, μ_r is the relative permeability, and μ_o is the free-space permeability of the material with:

$$\varepsilon_o = 8.854 \times 10^{-12} \text{Farads/m}$$

$$\mu_o = 4\pi \times 10^{-7} \text{Henrys/m}.$$

For typical optical materials, $\mu = \mu_o$, $\vec{J}$, and ρ are zero, and Maxwell's equations reduce to:

$$\nabla \times \vec{E} = -\mu_o \frac{\partial \vec{H}}{\partial t}$$

$$\nabla \times \vec{H} = \varepsilon \frac{\partial \vec{E}}{\partial t} \qquad (2.4)$$

$$\nabla \cdot \varepsilon \vec{E} = 0$$

$$\nabla \cdot \vec{H} = 0.$$

For optics, the electric field vector also indicates the polarization direction. In order to evaluate the basic nature of light propagation, several assumptions must be made and are true for most optical materials. First, significant propagation occurs in dielectric materials, and many of these materials are *isotropic* meaning that the material has the same properties regardless of the direction of propagation and polarization of the light. (Note that there are birefringent optical materials such as crystals that have different properties, but will not be considered at this time.) The material is also considered *homogeneous* in the sense that the dielectric constant does not change as a function of position along the light path. In addition, only light at a single optical frequency or light with a very narrow spectral bandwidth is considered such that the properties of the medium do not appreciably change as a function of optical frequency. In this sense, the material is *non-dispersive*.

With the above assumptions, it is possible to use Maxwell's equations to determine some of the basic propagation characteristics of light. Performing the operation $\nabla \times$ on the first of Equation 2.4 results in:

$$\nabla \times \left(\nabla \times \vec{E} \right) = -\mu_o \frac{\partial}{\partial t} \nabla \times \vec{H}$$

$$= -\mu_o \varepsilon \frac{\partial^2 \vec{E}}{\partial t^2}, \qquad (2.5)$$

then applying the vector operator identity:

$$\nabla \times \nabla \times \vec{E} = \nabla \nabla \cdot \vec{E} - \nabla^2 E, \qquad (2.6)$$

and noting that $\nabla \cdot \vec{E} = 0$ for homogeneous materials (i.e., $\nabla \varepsilon = 0$) produces:

$$\nabla^2 \vec{E} = \mu_o \varepsilon \frac{\partial^2 \vec{E}}{\partial t^2}, \qquad (2.7)$$

which is the wave equation for the electric field. In optics, it is more useful to use the refractive index rather than the permittivity, and the two are related through the relation:

$$n = \varepsilon_r^{1/2} = \left(\frac{\varepsilon}{\varepsilon_o} \right)^{1/2}. \qquad (2.8)$$

Using the refractive index, the wave equation then takes the form:

$$\nabla^2 \vec{E} = \mu_o \varepsilon_o n^2 \frac{\partial^2 \vec{E}}{\partial t^2} = \frac{n^2}{c^2} \frac{\partial^2 \vec{E}}{\partial t^2}, \tag{2.9}$$

where c is the speed of light in vacuum:

$$c = \left(\frac{1}{\varepsilon_o \mu_o} \right). \tag{2.10}$$

A similar result can be derived for the magnetic field vector $\vec{H}$. Equation 2.9 is the wave equation that is commonly used in optics to describe the propagation of the electric field in space and time.

The electromagnetic wave is transverse in the sense that $\vec{E}$ and $\vec{H}$ are perpendicular to each other with:

$$\vec{E} \cdot \vec{H} = 0, \tag{2.11}$$

and propagates in a direction given by the Poynting vector where:

$$\vec{S} = \vec{E} \times \vec{H}. \tag{2.12}$$

In optics, the Poynting vector specifies the direction of the optical ray or the propagation vector $\vec{k}$ with $\vec{k} = k_x \hat{x} + k_y \hat{y} + k_z \hat{z}$, and $\hat{x}, \hat{y}$, and $\hat{z}$ are unit vectors. The components of the propagation vector can also be expressed as:

$$k_x = \frac{2\pi n}{\lambda} \cos\alpha, \; k_y = \frac{2\pi n}{\lambda} \cos\beta, \; k_z = \frac{2\pi n}{\lambda} \cos\gamma, \tag{2.13}$$

where $\cos\alpha, \cos\beta$, and $\cos\gamma$ are direction cosines that indicate the components of a unit vector in the direction of $\vec{k}$ along the x, y, and z axes, respectively, as shown in Figure 2.1.

Since the direction cosines are related according to:

$$1 = \cos^2 \alpha + \cos^2 \beta + \cos^2 \gamma, \tag{2.14}$$

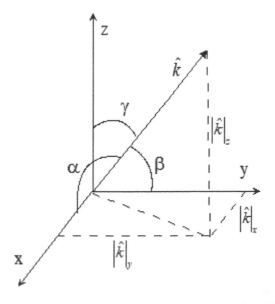

FIGURE 2.1 Propagation vector and components along the x-, y-, and z-directions corresponding to $\cos\alpha$, $\cos\beta$, and $\cos\gamma$.

the magnitude of the propagation vector is:

$$\left| \vec{k} \right| = \frac{2\pi n}{\lambda}. \tag{2.15}$$

One solution to the wave Equation (2.9) is a plane wave. Assuming Cartesian coordinates, a plane wave has the form:

$$\vec{E}\left(x, y, z, t \right) = E_o \hat{u} \cos\left(\vec{k} \cdot \vec{r} - \omega t + \varphi \right), \tag{2.16}$$

with E_o the amplitude of the field, $\hat{u}$ a unit polarization vector, ω the radian frequency, $\vec{k}$ the propagation vector, and φ is the optical phase of the field which in general is a function of position. $\vec{r}$ is the distance vector from the origin of the coordinate system to a point on the wavefront (x, y, z). If E_o is a constant, each field component x, y, and z can be represented as a scalar quantity with:

$$E\left(x, y, z, t \right) = E_o \cos\left(\vec{k} \cdot \vec{r} - \omega t + \varphi \right). \tag{2.17}$$

For many applications in optics, the scalar field representation is sufficient and provides a simpler solution and greater insight into the problem under investigation.

In many cases it is easier to use the complex description of the field to perform different mathematical operations and is given by:

$$E(x, y, z, t) = E_o \exp\left[j\left(\vec{k} \cdot \vec{r} - \omega t + \varphi \right) \right]. \tag{2.18}$$

The actual or physical field is the real part of Equation 2.18:

$$E(x, y, z, t) = E_o \operatorname{Re}\left\{ \exp\left[j\left(\vec{k} \cdot \vec{r} - \omega t + \varphi \right) \right] \right\}. \tag{2.19}$$

One advantage of the complex description is that the temporal and spatial factors can be separated. If the complex description is substituted into the wave equation (Equation 2.9), the time dependent component can be dropped to further simplify the expression. In this case, the time independent wave equation is:

$$\nabla^2 E + k^2 E = 0, \tag{2.20}$$

which is sometimes referred to as the Helmholtz equation [5], where:

$$E(x, y, z) = E_o \exp\left[j\left(\vec{k} \cdot \vec{r} - \varphi \right) \right]. \tag{2.21}$$

2.3 Polarization of an Optical Field

Knowing and controlling the state of polarization of the optical field is important for the formation of holograms. Therefore, a short summary of different polarization states of an optical field is presented in the following sections. A more comprehensive discussion of polarization can be found in references [1] and [6] at the end of this chapter.

As noted in the previous section, the transverse electric field (Equation 2.16) is a time and spatially varying vector and represents the polarization state of the electromagnetic field. The elliptical polarization state is the most general form with linear and circular polarization being important special cases. The description of the polarization state is based on the relative magnitudes of the vector components (i.e., E_x and E_y) and the phase difference (φ) that exists between these components.

2.3.1 Linear Polarization

For linearly polarized light, the tip of the electric field vector moves along a line in a plane transverse to the direction of propagation. If the propagation vector is along the *z*-axis, the transverse vectors are in the *x*- and *y*-directions. For this situation, the electric field vector can be expressed as:

$$\vec{E}(z,t) = E_{ox}\cos(kz - \omega t)\hat{x} + E_{oy}\cos(kz - \omega t + \varphi_{xy})\hat{y}, \tag{2.22}$$

where $E_{ox}\cos(kz - \omega t)$ is the magnitude of the field component in the *x*-direction and $E_{oy}\cos(kz - \omega t + \varphi_{xy})$ is the field component in the *y*-direction. If the phase $\varphi_{xy} = 2\pi m$ with $m = 0,1,2,\ldots$ then:

$$\vec{E}(z,t) = \left[E_{ox}\hat{x} + E_{oy}\hat{y}\right]\cos(kz - \omega t). \tag{2.23}$$

In this case, the tip of the polarization vector lies along a line that makes an angle ψ with the *y*-axis as shown in Figure 2.2 with $\tan\Psi = E_{ox}/E_{oy}$. This situation describes the linear polarization state. If $E_{oy} = 0$, then the field is polarized along the *x*-axis, and, conversely, if $E_{ox} = 0$, then the field is polarized along the *y*-axis.

2.3.2 Circular Polarization

Equation 2.22 can also be used to evaluate other important polarization states. In particular, if $E_{ox} = E_{oy}$ and $\varphi_{xy} = (2m-1)\pi/2$ with $m = 1,2,\ldots$, then Equation 2.22 can be rewritten as:

$$\begin{aligned}
\vec{E}_x(z,t) &= E_o\cos(kz - \omega t)x \\
\vec{E}_y(z,t) &= E_o\sin(kz - \omega t)\hat{y} .
\end{aligned} \tag{2.24}$$

When *z* is set equal to 0, the tip of the electric field vector will rotate in a clockwise direction when an observer at a +*z* position is looking back toward the oncoming beam as shown in Figure 2.3.

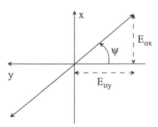

FIGURE 2.2 Diagram showing the path of the tip of the polarization vector for a linear polarized field described by Equation 2.21.

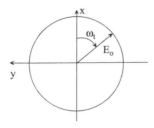

FIGURE 2.3 Rotation of the polarization vector at *z* = 0 when looking back toward the oncoming beam (i.e., looking in the −*z*-direction from an observation point at *z* = 0).

2.3.3 Elliptical Polarization

If no restrictions are placed on the magnitudes of E_{ox} and E_{oy} or the phase φ_{xy}, then the x and y field components can be written as:

$$E_x = E_{ox}\cos(kz - \omega t)$$

$$E_y = E_{oy}\cos(kz - \omega t + \varphi).$$

(2.25)

In order to determine the path traced by the tip of the electric field vector, it will be necessary to eliminate the time and z dependence from Equation 2.25. Rewriting the expression for E_y as:

$$\frac{E_y}{E_{oy}} = \cos(kz - \omega t)\cos\varphi - \sin(kz - \omega t)\sin\varphi,$$

(2.26)

and using $\cos(kz - \omega t) = E_x / E_{ox}$ from Equation 2.25 gives:

$$\frac{E_y}{E_{oy}} - \frac{E_x}{E_{ox}}\cos\varphi = -\sin(kz - \omega t)\sin\varphi.$$

(2.27)

Using the identity $\cos^2(\theta) + \sin^2(\theta) = 1$ in Equation 2.27 results in:

$$\left(\frac{E_x}{E_{ox}}\right)^2 + \left(\frac{E_y}{E_{oy}}\right)^2 - 2\left(\frac{E_x}{E_{ox}}\right)\cdot\left(\frac{E_y}{E_{oy}}\right)\cos\varphi = \sin^2\varphi,$$

(2.28)

which is the equation of an ellipse with the primary axis making an angle χ relative to E_x as shown in Figure 2.4 where the tilt angle χ is given by:

$$\tan 2\chi = \frac{2E_{ox}E_{oy}\cos\varphi}{E_{ox}^2 - E_{oy}^2}.$$

(2.29)

If $\chi = 0$, Equation 2.29 simplifies to:

$$\frac{E_x^2}{E_{ox}^2} + \frac{E_y^2}{E_{oy}^2} = 1,$$

(2.30)

and if $E_o = E_{ox} = E_{oy}$, then Equation 2.30 describes a circle with E_o equal to the radius.

$$E_o^2 = E_x^2 + E_y^2.$$

(2.31)

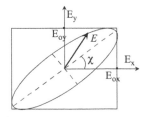

FIGURE 2.4   Diagram showing the path of the tip of the electric field vector $\vec{E}$ for elliptically polarized light.

2.4 Coherence

Holography is based on the recording of interference patterns formed by light originating from an optical source. The quality of the interference pattern in turn depends upon the coherence properties of light from the source. Therefore, in order to optimize the recording of holograms, it is important to understand at least the basic characteristics of coherence.

For most purposes two separate forms of coherence can be described. Temporal coherence indicates how well the phase difference between two different wavefronts remains a constant value. In contrast, spatial coherence indicates how constant the phase difference remains across a single wavefront. In this section, a review of the basic properties of temporal and spatial coherence are presented. A more complete review of coherence can be found in the text by Goodman on statistical optics [7].

2.4.1 Temporal Coherence

Temporal coherence can be expressed as either the length or time that the phase difference between the fields measured at two points along the propagation direction remains essentially constant. The temporal coherence time Δt_c and length Δl_c are simply related by a factor of the speed of light (c):

$$\Delta l_c = c \cdot \Delta t_c. \tag{2.32}$$

In addition, for a wave propagating in the z-direction as shown in Figure 2.5,

$$\Delta l_c = z_2 - z_1, \tag{2.33}$$

where z_1 and z_2 are the locations where the phase of the field are sampled and determined to be correlated.

The correlation between the fields can be quantified by analyzing the properties of the interference pattern between the fields at the two sampling positions. Assuming that the fields at the sampling locations are given by:

$$\tilde{E}_i = E_{o,i} \exp\left[j\varphi_i \right], \, i = 1,2; \tag{2.34}$$

the power or intensity of the two interfering fields becomes:

$$I = \left| \tilde{E}_1 + \tilde{E}_2 \right|^2 = \left| \tilde{E}_1 \right|^2 + \left| \tilde{E}_2 \right|^2 + \left\langle \tilde{E}_1 \tilde{E}_2^* + \tilde{E}_1^* \tilde{E}_2 \right\rangle$$

$$= I_1 + I_2 + 2\mathrm{Re}\left\langle \tilde{E}_1 \tilde{E}_2^* \right\rangle \tag{2.35}$$

$$= I_1 + I_2 + 2\sqrt{I_1 I_2} \, \cos\left(\varphi_1 - \varphi_2 \right),$$

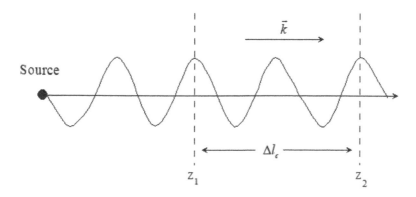

FIGURE 2.5 Diagram of a monochromatic wave with propagation vector $\vec{k}$. The wave is sampled at the points z_1 and z_2 along the direction of propagation.

where $\langle \tilde{E}_1 \tilde{E}_2^* \rangle$ is the time average of the product of the fields. The maximum intensity occurs when $(\varphi_1 - \varphi_2) = 2m\pi$ and a minimum when $(\varphi_1 - \varphi_2) = (2m-1)\pi$, where $m = 0, 1, 2,...$

Several important situations can now be examined. First, if the frequencies of the two fields are different $(\omega_1 \neq \omega_2)$, then the time average term in Equation 2.23 becomes:

$$\text{Re}\langle \tilde{E}_1 \tilde{E}_2^* \rangle = \frac{1}{2\Delta t} \int_{-\Delta t}^{\Delta t} \exp\left[-j(\omega_1 - \omega_2)t \right] dt$$

$$= \frac{\sin\left[(\omega_1 - \omega_2)\Delta t \right]}{\left[(\omega_1 - \omega_2)\Delta t \right]} \tag{2.36}$$

$$= \frac{\sin\left[2\pi \Delta v \Delta t \right]}{\left[2\pi \Delta v \Delta t \right]},$$

where $\Delta v = (\omega_2 - \omega_1)/(2\pi)$ is the effective spectral bandwidth of the source. Therefore, a large value for Re $\langle \tilde{E}_1 \tilde{E}_2^* \rangle$ will only occur when the frequencies are nearly equal. This expression also leads to the important result that the coherence will be maximum when the product:

$$\Delta v \Delta t = 1. \tag{2.37}$$

It implies that for sources with large spectral bandwidth, the coherence time will be shorter and conversely.

Quantitative measurements of the temporal coherence length can be made using a Michelson interferometer as shown in Figure 2.6. In this device, incident beam from the source is collimated and then divided by a beam splitter sending half the power to two mirrors M1 and M2. The light is reflected by the mirrors, passes again through the beam splitter, and is combined and displayed on a screen. The mirrors start out at equal distances from the beam splitter ensuring equal optical path lengths (i.e., $z_1 - z_2 = 0$). One of the mirrors is then moved to increase the difference in optical path length. If one of the mirrors is tilted, a set of fringes will appear across the screen allowing the maximum and minimum intensity of the interference pattern to be measured. The variation in intensity can be expressed in terms of the visibility defined as:

$$V = \frac{I_{max} - I_{min}}{I_{max} + I_{min}}. \tag{2.38}$$

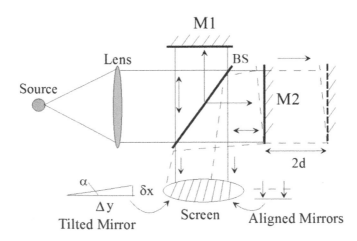

FIGURE 2.6 Diagram of a Michelson interferometer for measuring the visibility of the fringes resulting from sampling the field along the direction of propagation.

When $V = 1$, the source is completely coherent and when $V = 0$ completely incoherent. Typically, the temporal coherence length is specified at a value less than that corresponding to the distance between the $V = 1$ and $V = 0$ visibility locations (i.e., $V = 1/\sqrt{2}$). The distance $\Delta l_c = z_1 - z_2$ where the visibility equals the threshold value is considered the coherence length of the optical source.

Now consider the properties of optical sources. A purely monochromatic source emits light at a single wavelength or frequency without interruption. This type of source is assumed in describing many optical processes since it makes the analysis simpler. However, no optical source is truly monochromatic as there is always some spectral bandwidth associated with the emission process. The emission of photons from an actual source results from either spontaneous or stimulated processes as shown in Figure 2.7. In both of these processes an upper energy level is populated with electrons by some means such as optical, electrical, or chemical pumping. Spontaneous emitted photons are randomly released between an upper and lower energy level, while stimulated emitted photons have the same phase and polarization properties as the incident photon. Absorption is a third process related to the photon interaction between two energy bands and is the converse of stimulated emission. This provides a mechanism for populating the upper energy levels (Figure 2.7c).

The upper and lower energy levels actually consist of a range or band of energy values $\sum_i E_{1,i}$ and $\sum_i E_{2,i}$ and give rise to many possible energy values for spontaneous emitted photons. Therefore, sources governed by the spontaneous emission process such as light emitting diodes and thermal sources (i.e., incandescent filaments) have a broad spectral bandwidth. In contrast, lasers produce light primarily by stimulated emission and have a significantly smaller spectral spectral bandwidth which is primarily determined by the stability characteristics of the laser cavity.

Two of the most common spectral bandwidth profiles for describing light emission are Gaussian and Lorentzian functions [7]. For modeling purposes, it is also useful to consider a rectangular profile because it illustrates the main relationships source bandwidth and the coherence time and length. For example, Figure 2.8 shows the power spectrum for a Gaussian and a rectangular distribution. Most of the power is

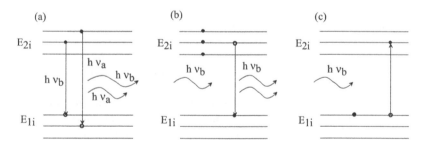

FIGURE 2.7 Diagram of the basic interactions between photons and a material system. (a) Spontaneous emission at multiple wavelengths; (b) stimulated emission at a wavelength with similar properties as the incident wavelength; and (c) absorption of an incident photon.

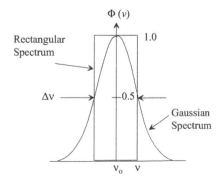

FIGURE 2.8 Power spectrum for a rectangular and Gaussian spectral distributions.

contained within the full width half maximum bandwidth (Δv) about the center frequency v_o. Therefore, it is possible to estimate the coherence length corresponding to a source with a spectral bandwidth Δv using the simpler rectangular spectral distribution. The analytical expression for the rectangular power spectrum can be written as:

$$\Phi(v) = \frac{1}{\Delta v} rect\left(\frac{v - v_o}{\Delta v}\right), \tag{2.39}$$

where

$$rect(z) = 1; \quad |z| < 1/2$$

$$= 1/2; |z| = 1/2 \tag{2.40}$$

$$= 0; \quad elsewhere.$$

As shown in Equation 2.36, a source of width Δv has a coherence time of Δt or a coherence length given by $\Delta l_c = c \cdot \Delta t_c$ (Equation 2.32). This provides a way to make a quick estimate of the coherence length of source simply by knowing the spectral bandwidth which is usually provided by the manufacturer.

2.4.2 Spatial Coherence

As described in the previous section, the temporal coherence of a source is a measure of the correlation of the optical field measured at two points in the direction of propagation. This represents the correlation between different wavefronts of the field propagating from the source. In contrast, the spatial coherence of a source indicates the correlation of the optical field at points across a single wavefront as shown in Figure 2.9. For an extended source (Δx_{src}), rays can reach sampling points P_1 and P_2 by different paths producing a variation in the wavefront. In addition, for a source dominated by spontaneous emission processes, the photons are emitted randomly in time. This causes the phase of the fields reaching the sampling plane to vary. These issues do not exist for a monochromatic point source, however, they must be considered for an extended source such as a light emitting diode or the emission region of an arc lamp.

The effects of spatial coherence can be evaluated using Young's interference experiment as shown in Figure 2.10 [7]. For this experiment, it is assumed that the source emits light with a relatively narrow spectral bandwidth at an average wavelength of $\bar{\lambda}$. Light propagates from each point on the extended source (S_o) to an intermediate screen that has two pinholes P_1 and P_2. The pinhole diameters are small in the sense that $\delta_{ph} << z_1\bar{\lambda}/x_{src}$, where δ_{ph} is the diameter of the pinholes. The length z_1 is the distance between the source and the intermediate screen, and x_{12} is the separation between the two pinholes and determines the locations where the wavefront is sampled. Light that passes through the pinholes propagates to an observation screen located a distance z_2 away and is measured at different

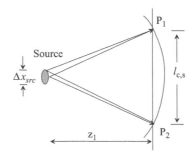

FIGURE 2.9 Diagram showing an extended source and sampling points P_1 and P_2 at a distance z_1 from the source.

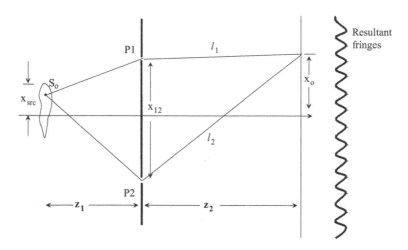

FIGURE 2.10 Diagram of Young's experiment showing sampling the wavefront from an extended source and formation of fringes on an observation plane.

positions, x_o, from the optical axis. The sampled light at z_2 results in an interference pattern and can be described by computing the intensity where:

$$I(x_o) = \left\langle \left(\tilde{v}_1 + \tilde{v}_2 \right)^2 \right\rangle$$

$$= \left\langle \tilde{v}_1 \cdot \tilde{v}_1^* \right\rangle + \left\langle \tilde{v}_2 \cdot \tilde{v}_2^* \right\rangle + \left\langle \tilde{v}_1 \cdot \tilde{v}_2^* + \tilde{v}_2 \cdot \tilde{v}_1^* \right\rangle \qquad (2.41)$$

$$= I_1 + I_2 + 2\,\mathrm{Re}\left\langle \tilde{v}_1 \cdot \tilde{v}_2^* \right\rangle,$$

where

$$\tilde{v}_1 = \tilde{v}\left(P_1, t - l_1/c \right) \text{ and } \tilde{v}_2 = \tilde{v}\left(P_2, t - l_2/c \right), \qquad (2.42)$$

and represent the fields that propagate through pinholes P_1 and P_2, respectively, on the intermediate screen. The time delay for light propagating through the two paths can be combined to form a single parameter:

$$\tau = \frac{l_2 - l_1}{c}, \qquad (2.43)$$

and used to simplify Equation 2.41 resulting in:

$$I\left(x_o, \tau \right) = I_1 + I_2 + 2\,\mathrm{Re}\left\langle \tilde{v}\left(P_1, t + \tau \right) \tilde{v}^*\left(P_2, t \right) \right\rangle. \qquad (2.44)$$

This result is similar to that found in Equation 2.35 for temporal coherence. In fact, both spatial and temporal coherence properties affect the fringes formed in Young's experiment. When $\tau = 0$ (i.e., zero path length difference), spatial coherence will dominate fringe visibility. However, when τ becomes larger, the fringe visibility will decrease due to temporal coherence effects.

The combined coherence effects are often expressed in terms of a generalized parameter called the "complex degree of coherence" [5,7,8] defined as:

$$\gamma_{12}\left(\tau \right) \equiv \frac{\Gamma_{12}\left(\tau \right)}{\left[\Gamma_{11}\left(0 \right)\Gamma_{22}\left(0 \right) \right]^{1/2}}, \qquad (2.45)$$

where

$$\Gamma_{12}(\tau) = \langle \tilde{v}(P_1, t + \tau) \cdot \tilde{v}^*(P_2, t) \rangle,$$

$$\Gamma_{11}(0) = \langle \tilde{v}(P_1, t) \cdot \tilde{v}^*(P_1, t) \rangle, \tag{2.46}$$

$$\Gamma_{22}(0) = \langle \tilde{v}(P_2, t) \cdot \tilde{v}^*(P_2, t) \rangle.$$

As can be seen, the complex degree of coherence is essentially the normalized modulation term in Equation 2.44 (i.e., $2\operatorname{Re}\langle \tilde{v}(P_1, t + \tau)\tilde{v}^*(P_2, t) \rangle$).

Equation 2.44 does not fully quantify the constraint that spatial coherence places on the observation region where an extended source may be considered to be spatially coherent and allow good visibility. This problem was addressed by Van Cittert [9] and Zernike [10] using diffraction propagation analysis and other coherence considerations. Their work resulted in a simple relationship [7]:

$$A_{coh} = \frac{(\bar{\lambda} z)^2}{A_{src}}, \tag{2.47}$$

where A_{src} is the area of the extended source, $\bar{\lambda}$ is the average emission wavelength, z is the distance between the source and the observer, and A_{coh} is the observation area that has high visibility. The result implies that a smaller extended source area forms a larger region of coherence when observed at a distance z. In addition, the further away the observation distance, the larger the coherence region. One example of this is that light from a star has a high degree of spatial coherence.

2.5 Geometrical Optics

Designing and conducting holography experiments require an understanding of basic optical components and imaging relations. In this section, a brief review of several useful principles and relationships of geometrical optics is presented. This discussion provides a basic framework for the design and layout of optical systems that are frequently used for holography. Geometrical optics treat the propagation of optical beams as a series of rays that pass through different components. The rays indicate the direction of the propagation and the Poynting vector of the optical field (Equations 2.12 and 2.13). While a geometrical optics evaluation does not consider physical effects such as diffraction, it provides a great deal of insight into the performance of an optical system. In particular, the distribution of rays indicates the optical power distribution and in many cases is sufficient for an application. For a more extensive treatment of geometrical optics, the reader is referred to the many texts on this subject, some of which are listed in the references [11–13].

2.5.1 Ray Propagation

There are three basic laws of geometrical optics that determine the propagation characteristics of rays through optical materials. These include:

1. *Transmission*: A ray trajectory remains unchanged in a medium with constant refractive index
2. *Reflection*: A ray making an angle θ_1 with respect to a surface normal is reflected at an angle θ_2 such that (Figure 2.11a):

$$\theta_1 = \theta_2 \tag{2.48}$$

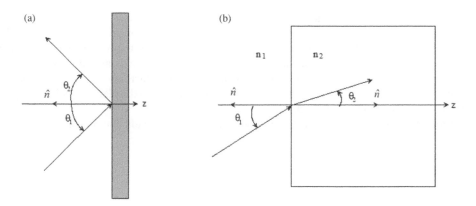

FIGURE 2.11 (a) Reflection of an incident ray at a planar interface and (b) Refraction at an interface with $n_2 > n_1$. z indicates the optical axis for the system.

3. *Refraction*: A ray propagating in a medium with refractive index n_1 at an angle θ_1 with respect to a surface normal will be refracted after entering a second material with refractive index n_2 at an angle θ_2 according to Snell's law such that (Figure 2.11b):

$$n_1 \sin\theta_1 = n_2 \sin\theta_2$$

$$\theta_2 = \sin^{-1}\left[\frac{n_1}{n_2}\sin\theta_1\right].$$

(2.49)

For both the laws of reflection and refraction, the rays and the normal vectors are coplanar. These simple laws are used to describe ray propagation through optical components such as lenses, prisms, beam splitters, and mirrors. These optical components are made by modifying either the refractive index or reflective properties of surfaces, the surface profile, or the thickness of the component as well as any combination of these parameters.

2.5.2 Reflection at Dielectric Interfaces

2.5.2.1 Fresnel Formulae

As shown in Figure 2.11b, Snell's law can be used to determine the direction of a transmitted ray that passes through an interface. However, a more complete description of the interaction of light at an interface is shown in Figure 2.12. The incident power is partially transmitted (T) and partially reflected (R)

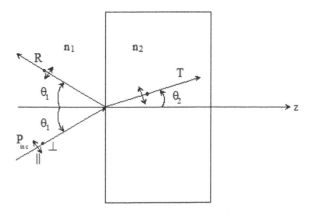

FIGURE 2.12 Diagram showing a ray incident from medium n_1 to n_2 with $n_2 > n_1$. The incident power P_{inc} is partially transmitted and reflected. $\perp$ and $\parallel$ indicate the polarization state of the incident beam.

at the interface with the angle that the transmitted ray makes with the optical axis (z) being described by Snell's law and the angle that the reflected component makes with the optical axis described by the law of reflection. The field amplitude and power of the reflected and transmitted components can be derived from the boundary matching conditions of the electric and magnetic fields and are called the Fresnel formulae [1]. Since the field components at the boundary are also a function of the polarization state of the incident beam, there will be a different relation for light polarized in the plane of incidence (p-,‖) and perpendicular (s-,⊥) to the plane of incidence.

The resulting Fresnel coefficients for the perpendicularly polarized field component (s-,⊥) are:

$$r_\perp = \frac{E_{R\perp}}{E_{inc\perp}} = \frac{n_1 \cos\theta_1 - n_2 \cos\theta_2}{n_1 \cos\theta_1 + n_2 \cos\theta_2}$$

$$t_\perp = \frac{E_{T\perp}}{E_{inc\perp}} = \frac{2n_1 \cos\theta_1}{n_1 \cos\theta_1 + n_2 \cos\theta_2},$$

(2.50)

where the angles and refractive index values are shown in Figure 2.12. The corresponding power of the reflected and transmitted components are:

$$R_\perp = \frac{P_{R\perp}}{P_{inc\perp}} = |r_\perp|^2$$

$$T_\perp = \frac{P_{T\perp}}{P_{inc\perp}} = \left(\frac{n_2 \cos\theta_2}{n_1 \cos\theta_1}\right)|t_\perp|^2.$$

(2.51)

Note that the transmittance coefficient $T_\perp$ includes an obliquity factor $\left(\frac{n_2 \cos\theta_2}{n_1 \cos\theta_1}\right)$ and is necessary to ensure power conservation, i.e., $1 = R_\perp + T_\perp$. The Fresnel coefficients for the parallel polarized (p-,‖) field components are:

$$r_\parallel = \frac{E_{R\parallel}}{E_{inc\parallel}} = \frac{n_2 \cos\theta_1 - n_1 \cos\theta_2}{n_1 \cos\theta_2 + n_2 \cos\theta_1}$$

$$t_\parallel = \frac{E_{T\parallel}}{E_{inc\parallel}} = \frac{2n_1 \cos\theta_1}{n_1 \cos\theta_2 + n_2 \cos\theta_1},$$

(2.52)

and the corresponding Fresnel coefficients for the optical power are:

$$R_\parallel = \frac{P_{R\parallel}}{P_{inc\parallel}} = |r_\parallel|^2$$

$$T_\parallel = \frac{P_{T\parallel}}{P_{inc\parallel}} = \left(\frac{n_2 \cos\theta_2}{n_1 \cos\theta_1}\right)|t_\parallel|^2.$$

(2.53)

These relations show that the amount of light reflected from a dielectric surface can be significant especially if the difference in refractive index between the two materials (n_2-n_1) or the angle of incidence θ_1 are large. If light in air is normally incident on silicon with a refractive index of 3.6, the reflected power for s- and p-polarized light are both equal to:

$$R = \left(\frac{n_2 - n_1}{n_2 + n_1}\right)^2 = 0.32.$$

(2.54)

In order to reduce the large reflection loss, anti-reflection coatings are often used. In this method, layers of dielectric films are deposited on the initial surface to change the phase of the reflected field component. The coatings have a refractive index value (n_c) that is between 1.0 (air) and n_2. Magnesium fluoride is often used that has a refractive index of 1.38 and can be deposited by evaporation methods. When the coating layer produces a phase shift of $\lambda_o/4$ (where λ_o is the free-space wavelength) between the incident and back surface of the film, destructive interference occurs and reduces the reflected intensity. The reflectance from a substrate (n_2) with a single coating layer (n_c) and light incident in medium with index n_1 is [3]:

$$R_{AR} = \frac{n_c^2\left(n_1 - n_2\right)^2 \cos^2\left(k_o d'\right) + \left(n_1 n_2 - n_c^2\right)^2 \sin^2\left(k_o d'\right)}{n_c^2\left(n_1 + n_2\right)^2 \cos^2\left(k_o d'\right) + \left(n_1 n_2 + n_c^2\right)^2 \sin^2\left(k_o d'\right)}, \tag{2.55}$$

where d' is the optical path length difference between light reflected from the front and back surface of the coating layer, and $k_o = 2\pi/\lambda_o$. If the optical path length difference is set to $d' = \lambda_o/4$, the Equation 2.55 reduces to:

$$R_{AR} = \frac{\left(n_1 n_2 - n_c^2\right)^2}{\left(n_1 n_2 + n_c^2\right)^2}, \tag{2.56}$$

and $R_{AR} = 0$ when

$$n_c = \sqrt{n_1 n_2}. \tag{2.57}$$

The optical path length difference d' depends on the angle of incidence and the refractive index of the layer as shown in Figure 2.13 with:

$$d' = n_c\left(AC + CD\right) - n_1 \cdot AB$$
$$= 2n_c d/\cos\theta_2 - 2n_1 d \tan\theta_2 \sin\theta_1. \tag{2.58}$$

Since the condition for zero reflectance is only true for a specific wavelength and angle of incidence, the reflectance will increase at other values. In order to extend the range of angles and wavelengths for minimum reflectance, multiple coatings are used. The performance of the anti-reflection coating can also

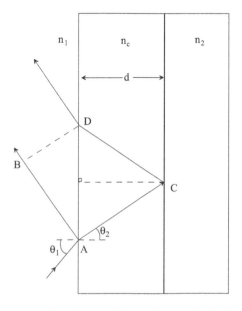

FIGURE 2.13   Diagram showing the optical path length difference in an anti-reflection coating.

be improved by setting the optical path length difference equal to an integral number of wavelengths. This is the basis for designing dichroic wavelength filters that reflect specific wavelength bands [14].

2.5.2.2 Brewster Angle

Figure 2.14 shows a plot of the reflectance as a function of angle for light incident from air with s- and p-polarization on a slab of glass with a refractive index of 1.50. The reflectance of the s-polarized light increases as a function of the angle of incidence. However, the reflectance for the p-polarized component passes through a zero or null value. This anomaly occurs because at the null angle, light within medium n_2 is polarized in the same direction of the reflected ray. Physically, the electrons in the material oscillate in the same direction as the polarization of the field within the medium and cannot radiate in the direction of oscillation. Therefore, a null in the reflectance for p-polarized incident light occurs. The null angle is called the Brewster angle [1] and is equal to:

$$\tan\theta_B = \frac{n_2}{n_1}.$$
(2.59)

For the case shown in Figure 2.14 with $n_1 = 1.00$ and $n_2 = 1.50$, the Brewster angle is 56.3° and only s-polarized light is reflected at this angle. It is possible to use this effect to make a polarizing beam splitter with the reflected light s-polarized and the transmitted light p-polarized (Figure 2.15).

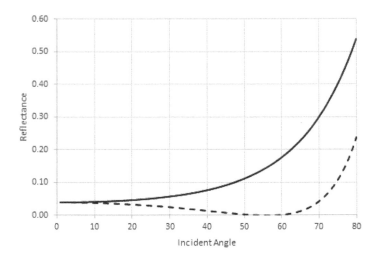

FIGURE 2.14   Plot of the reflectance for s- (solid line) and p- (dashed line) polarized light incident onto a dielectric with $n = 1.50$.

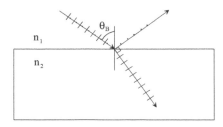

FIGURE 2.15   Diagram of randomly polarized light incident onto a material of refractive index n_2 from a refractive index of n_1.

2.5.2.3 *Total Internal Reflection*

If light passes from a higher to a lower refractive index medium, Snell's law (Equation 2.49) shows that θ_2 will reach 90° before θ_1 [1]. If θ_1 continues to increase, only reflection will occur at the boundary and is referred to as total internal reflection or TIR. The TIR angle can be determined directly from Snell's law and with $n_1 > n_2$:

$$\sin\theta_{TIR} = n_2/n_1. \tag{2.60}$$

TIR is the basis for making optical waveguides and will be discussed later in the design of substrate mode and edge-lit holograms. The TIR effect is also used to control beam propagation in different types of prisms. For example, Figure 2.16 shows light entering a prism at normal incidence and exiting another face after undergoing TIR.

2.5.3 Optical Lenses

Optical lenses are used to collect, transform, and focus light from an optical source or to image light from an object. These functions are accomplished by producing surface power either by shaping the surface of a material that has a constant refractive index, by varying the refractive index of a material that has a constant shape, or by changing both the shape and refractive index of a material along a direction perpendicular to the optical axis [11–13].

Surface power changes the phase of an incident optical field. To examine this, consider a cylindrical plano-convex lens (i.e., a lens that varies only in the x-direction) as shown in Figure 2.17. The lens has a

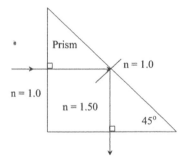

FIGURE 2.16 Light entering a glass prism with a refractive index of 1.50 from air and undergoing TIR.

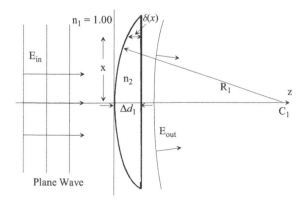

FIGURE 2.17 Planar wave incident on a plano-convex lens.

constant refractive index n_2, and the lens thickness varies in x. The operation of a lens can be expressed as a transmittance function $t_{lens}(x)$ that modifies an incident field $E_{in}(x)$ as shown in Equation 2.61 [2]:

$$E_{out}(x) = t_{lens}(x) E_{in}(x), \tag{2.61}$$

with

$$t_{lens}(x) = \exp\left[j\varphi_{lens}(x) \right], \tag{2.62}$$

and $\varphi_{lens}(x)$ is the phase change introduced by the lens as a function of position.

For a lens with a constant refractive index (n_2), $\varphi_{lens}(x)$ essentially represents the thickness variation of the lens with x and can be written as:

$$\varphi_{lens}(x) = \frac{2\pi n_2}{\lambda}\delta(x) + \frac{2\pi}{\lambda}\left[\Delta d_1 - \delta(x)\right], \tag{2.63}$$

where $\delta(x)$ is the thickness for different values of x, and Δd_1 is the lens thickness along the optical axis (i.e., $x = 0$). By convention, a surface with a positive radius of curvature, R_1, has the center, C_1, located to the right of the surface, and a negative curvature will have its center to the left of the surface. The thickness function $\delta(x)$ for this case is:

$$\delta(x) = \Delta d_1 - \left[R_1 - \left(R_1^2 - x^2\right)^{1/2} \right]$$

$$= \Delta d_1 - R_1\left[1 - \left(1 - \frac{x^2}{R_1^2}\right)^{1/2} \right]. \tag{2.64}$$

If it is also assumed that $x < R_1$, the binomial expansion:

$$\sqrt{1+a} = 1 + \frac{1}{2}a + \frac{1}{8}a^2 + \dots \quad ; a < 1, \tag{2.65}$$

can be applied in Equation 2.64 to simplify the radical so that:

$$\left(1 - \frac{x^2}{R_1^2}\right)^{1/2} \approx 1 - \frac{x^2}{2R_1^2}. \tag{2.66}$$

This in turn simplifies the thickness function:

$$\delta(x) = \Delta d_1 - \frac{x^2}{2R_1}, \tag{2.67}$$

and the phase function for the plano-convex lens becomes:

$$\varphi(x) = \frac{2\pi n_2}{\lambda}\Delta d_1 - \frac{2\pi}{\lambda}(n_2 - 1)\frac{x^2}{2R_1}. \tag{2.68}$$

Inserting this result into Equations 2.61 and 2.62 gives:

$$E_{out}(x) = t_{lens}(x) E_{in}(x)$$

$$= \exp\left[\frac{2\pi n_2}{\lambda} \Delta d_1\right] \exp\left[-j\frac{2\pi}{\lambda}(n_2-1)\frac{x^2}{R_1}\right] \cdot E_{in}(x). \tag{2.69}$$

Therefore, if a plane wave with $E_{in}(x) = 1$ is incident on the lens, the exiting wave has wavefront curvature given by $\exp[-j\frac{2\pi}{\lambda}(n_2-1)\frac{x^2}{R_1}]$. The term $(n_2-1)\frac{1}{R_1}$ is equal to the focal length of a lens through the lens makers formula:

$$\frac{1}{f} = (n_2-1)\left[\frac{1}{R_1} - \frac{1}{R_2}\right], \tag{2.70}$$

with $R_2 = \infty$ for the planar surface. Therefore, the resulting phase function for the lens is:

$$\varphi(x) = \frac{2\pi n_2}{\lambda} \Delta d_1 - \frac{2\pi}{\lambda}\frac{x^2}{2f}. \tag{2.71}$$

This relation can readily be extended to give the two-dimensional phase for a spherical lens:

$$\varphi(x,y) = \frac{2\pi n_2}{\lambda} \Delta d_1 - \frac{2\pi}{\lambda}\frac{(x^2+y^2)}{2f}. \tag{2.72}$$

When a plane wave is incident on a lens, the focal length f is also the radius of curvature of the field leaving the lens. If only geometrical optics are considered, an ideal spherical lens will focus an incident planar wave to a point. Later in this chapter it will be shown that due to diffraction by the finite aperture of the lens, light will not come to a point focus.

When the focal length of the lens is much greater than the lens thickness, the lens can be considered a *thin lens* and can be characterized simply by its focal length. The well-known thin lens imaging relation can be used to determine the position of an image when the object location is known and is given by [13]:

$$\frac{1}{f} = \frac{1}{d_o} + \frac{1}{d_i}, \tag{2.73}$$

where d_o is the distance of an axial object point to the left of the lens, and d_i is the distance of the corresponding axial image point to the right of the lens (Figure 2.18).

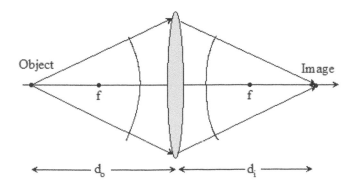

FIGURE 2.18 Imaging an on-axis object point to an image point.

2.5.4 Focusing Mirrors

Focusing elements can also be made with mirrors that have curved surfaces [1,11–13]. Using the law of reflection and a parallel incident ray (Figure 2.19), it can readily be seen that the focal length for a spherical mirror is:

$$f_{\text{Mirror}} = -R/2. \tag{2.74}$$

Mirrors are frequently used in large aperture optical systems such as telescopes since they can be supported with lower mass substrates than glass lenses. In addition, since mirrors do not refract light through glass, they are not subject to dispersion effects resulting from the wavelength dependence of the glass refractive index. This allows the use of mirrors with broad spectral band illumination without the need to correct for chromatic aberrations.

2.5.5 Paraxial Rays and Basic Image Analysis Methods

2.5.5.1 Paraxial Approximation and Ray Trace Relations

A very useful simplification for analyzing the first order properties of an optical system is the *paraxial approximation* [1,11–13]. In this approximation, the optical ray height is assumed to be close to the optical axis of the system, and the ray angles are assumed to be small such that:

$$u = \sin\theta = \tan\theta, \tag{2.75}$$

where u is the paraxial angle, and θ is the full angle of the ray with the optical axis. With these assumptions, ray propagation through an optical surface can be evaluated using the simple paraxial ray trace formula:

$$\phi = \left(n_2 - n_1\right)\frac{1}{R};$$

$$n_2 u_2 = n_1 u_1 - y_1\phi; \tag{2.76}$$

$$y_2 = y_1 + u_2 t_{12},$$

where ϕ is the surface power at a point where the ray intersects the surface, R is the radius of curvature of the surface, n_1 and n_2 are the refractive indices of the incident and second surface materials, u_1 and u_2 are the paraxial ray angles relative to the optical axis in the incident and secondary materials, y_1 and y_2 are the ray heights above the optical axis, and t_{12} is the propagation distance between planes 1 and 2.

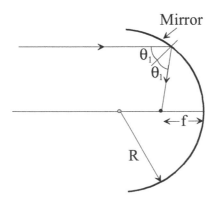

FIGURE 2.19   Diagram of rays reflected by a spherical surface mirror with radius of curvature R.

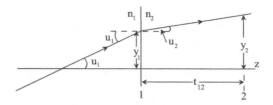

FIGURE 2.20 Diagram showing a paraxial ray propagate across an interface between materials with refractive indices n_1 and n_2.

The second relation in Equation 2.76 is the paraxial ray refraction at the boundary between materials n_1 and n_2, and the third relation determines the ray height y_2 after propagating to the transfer distance t_{12} to plane 2. A diagram of the paraxial ray parameters illustrating refraction at an interface between two materials and ray transfer to a second surface are shown in Figure 2.20.

2.5.5.2 Basic Image Analysis Methods

When analyzing the imaging properties of an optical system, some of the basic values that we would like to know include: the image location and magnification, the power that can be transferred through the system, and the field of view. Several metrics and special rays are useful in determining these values and assessing the overall performance of the optical system [11–13]. First, the basic premise of imaging with refractive and reflective components is that a point in object space corresponds to a single point in image space. This relation is often referred to as the conjugate imaging property.

The optical system has a set of cardinal points or planes that characterize the mapping between object and image points. These include the front and rear focal points, the front and rear principle points, and the front and rear nodal points. A ray passing from an object point through the front focal point and to the principle plane will emerge parallel to the optical axis as shown by rays b and b' in Figure 2.21. Similarly, a ray propagating parallel to the optical axis will emerge from the principle plane and pass through the focal point on the opposite side of the system. A ray passing through one principle plane (front or rear) will emerge from the other principle plane at the same ray height. This implies that the lateral magnification between the principle planes is 1. In general, the lateral magnification of an optical system is the ratio of the image to the object heights:

$$m = \frac{h_2}{h_1}. \tag{2.77}$$

In the case of the principle planes, the lateral magnification indicates the ratio of the ray height on the rear principle plane to the front principle plane and conversely.

The last set of cardinal points are the nodal points. A ray with angle u passing through one nodal point will emerge with angle u from the second nodal point. For a thin lens, the front and rear principle and nodal points are co-located at the vertex point on the optical axis representing the thin lens.

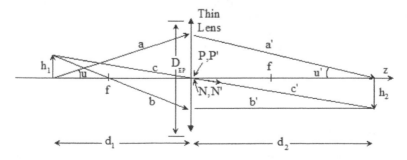

FIGURE 2.21 Characteristic rays for analyzing the imaging properties of a thin lens. P, P′ are the front and rear principle points, and N, N′ the front and rear nodal points of the thin lens.

Another set of parameters that are useful for analyzing an optical system are the aperture stop and the entrance and exit pupils. The aperture stop is the opening that limits the diameter of a beam that originates from an axial point on the object. The entrance pupil is the image of the aperture stop projected into object space, and the exit pupil is the image of the aperture stop in image space. For a system consisting of a single thin lens, the aperture stop and the pupils coincide at the lens. For more complicated systems, the aperture stop can be found by increasing the angle of the axial rays from the object, tracing the rays through the system, and noting the aperture that first blocks the rays.

An axial ray that makes the largest angle with the optical axis and makes it through the lens determines the light gathering capability of the lens. This angle is often expressed as the numerical aperture of the lens or optical system and is given by:

$$NA = n_1 \sin\theta_{1/2,\max} = n_1 u_{\max}, \tag{2.78}$$

where both the non-paraxial half angle ($\theta_{1/2,\max}$) and the paraxial half angle ($u_{\max}$) are given. For a thin lens with the limiting aperture at the lens and incident collimated light propagating parallel to the optical axis ($d_1 = -\infty$), the numerical aperture is also related to the f-number ($f/\#$) of the lens:

$$f/\# \simeq \frac{1}{2NA} = \frac{f}{D_{EP}}, \tag{2.79}$$

where D_{EP} is the diameter of the entrance pupil of the lens or optical system.

The marginal and chief rays are also used to obtain different imaging parameters of the system. The marginal ray is defined as an axial that passes through the edge of the entrance pupil of the system (rays a and a' in Figure 2.21), while a chief ray originates at the tip of the object and passes through the center of the entrance pupil (rays b and b' in Figure 2.21). The image of an object is located at the position where the marginal ray crosses the axis, while the height of the chief ray at this position gives the image height. Conversely, when a chief ray crosses the axis, it indicates the position of a stop or pupil, and the height of the marginal ray indicates the radius of the stop or pupil. The maximum chief ray that makes it through an optical system sets the *field of view* of the optical system. For a single thin lens, the field of view is usually limited by the aperture height of the detector at the image plane.

Another useful paraxial parameter that relates ray heights and angles at different locations in the optical system is the *Lagrange invariant* and is expressed as:

$$L = n\bar{u}y - nu\bar{y}, \tag{2.80}$$

where $\bar{u}$ and $\bar{y}$ are the angle and height of the chief ray and u and y the angle and height of the marginal ray. The value L is a constant at any point along the optical axis of the system and can be used to determine angles and ray heights at different locations in the system. At the object, $y = 0$ and $L = -nu\bar{y}$, while at the aperture stop and pupils $\bar{y} = 0$ and $L = n\bar{u}y$.

In geometrical optics, an ideal optical system can form a perfect point image and can be fully evaluated using paraxial optics. This first order is very useful, but does not provide the full story of the performance of actual systems. Image aberrations will form in real systems and can be modeled by expanding the analysis to include third and higher order terms to describe the wavefront of the optical beam that passes through the system. The analysis and correction of aberrations is beyond the scope of this book and the reader is referred to comprehensive treatment in the literature [1,11,12]. However, in Chapter 4, the aberrations of holographic lenses will be discussed and analyzed.

2.6 Diffraction Analysis

When an optical field propagates past or through an object, the field properties change. This interaction and effect of an aperture on the spatial properties of a wave can be analyzed using diffraction theory. In this section, a summary of the basic elements of scalar diffraction theory are presented since it is fundamental to understanding many aspects of holography. For a more thorough discussion of this subject, the reader is referred to several excellent and comprehensive treatments of diffraction including those by Goodman [2] and Born and Wolf [1].

2.6.1 Huygens-Fresnel Diffraction Relation

In 1678, Christiaan Huygens developed an intuitive explanation for the diffraction of light by an aperture and was based on the notion that light propagates as a wave. He suggested that a new wavefront can be described by placing a collection of point sources on the originating wavefront at an earlier time. The new wavefront is found by propagating the spherical wavefronts from the point sources by an equal distance and overlapping the spherical surfaces. This idea was later mathematically formulated by Augustin-Jean Fresnel in the early nineteenth century and allowed for quantitative prediction of the diffracted field.

Referring to Figure 2.22, the Huygens-Fresnel diffraction principle can be expressed as:

$$E(P_2) = \frac{1}{j\lambda} \iint_{A_1} E(P_1) \frac{\exp(jk s_{12})}{s_{12}} \cos\theta_{12} da, \tag{2.81}$$

where A_1 is the area of the aperture, r_{12} is the distance between point P_1 on the aperture and P_2 at the observation plane, da is an element of the aperture area, and θ_{12} is the angle from the point P_1 to P_2 relative to a surface normal $\check{n}$ at the aperture. The expression can be simplified by first noting that:

$$\cos\theta_{12} = z/s_{12}, \tag{2.82}$$

resulting in:

$$E(x_2, y_2) = \frac{z}{j\lambda} \iint_{A_1} E(x_1, y_1) \frac{\exp(jk s_{12})}{s_{12}^2} dx_1 dy_1. \tag{2.83}$$

In addition, if it is assumed that the observation point (x,y) is close to the optical axis, it is reasonable to let $s_{12} = z$ in the denominator within the integral resulting in:

$$E(x_2, y_2) = \frac{1}{j\lambda z} \iint_{A_1} E(x_1, y_1) \exp(jk s_{12}) dx_1 dy_1. \tag{2.84}$$

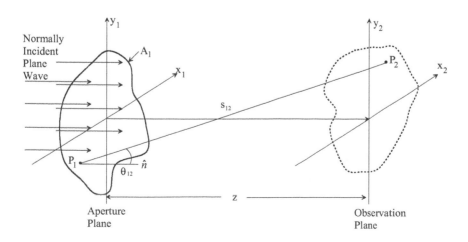

FIGURE 2.22 Diagram showing the geometry and parameters used for the Huygens-Fresnel principle.

Further simplifications to the integral can be made by making additional assumptions about the aperture size and the propagation distances. This process gives rise to the Fresnel (near-field) and Fraunhofer (far-field) diffraction approximations and is the subject of the next sections.

2.6.2 Fresnel or Near-Field Diffraction Region

Since the distance s_{12} in the exponential term is multiplied by $1/\lambda$, it cannot be simply set equal to z as phase can vary significantly. However, if it is assumed that $(\frac{x_2-x_1}{z})^2$ and $(\frac{y_2-y_1}{z})^2 < 1$, s_{12} can be approximated with sufficient accuracy using the binomial expansion (Equation 2.65). In this case s_{12} can be rewritten as:

$$s_{12} = \left[z^2 + (x_2 - x_1)^2 + (y_2 - y_1)^2 \right]^{1/2}$$

$$= z \left[1 + \left(\frac{x_2 - x_1}{z} \right)^2 + \left(\frac{y_2 - y_1}{z} \right)^2 \right]^{1/2}.$$

(2.85)

After substituting into Equation 2.84, the diffracted field becomes:

$$E(x_2, y_2) = \frac{\exp(jkz)}{j\lambda z} \int\limits_{-\infty}^{+\infty}\!\!\int E(x_1, y_1) \exp\left\{ j\frac{k}{2z} \left[(x_2 - x_1)^2 + (y_2 - y_1)^2 \right] \right\} dx_1 dy_1,$$

(2.86)

where the double integration is taken from $-\infty$ to $+\infty$ since it can be assumed that there is no light propagating to P_2 from the area outside of the clear aperture A_1.

If the quadratic terms in Equation 2.86 are expanded, the diffraction integral equation becomes:

$$E(x_2, y_2) = \frac{e^{jkz} \exp\left[j\frac{k}{2z}(x_2^2 + y_2^2) \right]}{j\lambda z} \times$$

$$\int\limits_{-\infty}^{\infty}\!\!\int \left[E(x_1, y_1) \exp\left(j\frac{k}{2z}(x_1^2 + y_1^2) \right) \right] \exp\left(-j\frac{2\pi}{\lambda z}(x_1 x_2 + y_1 y_2) \right) dx_1 dy_1.$$

(2.87)

This integral is essentially the Fourier transform of the term in brackets which consists of the product of the field at the aperture and a quadratic phase factor that varies with the distance from the aperture. As a result, the spatial distribution of the diffraction pattern in the "near-field" or "Fresnel region" will change with observation distance from the aperture (z). A representation of the changing diffraction pattern as a function of z is illustrated in Figure 2.23.

2.6.3 Fraunhofer or Far-Field Diffraction Region

If the observation distance from the aperture continues to increase, the magnitude of the quadratic phase factor:

$$\frac{k}{2z}(x_1^2 + y_1^2),$$

(2.88)

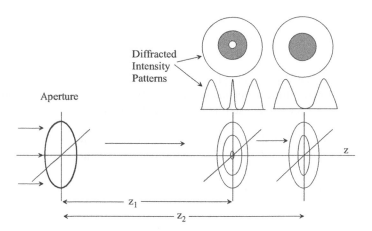

FIGURE 2.23 Illustration of the changing spatial distribution of the near-field diffraction pattern as a function of the distance from the aperture.

decreases. When this factor is <<1, the quadratic phase factor term in Equation 2.87 can be dropped resulting in:

$$E(x_2, y_2) = \frac{e^{jkz} \exp\left[j\frac{k}{2z}(x_2^2 + y_2^2) \right]}{j\lambda z} \int\int_{-\infty}^{\infty} E(x_1, y_1)\exp\left(-j2\pi(v_x x_1 + v_y y_1)\right)dx_1 dy_1, \qquad (2.89)$$

which is an exact Fourier transform of the field distribution at the aperture plane with spatial frequencies:

$$v_x = \frac{x_2}{\lambda z}; \quad v_y = \frac{y_2}{\lambda z}. \qquad (2.90)$$

In this case, the functional form of the spatial distribution of the diffraction pattern no longer changes with distance, z, and differentiates it from the near-field case.

The distance from the aperture required to be in the far-field diffraction region can be considerable. Using the condition that:

$$\frac{k}{2z}(x_1^2 + y_1^2) = \frac{\pi}{\lambda z}(x_1^2 + y_1^2)_{\max} << 1, \qquad (2.91)$$

a 1 cm aperture illuminated with 500 nm light requires that $z >> 600$ m. This makes it difficult to reach the far-field condition in normal laboratory settings by free-space field propagation alone.

2.6.4 Fourier Transform Properties of a Lens

In Section 2.5.3, it was shown that the operation of a lens can be viewed as a transmittance function that modifies the incident field:

$$E_{out}(x, y) = t_{lens}(x, y)E_{in}(x, y). \qquad (2.92)$$

It was also shown that the transmittance function of an ideal lens has the form:

$$t_{lens}(x, y) = \exp(-j\varphi(x, y)), \qquad (2.93)$$

with

$$\varphi(x,y) = -\frac{2\pi}{\lambda}\frac{\left(x^2+y^2\right)}{2f}, \tag{2.94}$$

and where $\varphi(x,y)$ is the spatially varying phase function of the lens. Note that in Equation 2.94 the constant phase factor related to the lens thickness Δd_l along the optical axis is not included since it does not change the functional form of the resulting field. In addition, the finite size of the lens aperture is not specified.

The effect of a lens on a diffracted field can now be evaluated by considering what happens when an object is placed directly in front of the lens and is illuminated with a normally incident plane wave as shown in Figure 2.24.

The field immediately past the object is:

$$E_L(x_l, y_l) = t_{obj}(x_l, y_l), \tag{2.95}$$

where the maximum transmittance of the object is assumed to be 1.0. The field distribution immediately past the lens is the product of the transmittance function of the object and the lens:

$$E_{L'}(x_l, y_l) = E_L(x_l, y_l)t_{lens}(x_l, y_l)$$

$$= t_{obj}(x_l, y_l)\exp\left(-j\frac{2\pi}{\lambda}\frac{\left(x_l^2+y_l^2\right)}{2f}\right). \tag{2.96}$$

If it is assumed that the focal length of the lens is long enough to satisfy the near-field diffraction condition, then the field distribution at the back focal plane of the lens can be written as:

$$E_f(x_f, y_f) = \frac{e^{j\frac{k}{2f}\left(x_f^2+y_f^2\right)}}{j\lambda f} \times$$

$$\int\limits_{-\infty}^{\infty}\int E_{L'}(x_l, y_l)\exp\left(j\frac{k}{2f}\left(x_l^2+y_l^2\right)\right)\exp\left[-j\frac{2\pi}{\lambda}\left(x_l x_f + y_l y_f\right)\right]dx_l dy_l. \tag{2.97}$$

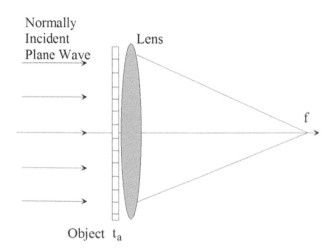

FIGURE 2.24 A transparent object (t_a) placed next to a lens and illuminated with a normally incident plane wave.

Substituting the field immediately past the lens, $E_{L'}$, (Equation 2.96) into Equation 2.97, cancels the quadratic phase term in the integral leaving:

$$E_f\left(x_f, y_f\right) = \frac{e^{j\frac{k}{2f}\left(x_f^2 + y_f^2\right)}}{j\lambda f} \int\int\limits_{-\infty}^{\infty} \left[t_o\left(x_l, y_l\right)e^{-j\frac{k}{2f}\left(x_l^2 + y_l^2\right)}\right] e^{j\frac{k}{2f}\left(x_l^2 + y_l^2\right)} e^{-j\frac{2\pi}{\lambda}\left(x_l x_f + y_l y_f\right)} dx_l dy_l$$

$$= \frac{e^{j\frac{k}{2f}\left(x_f^2 + y_f^2\right)}}{j\lambda f} \int\int\limits_{-\infty}^{\infty} t_o\left(x_l, y_l\right)e^{-j\frac{2\pi}{\lambda}\left(x_l x_f + y_l y_f\right)} dx_l dy_l,$$

(2.98)

which aside from the quadratic phase factor in front of the integral is a Fourier transform of the transmittance function of the object. It can further be shown [2] that by placing the object at the front focal plane of the lens, the phase factor in front of the integral can be eliminated as well leaving an exact Fourier transform of the object at the back focal plane of the lens. Therefore, the use of a lens is very helpful in performing Fourier transform operations within the confines of a laboratory.

One of the most useful configurations is a "4f" optical system shown in Figure 2.25. In this arrangement, an object is placed at the front focal plane of the first lens producing the Fourier transform of the object function in the back focal plane. Since the spectrum is in the front focal plane of the second lens, an exact Fourier transform is formed at the back focal plane of this lens. As will be shown in the next chapter, this arrangement is commonly used for hologram formation and reconstruction.

2.6.5 Diffraction by Apertures

It is now possible to compute the spatial diffraction patterns that result after light passes through different aperture shapes. Two of the more common shapes that occur in optical systems are the rectangular and circular apertures, and the diffraction patterns formed by these apertures are given in the following sections [2].

2.6.5.1 Rectangular Aperture

Consider a rectangular opening with dimensions a_x along the x_1 axis and a_y along the y_1-direction illuminated with a normally incident plane wave. The transmittance of the aperture can be represented by the product of two rect functions (see Equation A.9 in Appendix A) scaled by the dimensions a_x and a_y:

$$t_A\left(x_1, y_1\right) = rect\left(\frac{x_1}{a_x}\right) rect\left(\frac{y_1}{a_y}\right).$$

(2.99)

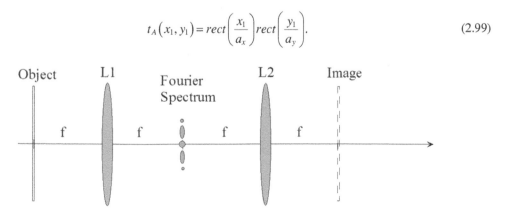

FIGURE 2.25 A 4f optical system with a transparent object placed in the front focal plane. The Fourier spectrum is formed at 2f and the inverse transform or image of the object is formed at 4f.

The far-field diffraction pattern can be found by using Equations 2.89 and A.10 resulting in:

$$E\left(x_2,y_2\right)=\frac{e^{jkz}\exp\left[j\dfrac{k}{2z}\left(x_2^2+y_2^2\right)\right]}{j\lambda z}a_xa_y\sin c\left(\frac{a_xx_2}{\lambda z}\right)\sin c\left(\frac{a_yy_2}{\lambda z}\right),\tag{2.100}$$

and the corresponding intensity distribution is found by:

$$I\left(x_2,y_2\right)=\left|E\left(x_2,y_2\right)\right|^2=\frac{\left(a_xa_y\right)^2}{\left(\lambda z\right)^2}\sin c^2\left(\frac{a_xx_2}{\lambda z}\right)\sin c^2\left(\frac{a_yy_2}{\lambda z}\right).\tag{2.101}$$

The intensity distribution has a peak value at $(x_2, y_2) = (0,0)$ and has side lobes with nulls when $a_xx_2/(\lambda z)$ or $a_yy_2/(\lambda z)$ are integers as shown in Figure 2.26.

2.6.5.2 Circular Aperture

The transmittance function of a circular aperture can be described with the circ function (see Appendix A, Equation A.12):

$$t_A\left(r_1\right)=circ(r_1/r_A),\tag{2.102}$$

where $r_1=\sqrt{x_1^2+y_1^2}$ in the aperture plane, and r_A is the radius of the circle. The far-field diffraction pattern can again be found using Equation 2.89, however, in this case, the Fourier Bessel transform (Equation A.13) is used with circular coordinates where:

$$B_F\left\{circ\left(\frac{r_1}{r_A}\right)\right\}=r_A^2\frac{J_1\left(\pi2r_A\rho_2\right)}{\pi r_A\rho_2}.\tag{2.103}$$

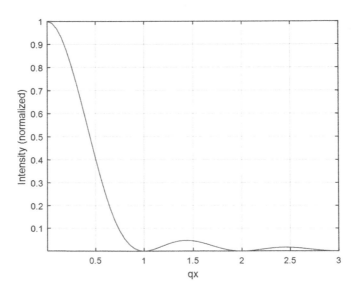

FIGURE 2.26 Normalized intensity for the far-field diffraction pattern from a rectangular aperture that is a_x on a side. $qx=a_xx_2/\lambda z$.

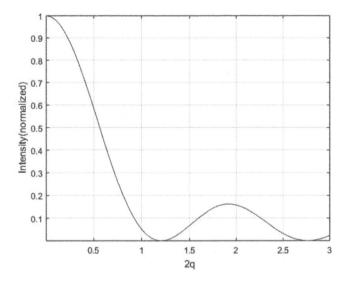

FIGURE 2.27 Normalized intensity of the far-field diffraction pattern from a circular aperture with radius r_A. $q = r_A r_2 / \lambda z$.

In this expression J_1 is a Bessel function of the first kind, $\rho_2 = \sqrt{\upsilon_{x2}^2 + \upsilon_{y2}^2}$ with $\upsilon_{x2} = x_2 / (\lambda z)$. and $\upsilon_{y2} = y_2 / (\lambda z)$ The resulting diffraction function for the circular aperture is:

$$E(r_2) = \frac{e^{jkz} \exp\left[j\dfrac{k}{2z}\left(r_2^2\right) \right]}{j\lambda z} \pi r_A^2 \frac{J_1\left(2\pi r_A r_2 / \lambda z\right)}{\pi r_A r_2 / \lambda z}, \tag{2.104}$$

and the corresponding intensity distribution is

$$I(r_2) = |E(r_2)|^2 = \left(\frac{A_A}{\lambda z}\right)^2 \left[2\frac{J_1\left(2\pi r_A r_2 / \lambda z\right)}{2\pi r_A r_2 / \lambda z} \right]^2, \tag{2.105}$$

This function has the familiar "Airy" pattern form as shown in Figure 2.27 with the width between the first nulls equal to:

$$\Delta r_2 = 1.22 \frac{\lambda z}{r_A}. \tag{2.106}$$

PROBLEMS

1. Consider a 100 W monochromatic point source. Determine the electric field amplitude at a distance 5 m from the source.
2. The average solar irradiance is $E = 970$ W/m². Determine the corresponding electric and magnetic field amplitudes by assuming that the average Poynting vector is equal to this irradiance.

3. Draw a sketch similar to Figure 2.4 illustrating the position of the polarization vector for the following waves:

$$\underline{E} = E_o v \left[\hat{x} \cos(\omega t - kz) + \hat{y} \cos\left(\omega t - kz + \frac{\pi}{4} \right) \right]$$

$$\underline{E} = E_o \left[\hat{x} \cos(\omega t - kz) + \hat{y} \cos\left(\omega t - kz + \frac{\pi}{2} \right) \right]$$

$$\underline{E} = E_o \left[\hat{x} \cos(\omega t - kz) + \hat{y} \cos\left(\omega t - kz - \frac{3\pi}{4} \right) \right].$$

4. Show that Equation 2.28 for elliptical polarization can be derived using Equations 2.25 through 2.27.

$$\left(\frac{E_x}{E_{ox}} \right)^2 + \left(\frac{E_y}{E_{oy}} \right)^2 - 2 \left(\frac{E_x}{E_{ox}} \right) \cdot \left(\frac{E_y}{E_{oy}} \right) \cos\varphi = \sin^2\varphi. \tag{2.28}$$

5. Silicon is transparent to light at 1.3 μm. Calculate the Fresnel power transmission coefficients for transverse electric (TE) and transverse magnetic (TM) fields at this wavelength incident at 20°, 40°, and 60° relative to the surface normal that passes through silicon ($n = 3.6$) with a layer of fused silica ($n = 1.46$) on the surface. The sample is in air.

6. A light beam with a flux density of 1 W/m^2 in air strikes a plane glass boundary ($n = 1.50$) at 45° relative to the normal. The beam is linearly polarized at 45° to the plane of incidence (i.e., equal TE and TM components). What fraction of the incident flux density is reflected at the boundary?

7. Consider a multi-layer dielectric as shown in Figure 2.28. A TE electric field is incident on the first surface at an angle of 32° relative to the surface normal.

   a. Compute the fraction of the incident light intensity that passes through slab n_1.

   b. Now assume that the light incident on medium n_1 is TM polarized. At what angle will the reflected light intensity go to zero?

8. TE polarized light is normally incident on the 45-90-45° glass prism at point A as shown in Figure 2.29. The prism has a refractive index of 2.25 and is surrounded by air.

   a. Where will the light emerge from the prism? Justify your answer.

   b. Determine the fraction of the incident light intensity that emerges from the prism at this location.

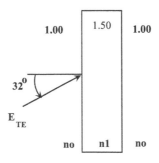

FIGURE 2.28   Problem 7.

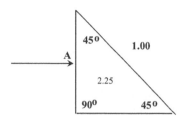

FIGURE 2.29 Problem 8.

9. A source emits an 8 pico-sec pulse of light at a wavelength of 532 nm. Assuming a rectangular spectral emission profile and that the spectral bandwidth of the source is determined by the full width half maximum of the spectrum:

   a. What is the coherence length of the source?

   b. What is the effective spectral bandwidth of the source?

10. A source with a nominal wavelength of 550 nm and a spectral bandwidth of 1.5 nm is 600 cm from a screen with two pinholes that are separated by 1 mm. If the spatial extent of the source and the diameter of the pinholes are <<1 mm, plot and label the interference pattern that forms on a second screen placed 100 cm from the screen with the pinholes. Label the important parameters in the figure.

11. A 1 cm high object is placed 2 cm from the focal point of a simple convex lens that has a focal length of 10 cm and diameter of 2 cm as shown in Figure 2.30. Determine:

   a. The image location using the thin lens imaging relations

   b. The image location and height using the paraxial ray trace relations

   c. The image magnification

   d. The f/# and numerical aperture of the lens.

12. A rectangular aperture with a width of 1 mm and a height of 3 mm is illuminated with normally incident plane wave light at a wavelength of 1.0 μm.

   a. At what distance from the aperture is the Fresnel approximation valid?

   b. At what distance from the aperture is the far-field approximation valid?

   c. The aperture is now placed against a lens with a focal length of 10 cm. Compute the intensity function in the back focal plane of the lens and sketch the pattern in the *x*- and *y*-directions showing the main features out to the second null.

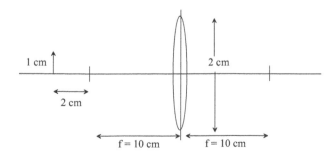

FIGURE 2.30 Problem 11.

REFERENCES

1. M. Born and E. Wolf, *Principles of Optics*, 5th ed., Pergamon Press, Oxford, UK (1975).
2. J. W. Goodman, *Introduction to Fourier Optics*, 4th ed., W. H. Freeman, New York (2017).
3. E. Hecht and A. Zajac, *Optics*, Addison Wesley, Menlo Park, CA (1974).
4. R. Guenther, *Modern Optics*, John Wiley & Sons, New York (1990).
5. M. Born and E. Wolf, *Principles of Optics*, p. 375, 5th ed., Pergamon Press, Oxford, UK (1975).
6. R. M. A. Azzam and N. M. Bashara, *Ellipsometry and Polarized Light,* North-Holland, New York (1977).
7. J. W. Goodman, *Statistical Optics*, Wiley Interscience, New York (1985).
8. R. J. Collier, C. B. Burckhardt, and L. H. Lin, *Optical Holography*, Academic Press, New York (1971).
9. P. H. Van Cittert, *Physica*, Vol. 1, 201–210 (1934).
10. F. Zernike, *Physica*, Vol. 5, 785–795 (1938).
11. D. C. O'Shea, *Elements of Modern Optical Design*, Wiley-Interscience, New York (1985).
12. W. J. Smith, *Modern Optical Engineering*, McGraw-Hill, New York (1966).
13. J. E. Greivenkamp, *Field Guide to: Geometrical Optics*, Vol. FG01, SPIE Press, Bellingham, WA (2004).
14. H. A. MacLeod, *Thin-Film Optical Filters*, 3rd ed., J. W. Arrowsmith, Bristol UK (2001).
15. R. W. Boyd, *Radiometry and the Detection of Optical Radiation*, John Wiley & Sons, New York (1983).

3

Introduction to the Basic Concepts of Holography

3.1 Introduction

This chapter introduces the basic principles, analytical methods, and terminology of holography. Concepts such as the holographic recording process, light scattering from a set of periodic scattering points and the grating equation, and Bragg matching are discussed. These ideas provide the framework for understanding the more in-depth discussion and analysis that follow in later chapters. Many of the hologram-recording geometries are introduced that have proven useful in many different applications. In addition, properties such as diffraction efficiency and dispersion are defined since they will be used throughout the remainder of the book to evaluate different holographic designs and optical elements.

3.2 Holographic Recording Process

The distinctive features of a holographic image result from recording both the optical phase and amplitude of the object field. As shown in Chapter 2, the optical field that satisfies the time independent wave equation (Equation 2.21) is:

$$E(x,y,z) = E_o \exp\left[j\left(\vec{k} \cdot \vec{r} + \varphi(x,y,z)\right) \right], \tag{2.21}$$

where $\varphi(x,y,z)$ is the phase. Typically, the phase cannot be measured directly and must be converted to an intensity distribution for detection with a camera, detector, or film recording material. Holography accomplishes this by adding a carrier or reference beam with the object beam and recording the resulting interference pattern. This is similar to coherent detection techniques that mix the detected signal with a local oscillator.

The holographic recording and reconstruction process consists of three steps: (1) superimposing the object and reference beams to form an interference pattern; (2) exposing a recording material to the interference pattern and convert it to a physical holographic grating; and (3) reconstructing the holographic image.

3.2.1 Step 1: Superimposing the Object and Reference Beams

For this step, let the reference and object fields be represented by the complex field functions:

$$\tilde{r} = a_r \exp\left[j\left(\vec{k}_r \cdot \vec{r} + \varphi_r(x,y,z)\right) \right]$$

$$\tilde{o} = a_o \exp\left[j\left(\vec{k}_o \cdot \vec{r} + \varphi_o(x,y,z)\right) \right], \tag{3.1}$$

where a_r and a_o are real amplitude constants, $\vec{r}$ is a position vector with respect to a coordinate system, $\vec{k}_r$ and $\vec{k}_o$ are propagation vectors, and φ_r and φ_o are the phase of the reference and object beams, respectively. Figure 3.1a shows the fringes formed by the interference between the object and reference beams assuming that these fields are plane waves. The intensity is found by taking the coherent sum of $\tilde{r}$ and $\tilde{o}$, and then squaring the result:

$$I = \left|\tilde{r} + \tilde{o}\right|^2$$
$$= \left|\tilde{r}\right|^2 + \left|\tilde{o}\right|^2 + \tilde{r}\tilde{o}* + \tilde{o}\tilde{r}*.$$

(3.2)

After substituting for the complex field functions from Equation 3.1, the intensity function for the interfering fields can be written as:

$$I = a_r^2 + a_o^2 + a_r a_o e^{j\left((\vec{k}_r - \vec{k}_o)\cdot\vec{r} + \varphi_r - \varphi_o\right)} + a_r a_o e^{-j\left((\vec{k}_r - \vec{k}_o)\cdot\vec{r} + \varphi_r - \varphi_o\right)}$$
$$= a_r^2 + a_o^2 + 2a_r a_o \cos\left((\vec{k}_r - \vec{k}_o)\cdot\vec{r} + \varphi_r - \varphi_o\right).$$

(3.3)

This expression shows that the intensity contains the phase difference between the reference and object fields $\varphi_r - \varphi_o$ and is a function of position.

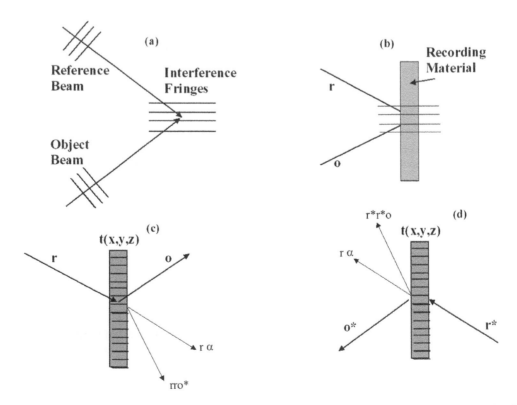

FIGURE 3.1 Diagram showing the main steps in constructing and reconstructing a holographic image: (a) forming an interference pattern; (b) exposing and forming a physical hologram; (c) reconstructing the original object wave; and (d) reconstructing the conjugate of the object wave.

3.2.2 Step 2: Expose the Recording Material and Convert to a Physical Holographic Grating

Figure 3.1b shows this step which is accomplished by placing a photosensitive material in the region where interference fringes occur. After processing, the hologram can be considered to be a transmittance function that is proportional to the intensity of the interference pattern:

$$t = \alpha I$$

$$= \alpha \left(\left| \tilde{r} \right|^2 + \left| \tilde{o} \right|^2 \right) + \alpha \left(\tilde{r}\tilde{o}* + \tilde{o}\tilde{r}* \right). \tag{3.4}$$

The first term represents a constant bias, and the second term is a modulation response of the material to the illumination and processing.

3.2.3 Step 3: Reconstructing the Holographic Image

The final step is the reconstruction of the image. This can be accomplished either with the original reference beam $\tilde{r}$ or with the conjugate of the reference beam $\tilde{r}^*$. When illuminated with the original reference beam (Figure 3.1c), the reconstructed field $\tilde{p}_r$ becomes:

$$\tilde{p}_r = t(x, y, z)\tilde{r}$$

$$= \tilde{r}\alpha \left(\left| \tilde{r} \right|^2 + \left| \tilde{o} \right|^2 \right) + \alpha \tilde{r}\tilde{r}\tilde{o}* + \alpha \tilde{r}\tilde{r}*\tilde{o}. \tag{3.5}$$

The first term is a constant bias, and the second term has a combined phase of $2\varphi_r - \varphi_o$. In the third term, the phase of the conjugate reference beam recorded in the hologram cancels the phase of the reconstruction beam. This leaves only the phase of the original object field and results in the reconstruction of the image of the object.

A useful reconstruction can also be achieved using the conjugate of the original reference beam. The conjugate reference is a beam that counter propagates with respect to the original reference beam and is designated as $\tilde{r}^*$ as shown in Figure 3.1d. In this case, the reconstructed field becomes:

$$\tilde{p}_{r^*} = t(x, y, z)\tilde{r}*$$

$$= \tilde{r}*\alpha \left(\left| \tilde{r} \right|^2 + \left| \tilde{o} \right|^2 \right) + \alpha \left(\tilde{r}\tilde{r}* \right)\tilde{o}* + \alpha \tilde{r}*\tilde{r}*\tilde{o}. \tag{3.6}$$

The first term is again a constant bias without modulation, and the last term has a combined phase of $(-2\varphi_r^* + \varphi_o)$. The second term, however, shows that the phase of the reference beam cancels leaving the phase of the conjugate of the object beam, φ_o^*. This results in an image that is the conjugate of the original object beam.

One application that uses the conjugate reconstruction is a holographic lens. Consider the recording of a point source (object beam) in combination with a plane wave (reference beam) as shown in Figure 3.2. When the resulting hologram is reconstructed with the original plane wave a virtual image is formed. However, when the reference wave direction is reversed to form a conjugate beam (r^*) a real image corresponding to o^* results. The resulting hologram acts like a lens that focuses incoming collimated light to a point. The holographic lens and its imaging properties will be described in more detail in Chapter 4 and also in Chapter 11 on holographic optical elements.

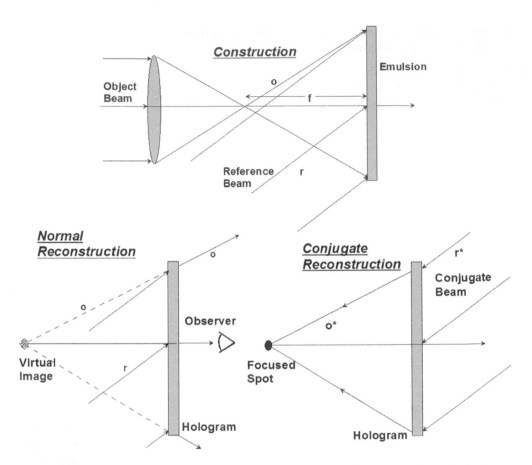

FIGURE 3.2   Top figure shows the formation of a holographic lens with a point source and an off-axis plane wave. The lower left figure shows reconstruction with the original object wave and the lower right figure with a conjugate wave.

3.3 Scattering from a Periodic Array of Scattering Points and the Grating Equation

The diffraction of an incident field by a grating also illustrates how phase properties can be embedded into an optical component. To illustrate this, consider a set of periodically spaced scattering points with period (Λ) arranged along the x-axis as shown in Figure 3.3. The array is illuminated with a coherent plane wave incident at an angle θ_i relative to the z-axis and with a free-space wavelength λ. In general, the refractive index on the incident side of the grating (n_1) can be different from the index on the diffracted side (n_2). Now consider two rays, incident at an angle θ_i intersecting adjacent scattering points at positions A and D on the grating as shown in the figure. The difference in optical path length for the upper and lower rays is equal to:

$$\Delta l_{opl} = n_2 AB - n_1 CD. \tag{3.7}$$

Whenever Δl_{opl} is equal to an integral number of wavelengths, the fields scattered from different sites *add in phase* and increase the intensity of the scattered field. This condition is typically written in the form:

$$n_2 AB - n_1 CD = q\lambda$$
$$\Lambda n_2 \sin\theta_{d,q} - \Lambda n_1 \sin\theta_i = q\lambda , \tag{3.8}$$

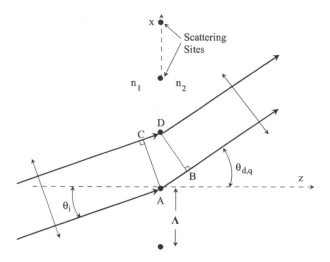

FIGURE 3.3 Light scattering by a periodic set of points placed along the *x*-axis.

where q is the diffraction order $(0, \pm 1, \pm 2,...)$. If the refractive indices on both sides of the grating are equal to n, the expression simplifies to the standard form of the *grating equation*:

$$\sin\theta_{d,q} - \sin\theta_i = q\frac{\lambda/n}{\Lambda}. \tag{3.9}$$

This is a very simple, but fundamental equation for explaining the direction of enhanced scattering or diffraction from a holographic grating.

It should be noted that a sign convention is used in this equation is such that counter clockwise angles relative to the z-axis are positive and clockwise angles are negative. In addition, a positive Δl_{opt} occurs when $q > 0$, and negative orders occur when $\Delta l_{opt} < 0$. As a result of this convention *positive (+) diffraction orders* appear counter clockwise to the zero order (ray that is not diffracted), and *negative (–) diffraction orders* occur clockwise to the zero order ray. When the illumination is at normal incidence to the grating surface as shown in Figure 3.4, the positive and negative diffraction

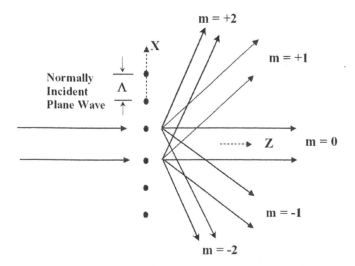

FIGURE 3.4 Diffraction of a normally incident plane wave by a set of point scattering sites.

orders emerge symmetrically about the z-axis which also coincides with the zero order. For the normally incident reconstruction, the grating equation simplifies to:

$$\sin\theta_{d,q} = q\frac{\lambda/n}{\Lambda},$$
(3.10)

and the largest propagating diffraction order occurs when $\sin\theta_{d,q} = 1$ or

$$q_{max} = \frac{\Lambda}{\lambda/n}.$$
(3.11)

Non-propagating or evanescent orders are also possible and will be discussed later in the chapter on coupled wave analysis.

3.4 Hologram Terminology

A variety of terms are useful in describing different types of holograms and their characteristics. In this section, the basic terminology that is frequently used in holography is reviewed.

3.4.1 Diffraction Efficiency

This term indicates the amount of optical power diffracted into a particular order relative to the power incident on the hologram and is equal to:

$$\eta_q = \frac{P_q}{P_{inc}},$$
(3.12)

where P_q is the optical power diffracted by the qth order and P_{inc} is the power incident on the hologram. This expression for the diffraction efficiency indicates the true efficiency of the hologram. However, in some instances, the diffraction efficiency is taken as the diffracted power in a particular diffracted order relative to the power emerging in all orders including the zero diffraction order. This latter form is sometimes used for modeling the performance of holograms with significant light absorption.

3.4.2 Linear, Computer Generated, and Digital Holography Recording

The type of material used to record the holographic grating has a significant impact on the efficiency of the reconstructed image. As mentioned previously a holographic grating consists of a periodic set of scattering sites that perturb the reconstruction field. If the scattering sites modify the conductivity, then part of the field will be absorbed as it passes through the hologram and forms an *absorption type* hologram. A grating can also be formed by modulating the refractive index, the thickness, or a combination of both factors to form a *phase grating*. Since an absorption type hologram reduces the power in the optical field, it introduces loss into the reconstruction process. On the other hand, a pure phase hologram can transfer all of the incident beam power into the diffracted beam and produces a more efficient hologram. Therefore, for many applications such as holographic optical elements, a phase type hologram is more desirable.

Absorption and phase holograms can be formed by optically exposing recording materials such as photographic film and photopolymers. However, it is also possible to form holograms using lithographic methods to produce digital encoding of phase and amplitude information at different points in aperture. These types of elements are called *computer-generated holograms* and can be used to form images of objects that are not realizable with conventional optical components. More discussion on computer-generated holograms is given in Chapter 6.

It is also possible to use a digital camera to record a holographic interference pattern. The information about the pattern is then stored in a computer and an image can be reconstructed mathematically. This process is referred to as a *digital holography* and is very useful for interferometry and microscopy. The topic of digital holography is covered in Chapter 7 in more detail and is also discussed in the applications sections later in the book.

3.4.3 Thin and Thick (Volume) Holographic Gratings

The grating thickness has a significant effect on the diffraction properties of the hologram. If the reconstruction field only interacts with a planar hologram as in a simple transmittance function, then the grating can be considered to be "thin." In this case, the following conditions generally hold true:

$$\Lambda \gg \lambda;$$
$$\Lambda \gg d, \tag{3.13}$$

where Λ is the grating period, d is the thickness, and λ is the wavelength. An illustration of these parameters for the thin grating case is shown in Figure 3.5.

In contrast, a field propagating through a "thick" or "volume" hologram continually interacts with it as it moves through the grating. This interaction allows the power of the diffracted field to increase by exchanging power with the reconstruction beam as it moves though the thickness of the grating. The conditions for a volume grating can be summarized as:

$$\Lambda \sim \lambda;$$
$$\Lambda \ll d; \tag{3.14}$$
$$d \gg \lambda,$$

and are illustrated in Figure 3.5. The volume grating can also be thought of as a cascade of thin gratings with each thin grating scattering light within the total thickness as shown in Figure 3.6. At each thin grating section, multiple diffraction orders can be formed, however, subsequent scattering within the thickness prevents enhancement of these orders unless the order satisfies the phase matching condition

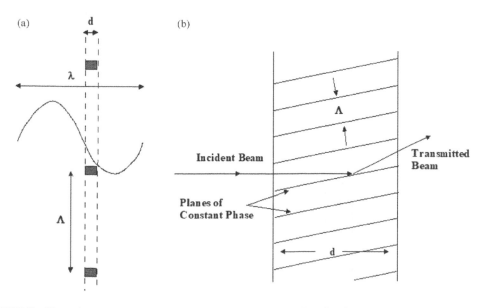

FIGURE 3.5 Illustration of a "thin" grating (a) and a "thick" or "volume" grating (b).

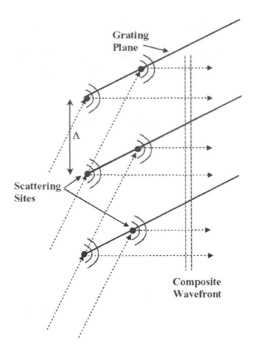

FIGURE 3.6   Scattering of light from grating planes within the thickness of the volume hologram.

throughout the entire thickness. However, when this matching condition is satisfied, nearly all the power in the incident beam can be transferred to the diffracted beam leading to very high diffraction efficiencies. This condition will be examined more thoroughly in the later chapter on coupled wave analysis and is very important for holographic optical elements and other components requiring high diffraction efficiency.

3.4.4 Transmission and Reflection Type Holograms

Holograms can also be categorized according to whether the diffracted light is directed through the hologram (transmitted) or back toward the incident beam (reflected). Figure 3.7 illustrates the orientation of the grating planes for an "unslanted" transmission and reflection hologram. For the transmission hologram, the grating planes are perpendicular to the grating surface (along the z-axis) and for a reflection hologram they are parallel to the surface (x-axis).

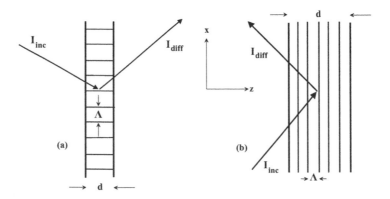

FIGURE 3.7   Illustration of an unslanted (a) transmission and (b) reflection hologram.

3.5 Hologram Geometries

Since the invention of holography, several useful geometrical configurations have been developed to overcome shortcomings in the optics or recording materials to optimize the recording arrangement for a specific application. This section gives a brief overview of several different hologram-recording geometries that have been found useful. Additional hologram geometries are presented in the later chapters on hologram applications.

3.5.1 "In-Line" (Gabor Type) Holograms

In 1948, when the concept of holography was first demonstrated by Gabor [1], optical sources with long coherence lengths and sufficient power did not exist. As a result, the first holograms were made using an "in-line" geometry as shown in Figure 3.8. In this approach the "object" is a transparency, $t(x,y)$, located a distance s_o from the recording material and is illuminated with a collimated beam from the optical source. Part of the incident light is scattered by the transparency and forms the object beam. The remaining light continues to propagate and serves as the reference beam for the holographic interference that exposes the film. After processing, the hologram is illuminated with the reconstruction beam as shown in Figure 3.9. The reconstruction simultaneously produces the original object beam o and the conjugate object beam o^*. An observer would see both of these fields and the reference beam which reduces the contrast of the reconstructed image.

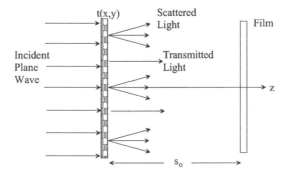

FIGURE 3.8 In-line (Gabor-type) hologram construction with an object transparency $t(x,y)$ illuminated with a plane wave.

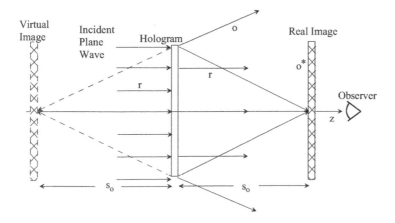

FIGURE 3.9 Reconstruction of an in-line hologram forming both the original object wave o, and the conjugate wave o^*.

3.5.2 "Off-Axis" Hologram

In order to eliminate the degradation caused by the overlapping fields for the Gabor type hologram, Leith and Upatnieks developed the "off-axis" geometry [2]. Their original experiments were conducted using a specially filtered mercury discharge lamp that provided adequate coherence length and power for the recording. With the advent of the laser, the off-axis configuration became easier to implement due to the longer coherence length and higher power of these sources. One configuration for constructing an off-axis hologram is shown in Figure 3.10. In this arrangement, the object is illuminated and scatters light that propagates to the recording material (film). In this case, the reference beam is an off-axis plane wave that makes an angle of θ_r with the optical axis. When the hologram is reconstructed as shown in Figure 3.11, a virtual image is formed as in the in-line hologram, however, the reference beam angle can be chosen so that during reconstruction it does not overlap with the object beam and leads to a higher contrast image. Figure 3.2 shown earlier is also an "off-axis" hologram with a plane wave reference beam at an angle to the optical axis and the object beam an on-axis point source formed by focusing an axial plane wave. As shown in Figure 3.2, when reconstructed with the conjugate reference beam, r^* light is focused to a point. When the hologram in Figure 3.11 is reconstructed with the conjugate reference beam, an extended real image is formed that can be projected on a screen for display.

3.5.3 Fourier Transform Hologram

The Fourier transform hologram, as first proposed by Vander Lugt [3,4], is very useful for applications such as holographic data storage. This is due to the ability to record most of the spatial frequency content of the object in a small region of the hologram. The Fourier transform hologram is recorded using an arrangement as shown in Figure 3.12. A transparent object is placed at the front focal plane of a lens and the recording material at the back focal plane. The object is illuminated with a plane wave propagating along the optical axis and forms the spatial Fourier transform of the object at the recording material plane. In addition, unit amplitude planar reference wave illuminates the recording material at an angle θ_r with the z-axis to form the exposing interference pattern.

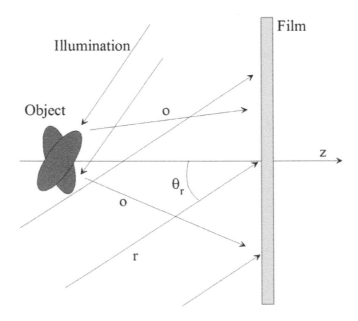

FIGURE 3.10   Construction of an off-axis hologram. The reference beam (r) is a plane wave incident at an angle θ_r.

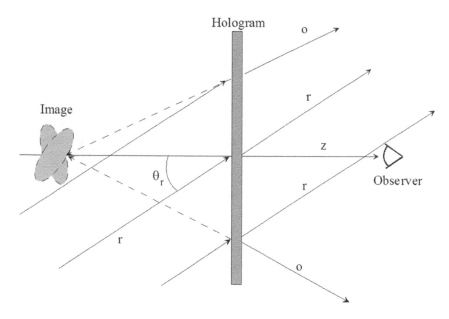

FIGURE 3.11 Reconstruction of an off-axis hologram and the formation of a virtual image.

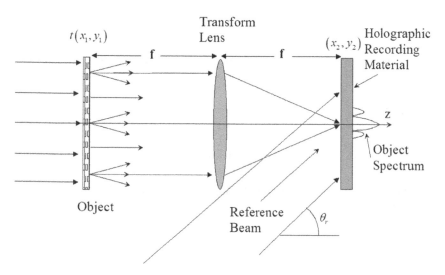

FIGURE 3.12 Arrangement for recording a Fourier transform hologram.

The recording beams at the film plane can be written as:

$$R(x_2, y_2) = \exp(jk_{rx}x_2)$$
$$O(x_2, y_2) = FT\{o(x_1, y_1)\},$$

(3.15)

where $k_{rx} = 2\pi \sin\theta_r / \lambda$, and the corresponding intensity at the recording plane (x_2, y_2) is:

$$I(x_2, y_2) = |R(x_2, y_2) + O(x_2, y_2)|^2$$
$$= 1 + |O(x_2, y_2)|^2 + O(x_2, y_2)e^{-jk_{rx}x_2} + O^*(x_2, y_2)e^{jk_{rx}x_2}.$$

(3.16)

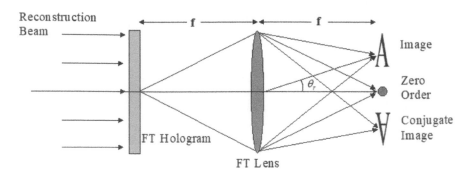

FIGURE 3.13   Reconstruction of a Fourier transform hologram.

After exposing and processing the film, the hologram is illuminated with an on-axis reconstruction beam. The field immediately past the hologram plane is:

$$w\left(x_2, y_2\right) \propto R \cdot I\left(x_2, y_2\right)$$

$$= R[1 + \left|O\left(x_2, y_2\right)\right|^2 + O\left(x_2, y_2\right)\exp\left(-jk_{rx}x_2\right) \tag{3.17}$$

$$+ O^*\left(x_2, y_2\right)\exp\left(jk_{rx}x_2\right)].$$

If the hologram is placed in the front focal plane of a lens, an image forms on the back focal plane as shown in Figure 3.13. The reconstructed field at the back focal plane of the lens is the Fourier transform of w:

$$E_f\left(x_2, y_2\right) \propto FT\left\{w\left(x_2, y_2\right)\right\}$$

$$= \left\{\delta\left(x, y\right) + o\left(x, y\right) \oplus o\left(x, y\right) + o\left(x - a, y\right) + o^*\left(-x + a, -y\right)\right\}, \tag{3.18}$$

where $\oplus$ represents the autocorrelation operation (Appendix A, Equation A.7). Use was also made of the transform relations:

$$FT\left\{O\left(x_2, y_2\right)e^{-jk_{rx}x_2}\right\} = o\left(x, y\right) \otimes \delta\left(x - a\right)$$

$$= o\left(x - a, y\right), \tag{3.19}$$

with $a = f \tan\theta_r \approx f \sin\theta_r$, and $\otimes$ being the convolution operation (Appendix A, Equation A.6). The resulting image consists of the δ and autocorrelation functions along the optical axis, a displaced image of the object, and a displaced image of the conjugate image of the object as shown in Figure 3.13.

3.5.4 Fraunhofer Hologram

Another hologram configuration that is useful for microscopy applications is the "far-field" or Fraunhofer geometry [5]. In this arrangement, one or more particles are illuminated with a reference beam as shown in Figure 3.14. Part of the reference beam is scattered by the particles becoming the object beam, and the remainder serves as the reference beam for exposing the film. The resulting situation is similar to the "in-line" or "Gabor" type hologram. However, the difference is that the object must be at a sufficient distance from the film plane so as to satisfy the "far-field" diffraction limit described in Chapter 2 and is restated here as:

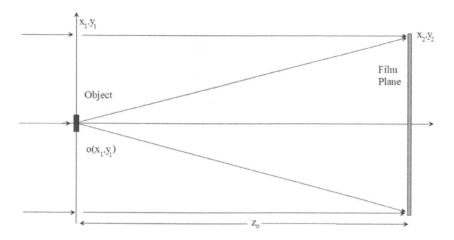

FIGURE 3.14 Geometry for forming a "Fraunhofer" hologram.

$$\frac{\left(x_1^2 + y_1^2\right)_{max}}{\lambda} \ll z_o, \tag{3.20}$$

where the subscript *max* indicates that the maximum size of the object is used to estimate the far-field distance.

Assuming that z_o satisfies the far-field condition and that $o(x_1, y_1)$ is the object field distribution just past the object, then the field distribution after propagating to the film plane becomes:

$$E_o\left(x_2, y_2\right) = \frac{c}{j\lambda z_o} \exp\left[\frac{j\pi}{\lambda z_o}\left(x_2^2 + y_2^2\right)\right] O\left(v_x, v_y\right), \tag{3.21}$$

where c is a constant and $O(v_x, v_y) = FT\{o(x_1, y_1)\}$, with $v_x = x_2/\lambda z_o, v_y = y_2/\lambda z_o$. The field at the film plane is essentially the Fourier transform of the object function times a quadratic phase function. If the particle is considered truly "small" such that $o(x_1, y_1) \rightarrow \delta(x_1, y_1)$, then

$$O(v_x, v_y) \rightarrow FT\{\delta(x_1, y_1)\} = 1, \text{ and in this case:}$$

$$E_o\left(x_2, y_2\right) = \frac{c}{j\lambda z_o} \exp\left[\frac{j\pi}{\lambda z_o}\left(x_2^2 + y_2^2\right)\right]. \tag{3.22}$$

The resulting field is a quadratic phase approximation to a spherical wave coming from the object point. After the film is exposed, processed, and illuminated with a planar reconstruction beam, a virtual image of the "point" object forms. When recorded with a pulsed laser, a "snapshot" of a microscopic scene is recorded that can later be observed in detail. Although this geometry was originally used in the early days of holography with fixed holographic recordings in photographic film, it is currently finding increasing use in using digital holographic techniques.

3.5.5 Hologram Geometry Diagram

A relative perspective of the different recording geometries can be obtained using the diagram in Figure 3.15 [6]. It shows the relative positioning of the recording material to form an in-line, off-axis, Fourier transform, and reflection hologram using the same object and reference beams. Note that the in-line, off-axis, and Fourier transform holograms are transmission type recordings.

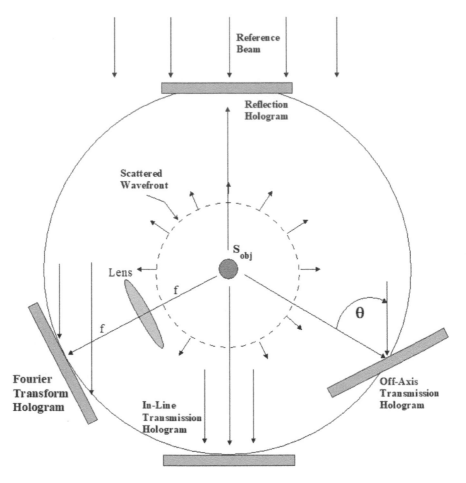

FIGURE 3.15 A hologram geometry diagram showing the formation of different types of hologram by inserting the recording material at different locations with respect to the reference and object beams. (From Collier, R.J. et al., *Optical Holography*, Academic Press, New York, 1971.)

3.6 Plane Wave Analysis of Holograms

Plane wave analysis is a very useful tool for examining the characteristics of interference fringe patterns and their resultant holographic grating structures. Using this construct, it is possible to evaluate even very complex interference patterns by restricting the observation region [7].

For this approach, assume that the reference and object waves are unit amplitude plane waves such that:

$$\tilde{r} = e^{-j\vec{k}_1 \cdot \vec{r}}$$

$$\tilde{o} = e^{-j\vec{k}_2 \cdot \vec{r}},$$

(3.23)

with $\left| \vec{k}_i \right| = \frac{2\pi}{\lambda_i}, i = 1, 2.$

For simplicity, assume that the refractive index is 1.00, and the wavelength for both fields is the same ($\lambda_1 = \lambda_2$). Therefore, the main difference between the two propagation vectors is their direction. An illustration of the interference pattern between two plane waves and the resulting interference

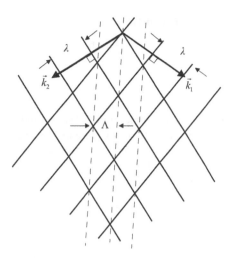

FIGURE 3.16 Interference between two plane waves showing propagation vectors and the resulting interference fringes.

pattern is shown in Figure 3.16. The propagation vectors $\vec{k}_1$ and $\vec{k}_2$ are perpendicular to the wavefronts. The interference fringes form along the bisector of the angle between the two propagation vectors and have a period Λ.

An expression for two interfering plane waves can be written as:

$$I = \left|E_1 + E_2\right|^2 = \left|E_1\right|^2 + \left|E_2\right|^2 + E_1 E_2^* + E_1^* E_2$$

$$= 1 + 1 + \exp\left[-j\left(\vec{k}_1 - \vec{k}_2\right)\cdot\vec{r}\right] + \exp\left[j\left(\vec{k}_1 - \vec{k}_2\right)\cdot\vec{r}\right] \tag{3.24}$$

$$= 2 + 2\cos\left[\left(\vec{k}_1 - \vec{k}_2\right)\cdot\vec{r}\right],$$

and shows that when the directional difference between the propagation vectors is large, the modulation term has a smaller period and conversely.

3.6.1 Grating Vector

A very useful parameter for evaluating the properties of gratings is the *grating vector* $\vec{K}$. It indicates the direction and spacing of the grating planes that result from the interference fringes that expose the recording material to form a hologram. This parameter has its roots in crystallography where it represents the "K-space" orientation and spacing of crystallographic planes. The magnitude of the grating vector is $\left|\vec{K}\right| = \frac{2\pi}{\Lambda}$, and the vector is perpendicular to the interference fringes as shown in Figure 3.17.

Interference Fringes

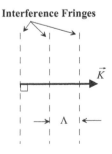

FIGURE 3.17 Relation between the interference fringes used to expose a recording material and form a hologram and the grating vector.

3.6.2 *K*-Vector Closure or Bragg Condition

The grating *K*-vector (magnitude and phase) is equal to the difference between the two propagation vectors that form the holographic grating:

$$\vec{K} = \vec{k}_1 - \vec{k}_2. \tag{3.25}$$

This relation is the Bragg condition for a volume grating [8] and indicates the reconstruction parameters that result in the maximum diffraction efficiency (Equation 3.12). In addition, as will be shown in Chapter 4 on imaging, it also represents the complete vector form of the grating equation (Equation 3.9). Computing the grating vector is one of the first steps required for analyzing hologram performance. The following is an example of the calculation of the grating vector for a reflection hologram. Other examples can be found in the exercises at the end of the chapter.

3.6.3 Reflection Hologram Example

As discussed previously, a reflection hologram has grating planes that are parallel or nearly parallel to the grating surface. Figure 3.18 shows grating vector ($\vec{K}$) for an "unslanted" reflection grating and the propagation vectors ($\vec{k}_1$ and $\vec{k}_2$) for constructing the hologram.

Typically, the magnitudes of the propagation vectors are the same since the same wavelength is used for both construction beams. Therefore, the magnitude of each propagation vector is:

$$k = \left|\vec{k}\right| = \frac{2\pi}{\lambda/n}. \tag{3.26}$$

The directions of the propagation vectors are defined by the angles θ_1 and θ_2 relative to the $+z$-direction. The resulting propagation vectors are:

$$\vec{k}_1 = k\sin\theta_1\,\hat{x} + k\cos\theta_1\,\hat{z}$$
$$\vec{k}_2 = k\sin\theta_2\,\hat{x} + k\cos\theta_2\,\hat{z}. \tag{3.27}$$

Using the Bragg condition, the grating vector $\vec{K}$ is found by combining x and z components of the propagation vectors:

$$\vec{K} = \vec{k}_1 - \vec{k}_2$$
$$= k\left[\left(\sin\theta_1 - \sin\theta_2\right)\hat{x} + \left(\cos\theta_1 - \cos\theta_2\right)\hat{z}\right] \tag{3.28}$$
$$= \frac{2\pi n}{\lambda}\left[\left(k_{1x} - k_{2x}\right)\hat{x} + \left(k_{1z} - k_{2z}\right)\hat{z}\right]$$

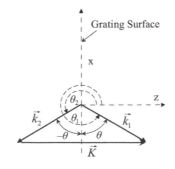

FIGURE 3.18 Construction of an unslanted reflection grating and the resultant grating *K*-vector.

For the case of an unslanted reflection grating, each propagation vector has an angle θ with respect to the x-axis as shown in Figure 3.18 and the $\underline{K}$ vector simplifies to:

$$\vec{K} = k2\cos\theta\hat{z}$$

$$= \frac{2\pi}{\Lambda}\hat{z} \qquad (3.29)$$

$$\frac{\lambda}{\Lambda} = 2n\cos\theta.$$

The resulting scalar expression is often stated as the Bragg condition, but is for the specific case of an unslanted grating [6,9]. The vector form given in Equation 3.17 is more general and can be used for any grating geometry.

3.6.4 Bragg Circle Diagram

The *Bragg circle diagram* [8] is a useful construction for evaluating grating characteristics (Figure 3.19). It is really just a graphical representation of the vector relation for the Bragg condition (Equation 3.25), but can be insightful when designing holograms for different applications. For example, it is often desired to form a grating at a wavelength where the recording material is sensitive and to reconstruct the grating at another wavelength. For instance, many holographic recording materials are sensitive at argon ion laser wavelengths (488 and 514.5 nm), however, it may be desired to use the grating at a telecommunication wavelength (1530–1550 nm). Another example are fiber Bragg gratings where the glass is sensitized for exposure in the ultra violet (UV) (244 nm), but is used at longer wavelengths (1530–1550 nm). The Bragg diagram can be very useful for evaluating these cases.

The radius of the Bragg circle is equal to the magnitude of the propagation vectors of the construction beams in the material $k_{1,i} = \frac{2\pi n_i}{\lambda_1}$; ($i = 1,2$). In general, the wavelengths of the two construction beams are the same as are the corresponding magnitudes of the propagation vectors. The resulting grating vector $\vec{K}$ connects the ends of the two formation vectors on the circle.

Once the hologram is recorded, the grating vector remains fixed. However, the reconstruction can take place with different wavelengths and will change the length of the propagation vectors. As a result, for Bragg matching, the orientation or angles of the propagation vectors must change. The new directions can readily be found with the Bragg circle by drawing a line perpendicular to the grating vector at the

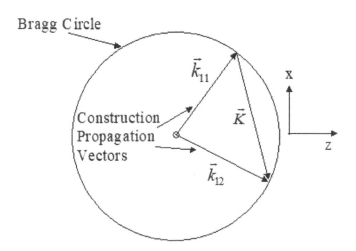

FIGURE 3.19 Bragg circle diagram showing the propagation vectors $\vec{k}_{11}$ and $\vec{k}_{12}$ and the resulting grating vector $\vec{K}$.

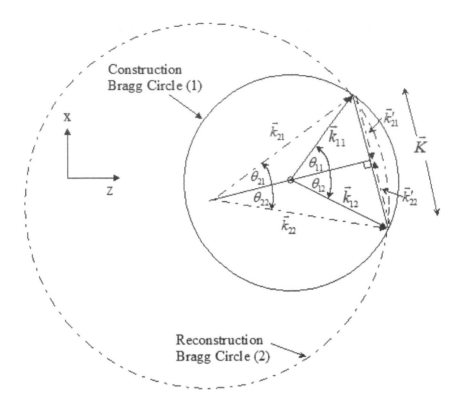

FIGURE 3.20   Bragg circle showing Bragg matched reconstruction at different wavelengths and propagation vectors.

midpoint as shown in Figure 3.20. In this Figure, the original construction vectors are $\vec{k}_{11}$ and $\vec{k}_{12}$, and the Bragg matched reconstruction propagation vectors are $\vec{k}_{21}$ and $\vec{k}_{22}$. A separate Bragg circle with radius equal to $2\pi n/\lambda_{i=1,2}$ is drawn for each wavelength with the circle intercepting the grating vector as shown. The directions for the reconstruction propagation vectors are then found by connecting a point from the center of circle 2 to the two ends of the grating vector. Notice as well that the shortest reconstruction vectors that are Bragg matched, $\vec{k}'_{21}$ and $\vec{k}'_{22}$, correspond to the longest wavelength that will produce maximum diffraction efficiency for a particular grating.

One might ask what happens when the vector relation Equation 3.25 is not satisfied? In this case, the diffraction efficiency of the volume grating decreases from the maximum value. For instance, if the two vectors $\vec{k}'_{21}$ and $\vec{k}'_{22}$ continue to decrease (i.e., wavelength increases), in Figure 3.20, the diffraction efficiency will continue to drop to a very low value. This situation is very similar to what happens when photons with different energies illuminate a semiconductor material with an energy bandgap as shown in Figure 3.21. Photons with sufficient energy are absorbed while those with lower energy pass through the material. The photons with higher energy correspond to propagation vectors that satisfy the Bragg matching condition (Equation 3.25), while those with lower energy that are

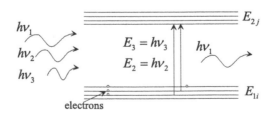

FIGURE 3.21   Energy bandgap for a semiconductor illuminated with photons with different energies. Only those photons with sufficient energy (E_2, and E_3) are absorbed. The remaining photons pass through the material.

not absorbed correspond to propagation vectors that are too short, do not satisfy the Bragg condition, and have lower diffraction efficiency. In this sense a volume grating is similar to a one-dimensional "photonic bandgap" device [10].

3.7 Dispersion of Thin Gratings

Examination of the *grating equation* (Equation 3.9) shows that there is a connection between the diffraction angle and the wavelength of the diffracted beam. This effect is illustrated in Figure 3.22 and is called dispersion. A dispersion relation or dispersion factor can be used to quantify the change in angle of a ray emerging from a grating as a function of the change in wavelength:

$$DF = \frac{\Delta\theta}{\Delta\lambda}. \tag{3.30}$$

To derive the dispersion factor (DF) for a grating, consider the grating equation (Equation 3.9) with a fixed reconstruction angle, $\theta_{r=i}$ and a small difference between the reconstruction and construction wavelengths, $\Delta\lambda$. This results in a shift in the diffraction angle given by:

$$\sin(\theta_d + \Delta\theta) - \sin\theta_i = \frac{\lambda + \Delta\lambda}{\Lambda}. \tag{3.31}$$

Note that in this case, it is assumed that the refractive index surrounding the grating is $n = 1.00$, only the $m = 1$ diffraction order is considered, and $\Delta\theta \ll \theta_{i,d}$. Expanding the sine functions in Equation 3.31 results in:

$$\begin{aligned}
\sin\theta_d \cos\Delta\theta - \sin\Delta\theta \cos\theta_d - \sin\theta_i &= \frac{\lambda}{\Lambda} + \frac{\Delta\lambda}{\Lambda} \\
\sin\theta_d - \sin\Delta\theta \cos\theta_d - \sin\theta_i &\approx \frac{\lambda}{\Lambda} + \frac{\Delta\lambda}{\Lambda} \\
\sin\Delta\theta \cos\theta_d &\approx \frac{\Delta\lambda}{\Lambda} \\
\Delta\theta \cos\theta_d &\approx \frac{\Delta\lambda}{\Lambda} \\
\frac{\Delta\theta}{\Delta\lambda} &\approx \frac{1}{\cos\theta_d\Lambda},
\end{aligned} \tag{3.32}$$

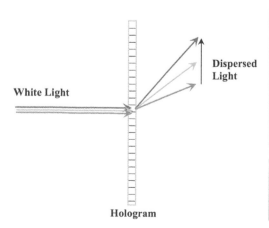

FIGURE 3.22 (See color insert.) A schematic and experimental illustration of dispersion from a holographic grating.

where simplifications for small angle approximations have been made. The result shows that dispersion is inversely proportional to the grating period Λ and is a direct result of the phase matching condition inherent in the grating equation.

3.7.1 Example of a Spectrometer with a Holographic Grating

The dispersive effect of a hologram can be used in a spectrometer to resolve different spectral components of an incident optical beam. As an example, consider a spectrometer consisting of a holographic grating and a lens as shown in Figure 3.23. There are two factors that determine the spectral resolving power of this spectrometer: (1) the diffraction limited spot diameter formed by the lens and (2) the dispersion factor for the holographic grating. The width of the diffraction limited spot produced by the lens can be approximated by using the distance between the first two nulls of the Airy pattern that is formed on the back focal plane of the lens (Chapter 2). This value is equal to:

$$2\Delta x_d = 2.44 \frac{\lambda}{D} f, \tag{3.33}$$

where Δx_d is the half width of the Airy pattern, λ is the nominal wavelength, D is the diameter of the lens, and f is the focal length of the lens. The second value of importance is the separation distance between two wavelengths (λ_1, λ_2) at the back focal plane of the lens that results from the dispersion produced by the hologram. This distance is given by:

$$\Delta x_H = \left(\frac{d\theta}{d\lambda} \Delta\lambda \right) \cdot f, \tag{3.34}$$

where $\Delta\lambda$ is the wavelength difference between λ_1 and λ_2. In order to distinguish two different wavelengths (λ_1, λ_2) illuminating the spectrometer:

$$\Delta x_H \geq \Delta x_d. \tag{3.35}$$

For perspective, consider a lens with $D = 5$ cm, $f = 10$ cm, and $\lambda = 1.0$ μm and a hologram formed with construction beams at $\theta_1 = 30°, \theta_2 = 0°$ and $\lambda = 1.0$ μm. In this case, $\Delta\lambda = 0.02$ nm.

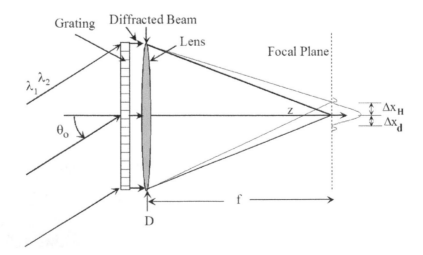

FIGURE 3.23   Spectrometer formed with a holographic grating and a lens.

PROBLEMS

1. Compute the grating vector direction and magnitude for the construction situations a, b, and c shown in Figure 3.24. Assume that the wavelength is 514.5 nm (argon laser line) and that the refractive index of the recording material is 1.51.

2. Problem related to a holographic lens and the grating vector. Refer to Figure 3.25:

    a. Compute the grating vector (magnitude and orientation) at positions (A, B, and C) on the aperture of the holographic lens. The hologram is recorded with a wavelength of 514.5 nm, and the refractive index of the recording material is 1.52.

    b. Determine the angles for the diffraction orders that emerge from the grating at the location B on the hologram aperture when the hologram is illuminated with a reference beam at 20° (same as the construction), but at a wavelength of 532 nm. Show a diagram illustrating the diffracted beams.

3. It is desired to form a multiplexed hologram that is reconstructed with a HeNe laser (632.8 nm) at the same angle of incidence (−15°). The propagation vectors for the beams diffracted from the two holograms are shown in Figure 3.26.

    Determine the construction angles for the two gratings formed with an exposing wavelength of 514.5 nm. It is useful to use Bragg circle diagrams for the construction and reconstruction processes. Assume a refractive index of 1.50 for both the medium and the surrounding regions (i.e., no need to consider refraction at the medium boundary). The angles shown are in the medium.

4. A hologram is recorded with two plane waves at ±θ relative to the z-axis at a free-space wavelength of 670 nm in a material that has a maximum spatial frequency recording capability of

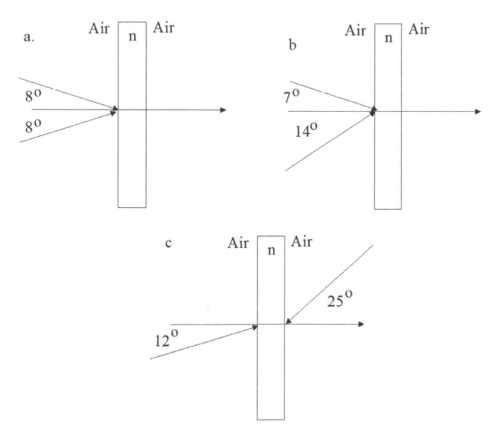

FIGURE 3.24 Configurations a, b, and c for Problem 3.1.

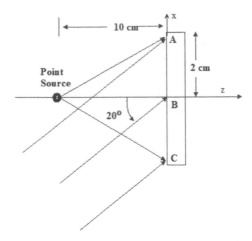

FIGURE 3.25   Holographic lens for Problem 3.2.

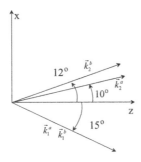

FIGURE 3.26   Configurations for Problem 3.3.

2000 lines/mm (0.5 μm resolution) and a refractive index of 1.50. What is the *largest angle* between the object and reference beams *in air* that can be recorded with these conditions?

5. Describe the *steps* involved for *recording and reconstructing* a holographic image. Show a *sketch* and the *relevant equation* for each step. Assume that the object wave is $\tilde{o} = a_o e^{-j\varphi_o}$ and the reference wave is: $\tilde{r} = a_r e^{-j\varphi_r}$. Give the reconstruction conditions and the expressions for the reconstruction of the original object wave and the conjugate object wave. It will be easiest to describe the steps with simple sketches.

6. For the diagram of the holographic lens shown in Figure 3.27:

   a. Compute the magnitude and direction of the grating vector at point B. The wavelength is 0.55 μm and you can assume that the refractive index is 1.0 both in and out of the material. How does the magnitude of the grating vector at point B compare to the magnitude of the grating vector at point A (no need to compute for point A, just indicate which grating vector is larger based on the graphical information)?

   b. Using the grating vector found in part (*a*) at point B, what is the longest wavelength that can be Bragg matched to the grating at this location?

7. a. Draw a diagram showing the components and their positions in the system required for the formation and reconstruction of a Fourier transform hologram. Indicate what happens to the object field at the recording plane during construction and the image plane during reconstruction. Assume that the reference beam $R(x_2) = \exp(j2\pi x_2 \sin\theta_R)$ illuminates the film at a small angle θ_R with the z-axis (optical axis) during hologram formation and that during reconstruction a normally incident plane wave illuminates

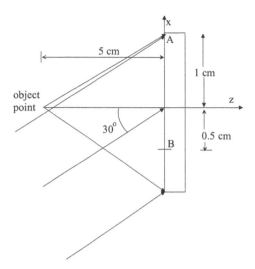

FIGURE 3.27 Holographic lens for Problem 3.5.

the hologram (as described in class). What is the lateral separation of the images at the image plane in terms of the lens parameters and hologram parameters (i.e., reference beam angle)?

b. Show the geometry for recording and reconstructing a "Gabor type" hologram. What is the problem with this recording geometry?

REFERENCES

1. D. Gabor, "A new microscope principle," *Nature*, Vol. 161, 777–778 (1948).
2. E. Leith and J. Upatnieks, "Reconstruction wavefronts and communication theory," *J. Opt. Soc. Am.*, Vol. 52, 1123 (1962).
3. A. Vander Lugt, "Signal detection by complex spatial filtering," *IEEE Trans. Inform. Theory*, IT-10, 139–145 (1964).
4. A. Vander Lugt, "Design relationships for holographic memories," *Appl. Opt.*, Vol. 12, 1675–1685 (1973).
5. B. J. Thompson, J. H. Ward, and W. Zinky, "Application of hologram techniques for particle-size analysis," *Appl. Opt.*, Vol. 6, 519–526 (1967).
6. R. J. Collier, C. B. Burckhardt, and L. H. Lin, *Optical Holography*, Academic Press, New York (1971).
7. R. R. A. Syms and L. Solymar, "Localized one-dimensional theory for volume holograms," *Opt. Quant. Electron.*, Vol. 13, 415 (1981).
8. H. Kogelnik, "Coupled wave theory for thick hologram gratings," *Bell Syst. Tech. J.*, Vol. 48, 2909–2947 (1969).
9. E. Hecht and A. Zajac, *Optics*, Chapter 10, Addison-Wesley, Menlo Park, CA (1974).
10. J. D. Joannopoulos, R. D. Meade, and J. N. Winn, *Photonic Crystals*, Princeton University Press, Princeton, NJ (1995).

4

Holographic Image Formation

4.1 Introduction

The performance of a hologram in an optical system requires separate analysis of the imaging characteristics and the diffraction efficiency. The imaging characteristics are determined by the electromagnetic boundary matching conditions at the hologram interface with the surrounding medium. In contrast, the diffraction efficiency results from the exchange of power between the incident beam and the diffracted field by coupling with the grating. While it is possible to treat these two aspects as one general problem, it is often more intuitive and expedient to evaluate them separately. In addition, a separate approach to image analysis provides a more direct interface with many optical design programs and allows the use of the powerful optimization techniques developed for these programs. Therefore, in this chapter, methods for analyzing the imaging properties of holographic elements are developed and is followed by an analysis of hologram diffraction efficiency in Chapter 5.

In refractive and reflective systems, the laws of refraction and reflection are used repeatedly to describe the trajectory of light through the optical system and the eventual formation of an image. Ray tracing methods based on these relations are used in sophisticated simulation programs to optimize designs. Paraxial imaging relations and aberration coefficients are also used to obtain the basic design parameters and the starting point for more in-depth optical system analysis. Not surprisingly the methods used to evaluate the imaging characteristics of holographic optical elements follow those developed for systems using refractive and reflective optical components [1–4]. In this chapter, methods for exact ray tracing, paraxial imaging relations, and aberration analysis of holographic optical elements will be developed. Techniques for analyzing the properties of holographic components with optical design simulation programs and approaches to improve the performance of holographic elements are also reviewed.

4.2 Exact Ray Tracing

4.2.1 Exact Ray Tracing Algorithm

Ray tracing through an optical system involves calculating the change in the direction of a ray at a surface and determining the ray transfer and point of intersection of the ray at the next surface in the system. For refractive and reflective systems, the change in ray direction is determined by applying Snell's law (Equation 2.47) and the law of reflection (Equation 2.46) at the point where the ray hits the surface. The coordinates of the point where a ray intersects the next surface is then determined by applying a geometric transfer relation to the subsequent plane. The primary difference for ray tracing through a hologram is that the ray direction after encountering the hologram is determined by diffraction. In Chapter 3, the two-dimensional grating equation (Equation 3.9) was derived using phase matching to determine the direction of a diffracted beam. However, for ray tracing it is more useful to generalize the diffraction direction to three dimensions. A straightforward method to accomplish this follows in the remainder of this section along with corresponding transfer relation between optical surfaces.

A three-dimensional form for diffraction can be derived using the geometry in Figure 4.1. The volume grating has a grating vector $\vec{K}$, volumetric grating period Λ, and surface grating period Λ_T. A unit vector normal to the surface is $\hat{n}$, and tangent to the surface is $\hat{t}$. This coordinate system allows the grating

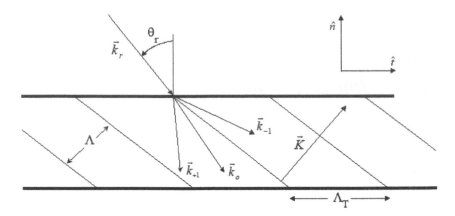

FIGURE 4.1   Geometry for diffraction by a volume grating with period Λ and surface period Λ_T.

surface to have arbitrary shape. The Bragg condition (Equation 3.25) gives the relation between the construction and reconstruction propagation vectors and the grating vector:

$$q\left(\vec{k}_o - \vec{k}_r\right) = q\vec{K} = \vec{k}_{i,q} - \vec{k}_p,\tag{4.1}$$

where $\vec{k}_r$ and $\vec{k}_o$ are, respectively, the reference and object propagation vectors used to form the hologram, $\vec{k}_p$ and $\vec{k}_i$ are the reconstruction and image propagation vectors, and q is the diffraction order. In terms of unit vectors, the relation between construction and reconstruction propagation vectors can be written as:

$$q\frac{2\pi n_1}{\lambda_1}[\,\hat{k}_o - \hat{k}_r\,] = \frac{2\pi n_2}{\lambda_2}[\,\hat{k}_{i,q} - \hat{k}_p\,],\tag{4.2}$$

where n_1 and n_2 are, respectively, the refractive index of the hologram material during construction and reconstruction, and λ_1 and λ_2 are the wavelengths during construction and reconstruction. Taking the "dot" product of a unit tangent vector on both sides of this relation gives the grating equation:

$$q\frac{2\pi n_1}{\lambda_1}\left(\hat{k}_o - \hat{k}_r\right)\cdot\hat{t} = \frac{2\pi n_2}{\lambda_2}\left(\hat{k}_{i,q} - \hat{k}_p\right)\cdot\hat{t} = q\vec{K}\cdot\hat{t}$$

$$\frac{2\pi n_2}{\lambda_2}\sin\theta_{i,q} - \frac{2\pi n_2}{\lambda_2}\sin\theta_p = q\frac{2\pi}{\Lambda_T},\tag{4.3}$$

$$\sin\theta_{i,q} - \sin\theta_p = q\frac{\lambda_2}{n_2\Lambda_T}$$

where Λ_T is the tangential component of the grating period. The first term can also be written as:

$$\left[q\frac{2\pi n_1}{\lambda_1}\left(\hat{k}_o - \hat{k}_r\right) - \frac{2\pi n_2}{\lambda_2}\left(\hat{k}_{i,q} - \hat{k}_p\right)\right]\cdot\hat{t} = 0\tag{4.4}$$

Therefore, the term in brackets is a vector that is normal to the surface since in general the operation $[\quad]\cdot\hat{t} = 0$ implies that the non-zero vector in the brackets is in the direction of $\hat{n}$. In addition, according to vector calculus, a vector $\hat{n}$ also follows the relation $\hat{n}\times a\,\hat{n} = 0$, where a is a constant. Combining these two identities and assuming that the refractive indices $n_1 = n_2$ yields:

$$\frac{\lambda_1}{\lambda_2}\,\hat{n}\times(\hat{k}_{i,q} - \hat{k}_p) = q\hat{n}\times(\hat{k}_o - \hat{k}_r).\tag{4.5}$$

If it is also assumed that $\hat{n}$ is in the $\hat{z}$ direction, then:

$$\frac{\lambda_1}{\lambda_2}\hat{z}\times(\hat{k}_{i,q}-\hat{k}_p)=q\hat{z}\times(\hat{k}_o-\hat{k}_r).$$ (4.6)

This expression is the vector form of the scalar grating equation found in Equation 4.3 that describes diffraction from the grating surface in three dimensions.

The three-dimensional grating equation can be placed in a more workable form for ray tracing by using direction cosines. As described in Chapter 2, the direction cosines represent the x, y, and z components of a unit vector and is illustrated in Figure 4.2. Direction cosines along the x- and y-axis for a vector between points (x_1,y_1,z_1) and (x_2,y_2,z_2) can be found using the construction (Figure 4.3):

$$\cos\alpha = \frac{x_2-x_1}{r_{12}};\cos\beta = \frac{y_2-y_1}{r_{12}};\text{with}$$
$$r_{12} = \sqrt{(x_2-x_1)^2+(y_2-y_1)^2+(z_2-z_1)^2}.$$ (4.7)

Since the direction cosines must satisfy the relation:

$$\cos^2\alpha + \cos^2\beta + \cos^2\gamma = 1,$$ (4.8)

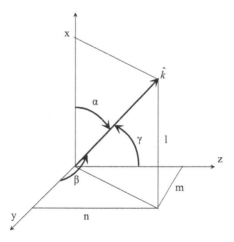

FIGURE 4.2 Illustration of the relation between the propagation vector and angles relative to an x-y-z coordinate system.

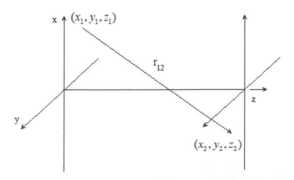

FIGURE 4.3 Formation of the direction cosines between two sets of coordinates (x_1, y_1, z_1) and (x_2, y_2, z_2).

the direction cosine along the z-axis is:

$$\cos\gamma = \pm\sqrt{1 - \cos^2\alpha - \cos^2\beta}, \tag{4.9}$$

where the $+$ or $-$ sign is used depending on the relative values for z_1 and z_2.

The cross-product terms $\hat{z} \times \hat{k}_j$ in Equation 4.6 expressed in the form of direction cosines are:

$$\hat{z} \times \hat{k}_j = \begin{vmatrix} \hat{x} & \hat{y} & \hat{z} \\ 0 & 0 & 1 \\ \cos\alpha_j & \cos\beta_j & \cos\gamma_j \end{vmatrix} = \begin{vmatrix} \hat{x} & \hat{y} & \hat{z} \\ 0 & 0 & 1 \\ l_j & m_j & n_j \end{vmatrix} \tag{4.10}$$

$$= -m_j\hat{x} + l_j\hat{y},$$

where $j = o, r, p, i$. Equating the y-vector components yields the relation for the direction cosines along the x-direction:

$$l_{i,q} = q\frac{\lambda_2}{\lambda_1}(l_o - l_r) + l_p. \tag{4.11}$$

Similarly, equating the x-vector components gives the direction cosines along the y-direction:

$$m_{i,q} = q\frac{\lambda_2}{\lambda_1}(m_o - m_r) + m_p, \tag{4.12}$$

and using Equation 4.9, the direction cosine of the diffracted ray in the z-direction is:

$$n_{i,q} = \pm\left[1 - l_{i,q}^2 - m_{i,q}^2\right]^{1/2}. \tag{4.13}$$

The $\pm$ sign in this case indicates whether the hologram is a transmission $(+)$ or a reflection type $(-)$.

The direction cosines $(l_{i,q}, m_{i,q}, n_{i,q})$ give the direction of the diffracted ray similar to Snell's law, giving the angle for a ray refracted from an optical interface between two dielectric materials.

The next step is to determine the ray transfer relations between two planes in the optical system. For simplicity, only the first diffraction order $q = 1$ will be used, and the geometry for construction and reconstruction of the hologram is shown in Figure 4.4. The hologram is formed with light from two point sources located at O and R and the reconstruction source point is P. The interference of light from points

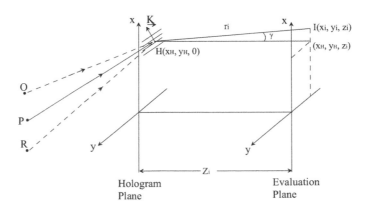

Hologram
Plane

Evaluation
Plane

FIGURE 4.4 Diagram showing ray transfer between the hologram plane ($zh = 0$) and image evaluation plane located at $z = z_i$. (From Andrews, D.L.: *Photonics Technology and Instrumentation.* 2015. Copyright Wiley-VCH Verlag GmbH & Co. KGaA. Reproduced with permission.)

O and R results in a grating at a point $H(x_H,y_H,0)$ with a grating vector $\underline{K}$, and the diffracted ray at location H makes an angle γ with respect to the z-axis. The angle γ_i is computed from the cosine relation:

$$n_i = \frac{z_i}{r_i} = \cos\gamma_i, \qquad (4.14)$$

where r_i is the length of the vector from points H to I with $r_i = [(x_i - x_H)^2 + (y_i - y_H)^2 + z_i^2]^{1/2}$, and z_i the axial distance between the hologram and evaluation planes. Using these values, the diffracted ray intercepts a point on the image plane (x_i, y_i) with:

$$
\begin{aligned}
x_i &= r_i l_i + x_H \\
y_i &= r_i m_i + y_H,
\end{aligned}
\qquad (4.15)
$$

and are the transfer relations for the hologram ray trace operation.

Combining these results and referring to Figure 4.4, the ray tracing procedure for evaluating an image formed by a hologram can be summarized in the following algorithm:

1. Divide the hologram aperture into a set of equally spaced coordinates
2. Determine the direction cosines for rays from source points O, R, P to a coordinate in the hologram aperture
3. Compute the image ray direction cosines (l_i, m_i, n_i) at the coordinate in the hologram aperture
4. Determine the path distance r_i using the relation $r_i = z_i / n_i$
5. Use the transfer relations (Equation 4.15) to determine the intercept location on the evaluation plane
6. Repeat steps 1–5 for each coordinate on the hologram aperture and construct an intercept plot on the image/evaluation plane.

The set of ray intercept coordinates can then be used to generate a spot diagram for the hologram reconstruction configuration. Notice that in this procedure the evaluation plane, located at a distance z_i from the hologram plane, may or may not be the "image" plane. As with ray tracing through refractive and reflective optical systems, the location of the image plane is usually determined by other criteria such as the paraxial image location or the distance at which the point spread function is a minimum. Some of these criteria for imaging with a hologram are discussed later in this chapter.

4.2.2 Primary and Secondary Image Formation

As previously shown in Chapter 3, when the hologram is reconstructed with the original reference beam, both rr^*o and rro^* are formed:

$$
\begin{aligned}
I_r &= t(x,y,z)r \\
&= r\alpha\left(|r|^2 + |o|^2\right) + \alpha r r o^* + \alpha r r^* o.
\end{aligned}
\qquad (3.5)
$$

In this case, the image rr^*o corresponds to the $q = +1$ order and results in an undistorted reconstruction of the object beam. The term rro^* corresponds to $q = -1$ and since the combined phase is $2\varphi_r - \varphi_o$, it results in a distorted image of the conjugate object beam. Oftentimes the image rr^*o is referred to as the primary image and rro^* the secondary image. The two images are illustrated in Figure 4.5.

4.2.3 Forming a Real Image with a Conjugate Reconstruction Beam

Frequently, it is desired to reconstruct a non-distorted real image with a holographic lens. In this case, careful consideration of the reconstruction conditions should be made. As an example, consider the

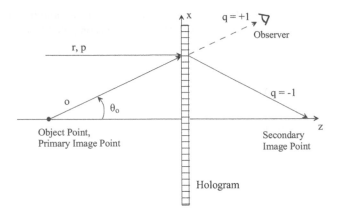

FIGURE 4.5   Reconstruction of a hologram with the reference beam r (i.e., $p = r$) and the formation of the primary ($q = +1$) and secondary ($q = -1$) images.

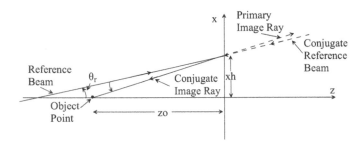

FIGURE 4.6   Geometry showing hologram reconstruction with the original reference and the conjugate reference beams.

simple case of forming a holographic lens with an on-axis point source and an off-axis plane wave as shown in Figure 4.6. The object is a point source located at $(x_o, y_o, -z_o)$, and the reference beam is a plane wave making an angle of θ_r relative to the z-axis. The primary image ($q = +1$) is formed by reconstructing the hologram with the original reference beam. This diffracts a beam as shown in the figure and results in a ray diverging from the object point forming a virtual image. If the hologram is reconstructed with a conjugate reference beam, the hologram is illuminated from the $+z$ side and diffracts the beam toward the original object point. This produces a real image with $q = -1$. Therefore, when using the ray tracing algorithm developed in Section 4.2.1, the correct sign for the diffraction order should be used in combination with the original or conjugate reconstruction beam to obtain the desired primary or secondary image.

4.3 Hologram Paraxial Imaging Relations

As noted in the previous section, exact ray tracing does not in itself establish an image plane, but only evaluates the ray intercepts at an evaluation plane. In refractive and reflective systems, the paraxial image location is often used as a starting point for the location of the image. It is also possible to form a set of paraxial imaging relations for holographic optical elements to predict first order imaging properties [1]. These relations are developed using point sources for constructing and reconstructing the holographic image. Just as in refractive and reflective systems, the paraxial relations for holographic imaging result from restricting the analysis to regions near the optical (z-axis) of the optical system. This allows the phase of the spherical waves emitted from the point sources to be expressed as quadratic phase functions and simplifies the analysis.

4.3.1 Analysis of the Phase Distribution from a Point Source to a Hologram Plane

Consider a hologram formed by exposing a recording material to reference point source located at $R(x_r,y_r,z_r)$ and an object point source at $O(x_o,y_o,z_o)$. The hologram is reconstructed with a third point source located at $P(x_p,y_p,z_p)$. The hologram is constructed with light at wavelength λ_1 and reconstructed at λ_2, and the recording material is located at $z = 0$. A spherical wave from one of the point sources can be written as:

$$E_q(r_q) = \frac{\exp(jkr_q)}{r_q}, \tag{4.16}$$

where $k = 2\pi/\lambda$, the subscript q represents one of the source points R, O, or P (Figure 4.7), and r_q is the distance from a point $Q(x_q,y_q,z_q)$ to a point $H(x,y,0)$ on the hologram. Instead of computing the total phase from points Q to H, consider the phase difference with a path from point Q to a point C located at $(0,0,0)$ at the center of the hologram:

$$\varphi_q(x,y) = \frac{2\pi}{\lambda}d = \frac{2\pi}{\lambda}(QH - QO)$$

$$= \frac{2\pi}{\lambda}\left\{\left[(x-x_q)^2 + (y-y_q)^2 + z_q^2\right]^{1/2} - \left(x_q^2 + y_q^2 + z_q^2\right)^{1/2}\right\} \tag{4.17}$$

$$= \frac{2\pi}{\lambda}z_q\left(\left\{1 + \left[(x-x_q)^2 + (y-y_q)^2\right]/z_q^2\right\}^{1/2} - \left[1 + \left(x_q^2 + y_q^2\right)/z_q^2\right]^{1/2}\right),$$

where d represents the path length from the hologram plane to the wavefront as shown in Figure 4.7. Applying the binomial expansion to the square root terms reduces the relation for φ_q to:

$$\varphi_q(x,y) = \frac{2\pi}{\lambda}\left[\frac{1}{2z_q}\left(x^2 + y^2 - 2xx_q - 2yy_q\right) - \frac{1}{8z_q^3}\left(x^4 + y^4 - 2x^2y^2 - 4x^3x_q - 4y^3y_q\right.\right.$$

$$- 4x^2yy_q - 4xy^2x_q + 6x^2x_q^2 + 6y^2y_q^2 + 2x^2y_q^2 + 2y^2x_q^2 + 8xyx_qy_q$$

$$\left.\left. - 4xx_q^3 - 4yy_q^3 - 4xx_qy_q^2 - 4yy_qx_q^2 + H.O.T.\right)\right], \tag{4.18}$$

where H.O.T. refers to higher order terms in the expansion.

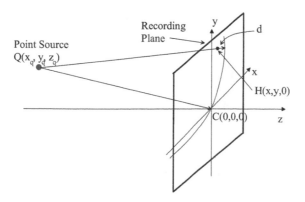

FIGURE 4.7 Geometry for an off-axis reference wave propagating from a point source to the hologram plane.

The holographic paraxial imaging relations are obtained from the coefficients of $1/z$ in the phase function. The coefficients for the fields R, O, and P are:

$$E_r\left(r_r\right) \approx \frac{1}{z_r}\exp\left(jk_1 z_r\right)\exp\left(-jk_1\left(x_r^2 + y_r^2\right)z_r\right)\exp\left\{j\frac{k_1}{2}\left[\frac{\left(x-x_r\right)^2}{z_r} + \frac{\left(y-y_r\right)^2}{z_r}\right]\right\}$$

$$= A_r\exp\left\{j\frac{\pi}{\lambda_1}\left[\frac{\left(x-x_r\right)^2}{z_r} + \frac{\left(y-y_r\right)^2}{z_r}\right]\right\} = A_r\exp\left(j\varphi_r\right)$$

$$E_o\left(r_o\right) \approx A_o\exp\left\{j\frac{\pi}{\lambda_1}\left[\frac{\left(x-x_o\right)^2}{z_o} + \frac{\left(y-y_o\right)^2}{z_o}\right]\right\} = A_o\exp\left(j\varphi_o\right)$$

$$\tag{4.19}$$

$$E_p\left(r_p\right) \approx A_p\exp\left\{j\frac{\pi}{\lambda_2}\left[\frac{\left(x-x_p\right)^2}{z_p} + \frac{\left(y-y_p\right)^2}{z_p}\right]\right\} = A_p\exp\left(j\varphi_p\right).$$

The paraxial imaging relations are found by using the first order field terms in the holographic recording and image reconstruction process. The intensity interference pattern at the hologram plane with the first order field terms are:

$$I(x,y) = \left|E_r\left(x,y\right) + E_o\left(x,y\right)\right|^2$$

$$= A_r^2 + A_o^2 + A_r A_o^*\exp\left(j\frac{\pi}{\lambda_1}\left\{\frac{1}{z_r}\left[\left(x-x_r\right)^2 + \left(y-y_r\right)^2\right] - \frac{1}{z_o}\left[\left(x-x_o\right)^2 + \left(y-y_o\right)^2\right]\right\}\right)$$

$$+ A_o A_r^*\exp\left(j\frac{\pi}{\lambda_1}\left\{-\frac{1}{z_r}\left[\left(x-x_r\right)^2 + \left(y-y_r\right)^2\right] + \frac{1}{z_o}\left[\left(x-x_o\right)^2 + \left(y-y_o\right)^2\right]\right\}\right)$$

$$\tag{4.20}$$

$$= A_r^2 + A_o^2 + A_o A_r^*\exp\left[j\left(\varphi_o - \varphi_r\right)\right] + A_r A_o^*\exp\left[j\left(\varphi_r - \varphi_o\right)\right].$$

The intensity pattern is then converted into the hologram which is expressed as a transmittance function:

$$t(x,y) = \gamma I(x,y), \tag{4.21}$$

where γ is a constant related to the exposure time and processing of the recording material. A holographic image field is formed by illuminating with a point source located at a point P:

$$E_i(x,y) = E_p(x,y)t(x,y)$$

$$= E_p\gamma\left\{A_r^2 + A_o^2 + A_o A_r^*\exp\left[j\left(\varphi_o - \varphi_r\right)\right] + A_r A_o^*\exp\left[j\left(\varphi_r - \varphi_o\right)\right]\right\}$$

$$= \gamma A_p\exp\left(j\varphi_p\right)\left\{A_r^2 + A_o^2 + A_o A_r^*\exp\left[j\left(\varphi_o - \varphi_r\right)\right] + A_r A_o^*\exp\left[j\left(\varphi_r - \varphi_o\right)\right]\right\} \tag{4.22}$$

$$= \gamma A_p\left\{\left(A_r^2 + A_o^2\right)\exp\left(j\varphi_p\right) + A_r A_o^*\exp\left[j\left(\varphi_r - \varphi_o - \varphi_p\right)\right]\right.$$

$$\left. + A_o A_r^*\exp\left[j\left(\varphi_o - \varphi_r - \varphi_p\right)\right]\right\},$$

and consists of four terms. The first two terms are constants with the same phase as E_p, however, the third and fourth terms contain a combination of the phase of E_p plus that of E_r and E_o. The resulting image field can be simplified to:

$$E_i = (c_1 + c_2)\exp(j\varphi_p) + c_3 \exp\left[j(\varphi_p - \varphi_r + \varphi_o)\right] + c_4 \exp\left[j(\varphi_p + \varphi_r - \varphi_o)\right]$$

$$= U_1 + U_2 + U_3 + U_4,$$

(4.23)

where c_{1-4} are constants and U_{1-4} are the complex field terms. The first two terms (U_1 and U_2) are fields with the same phase as the reconstruction beam. However, as pointed out in Chapter 3, the terms U_3 and U_4 are of more interest since when $\varphi_p = \varphi_r$ the original object phase (φ_o) is reconstructed, and when $\varphi_p = -\varphi_r$ the conjugate object wave is formed and are illustrated in Figure 4.8. As shown, the reconstruction of U_3 with $\varphi_p = \varphi_r$ will in general form a virtual image, while the reconstruction of U_4 with $\varphi_p = -\varphi_r$ forms a real image. The terms U_3 and U_4 can be simply expressed as:

$$U_3 = c_3 \exp\left(j\varphi_3^{(1)}\right) = c_3 \exp j(\varphi_p - \varphi_r + \varphi_o)$$

$$U_4 = c_4 \exp\left(j\varphi_4^{(1)}\right) = c_4 \exp j(\varphi_p + \varphi_r - \varphi_o),$$

(4.24)

where the superscript (1) in φ_3 and φ_4 indicates the phase of these field terms as expressed with Equation 4.24.

Another representation of the image fields U_3 and U_4 is also possible. If the object field used to form the hologram is a point source, the phase of the resulting image field is a spherical wave. During the reconstruction process, the fields will either diverge (U_3) or converge (U_4) to a point. Using this result and Equations 4.19 and 4.20, the phase for the field terms U_3 and U_4 can be expressed as:

$$\varphi_3^{(2)} = \frac{\pi}{\lambda_2 z_3}\left[x^2 + y^2 - 2xx_3 - 2yy_3\right]$$

$$\varphi_4^{(2)} = \frac{\pi}{\lambda_2 z_4}\left[x^2 + y^2 - 2xx_4 - 2yy_4\right],$$

(4.25)

where the subscript (2) indicates the point source interpretation for the phase of U_3 and U_4. Note, however, that the two representations for the phase of the fields are equal, therefore:

$$\varphi_3^{(1)} = \varphi_3^{(2)}$$

$$\varphi_4^{(1)} = \varphi_4^{(2)}.$$

(4.26)

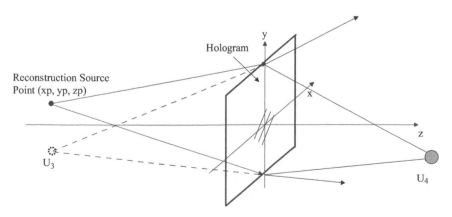

FIGURE 4.8 Reconstruction of a hologram showing the formation of the primary (U_3) and secondary (U_4) images. (From Andrews, D.L.: *Photonics Technology and Instrumentation*. 2015. Copyright Wiley-VCH Verlag GmbH & Co. KGaA. Reproduced with permission.)

Matching the two-phase expressions for U_3 results in the following relation:

$$\varphi_3^{(1)} = \varphi_3^{(2)}$$

$$\frac{\pi}{\lambda_2}\left[\left(x^2+y^2\right)\left(\frac{1}{z_p}+\frac{\mu}{z_o}-\frac{\mu}{z_r}\right)-2x\left(\frac{x_p}{z_p}+\frac{\mu x_o}{z_o}-\frac{\mu x_r}{z_r}\right)\right]-2y\left(\frac{y_p}{z_p}+\frac{\mu y_o}{z_o}-\frac{\mu y_r}{z_r}\right) \tag{4.27}$$

$$=\frac{\pi}{\lambda_2}\frac{x^2+y^2-2xx_3-2yy_3}{z_3}.$$

Equating the first and second order terms in x and y in Equation 4.27 results in the coordinates for (x_3,y_3,z_3):

$$x_3 = \frac{x_p z_o z_r + \mu x_o z_p z_r - \mu x_r z_p z_o}{z_o z_r + \mu z_p z_r - \mu z_p z_o}$$

$$y_3 = \frac{y_p z_o z_r + \mu y_o z_p z_r - \mu y_r z_p z_o}{z_o z_r + \mu z_p z_r - \mu z_p z_o}, \tag{4.28}$$

$$z_3 = \frac{z_p z_o z_r}{z_o z_r + \mu z_p z_r - \mu z_p z_o}$$

where $\mu = \lambda_2/\lambda_1$. The point (x_3,y_3,z_3) is the paraxial image point of the primary image formed by the hologram. Similarly, matching the phase in the expressions for U_4 results in:

$$\varphi_4^{(1)} = \varphi_4^{(2)}$$

$$\frac{2\pi}{\lambda_2}\frac{1}{2}\frac{x^2+y^2-2xx_p-2yy_p}{z_p}-\frac{2\pi}{\lambda_1}\frac{1}{2}\frac{x^2+y^2-2xx_o-2yy_o}{z_o}+\frac{2\pi}{\lambda_1}\frac{1}{2}\frac{x^2+y^2-2xx_r-2yy_r}{z_r} \tag{4.29}$$

$$=\frac{2\pi}{\lambda_2}\frac{1}{2}\frac{x^2+y^2-2xx_4-2yy_4}{z_4},$$

and solving for the coordinate points (x_4,y_4,z_4) yields:

$$x_4 = \frac{x_p z_o z_r - \mu x_o z_p z_r + \mu x_r z_p z_o}{z_o z_r - \mu z_p z_r + \mu z_p z_o}$$

$$y_4 = \frac{y_p z_o z_r - \mu y_o z_p z_r + \mu y_r z_p z_o}{z_o z_r - \mu z_p z_r + \mu z_p z_o}, \tag{4.30}$$

$$z_4 = \frac{z_p z_o z_r}{z_o z_r - \mu z_p z_r + \mu z_p z_o},$$

which gives the location of the real paraxial image point.

The connection between the reconstruction point coordinates (x_p,y_p,z_p) and the image point coordinates (x_3,y_3,z_3) provides a set of paraxial imaging relations for a holographic lens in the same manner as the object point and conjugate image points are used in the paraxial imaging relations for a refractive or reflective imaging system.

In many cases the image point (x_3,y_3,z_3) is a virtual image point formed in object space and (x_4,y_4,z_4) is a real image point. However, this will depend on the geometry of the hologram and the reconstruction conditions. As a result, it is often less confusing to refer to the image formed by U_3 as the primary image and that formed by U_4 as the secondary image. As shown in Equation 4.23, both the primary and secondary images are formed during reconstruction, however, in general, one image is much more resolved than the other and is of more significance.

4.3.2 Image Magnification Effects

The holographic paraxial imaging relations can be used to determine the lateral, longitudinal, and angular magnification in optical systems that contain holographic optical elements and are very useful in the initial optical design process.

The lateral image magnification is defined as the change in lateral image height with the change in the object point height and is illustrated in Figure 4.9 for the x coordinate. Notice that unlike a conventional refractive lens, the image height is not necessarily inverted. The expressions for lateral magnification can be found by differentiating the paraxial image location x_3 with respect to the object position x_o:

$$M_{Lat,3} = \frac{dx_3}{dx_o} \qquad \text{or} \qquad M_{Lat,4} = \frac{dx_4}{dx_o}$$

$$M_{Lat,3} = \frac{1}{\left[1 + z_o \left(\dfrac{1}{\mu z_p} - \dfrac{1}{z_r}\right)\right]} \quad \text{or} \quad M_{Lat,4} = \frac{1}{\left[1 - z_o \left(\dfrac{1}{\mu z_p} + \dfrac{1}{z_r}\right)\right]}. \tag{4.31}$$

The angular magnification indicates the change in the angle subtended by the image with a change in object angle. The angles subtended by the object and image are with respect to the hologram location as shown in Figure 4.10. The angular magnification is found by differentiating the image angle with respect to the object angle, and for the primary image in the x–z plane is given by:

$$M_{Ang,3} = \frac{d(x_3/z_3)}{d(x_o/z_o)} \rightarrow M_{Ang,3} = \mu = \frac{\lambda_2}{\lambda_1}. \tag{4.32}$$

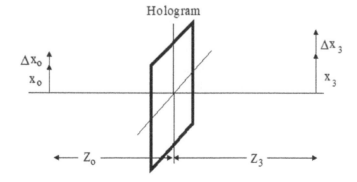

FIGURE 4.9 Lateral magnification by a hologram in the x-direction.

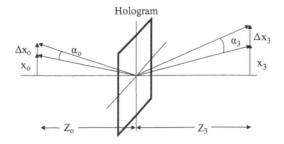

FIGURE 4.10 Illustration of angular magnification by a holographic optical element.

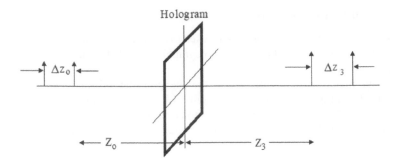

FIGURE 4.11 Illustration of paraxial longitudinal magnification of a holographic optical element.

Similarly, the longitudinal magnification (Figure 4.11) indicates the change in axial position of an image with a change in the axial position of the object and is found by differentiating the paraxial expression for z_3 with respect to z_o:

$$M_{Long,3} = \frac{dz_3}{dz_o} = \frac{1}{\mu}\frac{d}{dz_o}\left\{\frac{z_o}{1 + z_o\left[1/(\mu z_p) - 1/z_r\right]}\right\}$$

$$= \frac{1}{\mu}\frac{1}{\left[1 + z_o\left[1/(\mu z_p) - 1/z_r\right]\right]^2} = \frac{1}{\mu}M_{Lat,3}^2,$$

(4.33)

and for the secondary image is:

$$M_{Long,4} = -\frac{1}{\mu}M_{Lat,4}^2.$$

4.3.3 Effect of Spectral Bandwidth on Hologram Image Resolution

The effect of spectral bandwidth on the holographic image quality can also be evaluated using the paraxial imaging relations. For this analysis, it is assumed that the hologram is formed in the usual manner with a narrow bandwidth optical source at wavelength λ_1 and reconstructed with a source having a nominal wavelength of λ_2 and a spectral bandwidth of $\Delta\lambda_2$. The object point is located at $O(x_o, y_o, z_o)$, and both the reference and reconstruction points are located at infinity $z_p = z_r = -\infty$ (Figure 4.12). The source points at $-\infty$ can be described as the tangent of the respective angles with respect to the optical axis:

$$\tan\theta_r = \frac{x_r}{z_r}$$

(4.34)

$$\tan\theta_p = \frac{x_p}{z_p},$$

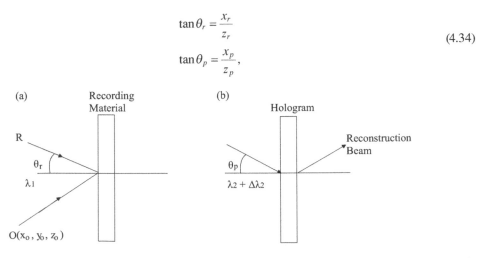

FIGURE 4.12 Schematic diagram of the construction and reconstruction conditions for analyzing the effects of spectral bandwidth of the reconstruction source. (a) shows the construction geometry and (b) shows the reconstruction.

where it is assumed that $y_r = y_p = 0$.

With these construction and reconstruction conditions, the paraxial image position becomes:

$$x_3 = x_o + \frac{x_p}{z_p}\frac{z_o}{\mu} - \frac{x_r}{z_r}z_o,$$

(4.35)

$$z_3 = z_o/\mu$$

with $\mu = \lambda_2/\lambda_1$. The change in image position with λ_2 can be determined taking the derivative of x_3 with respect to λ_2:

$$\frac{dx_3}{d\lambda_2} = -\left(\frac{x_p}{z_p}\right)\left(\frac{z_o}{\mu}\right)\frac{1}{\lambda_2}$$

(4.36)

$$\frac{dz_3}{d\lambda_2} = -\left(\frac{z_o}{\mu}\right)\frac{1}{\lambda_2}.$$

If it is also assumed that the nominal construction and reconstruction wavelengths are nearly equal $\lambda_2 \simeq \lambda_1$, then $\mu \simeq 1$. It can then be seen that if the reconstruction source has a bandwidth of $\Delta\lambda_2$, the image location will spread:

$$|\Delta x_3| = \left(\frac{x_p}{z_p}\right)z_o\frac{\Delta\lambda_2}{\lambda_2}$$

(4.37)

$$|\Delta z_3| = z_o\frac{\Delta\lambda_2}{\lambda_2}.$$

Notice that if $z_o = 0$ then $|\Delta x_3| = |\Delta z_3| = 0$. This is the condition for an image plane hologram and is one of the advantages of using this hologram geometry.

Consider an example where a hologram is formed with an argon laser at 514.5 nm and reconstructed with a light emitting diode with $\lambda_2 = 535$ nm and $\Delta\lambda = 20$ nm. The construction and reconstruction beams are plane waves at an angle of 30° to the optical axis. Under what conditions can the image blur in the x-direction be restricted to less than 0.5 mm? Examination of Equation 4.37 shows that the resolution is proportional to z_o, therefore:

$$z_o = \frac{\lambda_2}{\Delta\lambda_2}\frac{z_p}{x_p}\Delta x_3 = \frac{535\,\text{nm}}{20\,\text{nm}}\frac{1}{\tan(30°)}(0.5\,\text{mm}) = 23.166 \text{ mm}.$$

(4.38)

And shows that the object point used during the recording should not be any closer than 23.166 mm from the hologram plane. The corresponding spread in the longitudinal position of the image is:

$$|\Delta z_3| = 23.166\,\text{mm}\left(\frac{20\,\text{nm}}{535\,\text{nm}}\right) = 0.866\,\text{mm}.$$

(4.39)

4.4 Aberrations in Holographic Imaging

As shown in the previous section, the holographic paraxial imaging relations give the ideal image location for holograms formed with point sources. However, when the location or wavelength of the reconstruction source differs from the construction source the image quality degrades. Holographic image aberrations have similar properties and characteristics to images formed with refractive lenses [5].

Therefore, it is useful to build a framework for analyzing hologram image aberrations upon the vast body of information available on aberration theory and reduction for refractive and reflective focusing elements. For example, relations for third, fifth, and higher order Seidel aberration coefficients [6] can be formed for holographic optical elements and used to balance aberrations and reduce image degradation. In the remainder of this section, the third order aberration coefficients for holographic lenses formed with point sources are developed.

To see where the third order aberration coefficients arise for a holographic image, consider Equation 4.18 for the phase of a wavefront. This expression contains factors that are powers of $1/z_q$ and $1/z_q^3$. The terms with factors of $1/z_q$ were used to develop the holographic paraxial imaging relations. In a similar manner, the third-order aberration coefficients are derived using terms with factors of $1/z_3^3$ or $1/z_4^3$, where the subscripts 3 and 4 refer to the primary and secondary paraxial images. The third order expression for the phase of the wavefront converging to the paraxial image point (x_4, y_4, z_4) are:

$$
\begin{aligned}
\varphi_4^{(3)} = \frac{2\pi}{\lambda_2} \Bigg\{ &-\frac{1}{8z_4^3} [x^4 + y^4 + 2x^2y^2 - 4x^3x_4 - 4y^3y_4 - 4x^2yy_4 - 4xy^2x_4 \\
&+ 6x^2x_4^2 + 6y^2y_4^2 + 2x^2y_4^2 + 2y^2x_4^2 + 8xyx_4y_4 - 4xx_4^3 - 4yy_4^3 \\
&- 4xx_4y_4^2 - 4yx_4^2y_4] \Bigg\},
\end{aligned}
$$

(4.40)

where x and y refer to the coordinates in the hologram plane. Replacing the rectangular coordinates x and y with radial coordinates $\rho = \sqrt{x^2 + y^2}$ and $\theta = \tan^{-1}(y/x)$ as shown in Figure 4.13 results in an expression for the phase of the wavefront corresponding to the third order aberrations:

$$
\begin{aligned}
W = \frac{2\pi}{\lambda_2} \Bigg[&-\frac{1}{8}\rho^4 S + \frac{1}{2}\rho^3 (C_x \cos\theta + C_y \sin\theta) \\
&-\frac{1}{2}\rho^2 (A_x \cos^2\theta + A_y \sin^2\theta + 2A_{xy} \cos\theta \sin\theta) \\
&-\frac{1}{4}\rho^2 F + \frac{1}{2}\rho (D_x \cos\theta + D_y \sin\theta), \Bigg]
\end{aligned}
$$

(4.41)

where S is the spherical aberration coefficient, C_x and C_y are the coma aberration coefficients, A_x, A_y, and A_{xy} are the astigmatism aberration coefficients, F is the field curvature coefficient, and D_x and D_y are the distortion aberration coefficients. The following expressions are the third order aberration coefficients for the primary (x_3, y_3, z_3) and secondary image (x_4, y_4, z_4) formed by a point source hologram.

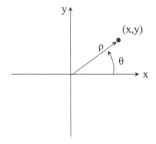

FIGURE 4.13   Radial coordinate system for the aberration coefficients.

4.4.1 Spherical Aberration Coefficient

The spherical aberration coefficients for the primary and secondary images are [1]:

$$S_3 = \frac{1}{z_p^3} + \frac{\mu}{z_o^3} - \frac{\mu}{z_r^3} - \frac{1}{z_3^3}$$

$$S_4 = \frac{1}{z_p^3} - \frac{\mu}{z_o^3} + \frac{\mu}{z_r^3} - \frac{1}{z_4^3},$$

(4.42)

where $\mu = \lambda_2/\lambda_1$. This relation shows that spherical aberration primarily depends on the axial position of the construction and reconstruction point sources.

When z_r and $z_p = \infty$, i.e., the reference and reconstruction beams are plane waves and the spherical aberration coefficient reduces to [1]:

$$S = \mu\left[\mu^2 - 1\right]\frac{1}{z_o^3}.$$

(4.43)

In this case, it can also be seen that when $\lambda_1 = \lambda_2 \rightarrow S = 0$. Another situation of interest occurs when the reference and object point sources are located at the same distance from the hologram plane, i.e., when $z_r = z_o$. In this case, $S = 0$ regardless of the values of z_p and μ.

4.4.2 Coma Aberration Coefficient

The aberration coefficients for coma in an image formed by a hologram are [1]:

$$C_{x3} = \frac{x_p}{z_p^3} + \frac{\mu x_o}{z_o^3} - \frac{\mu x_r}{z_r^3} - \frac{x_3}{z_3^3}$$

$$C_{x4} = \frac{x_p}{z_p^3} - \frac{\mu x_o}{z_o^3} + \frac{\mu x_r}{z_r^3} - \frac{x_4}{z_4^3},$$

(4.44)

This aberration depends on the off-axis skewness (x/z) of the source points used to form and reconstruct the hologram and is analogous to coma for refractive elements. Coma in the holographic image can be reduced by making $x_p/z_p^3 = -\mu x_r/z_r^3$ or by minimizing the off-axis displacement of the construction and reconstruction source points.

4.4.3 Astigmatism and Field Curvature Aberration Coefficients

The coefficients for astigmatism and field curvature have the same functional dependence. They depend on the square of the lateral coordinates of the construction and reconstruction source points divided by the cube of the longitudinal coordinates for the sources. The expressions for the astigmatism and field curvature coefficients in terms of the source point coordinates are [1]:

$$A_{x3} = \frac{x_p^2}{z_p^3} + \frac{\mu x_o^2}{z_o^3} - \frac{\mu x_r^2}{z_r^3} - \frac{x_3^2}{z_3^3}$$

$$A_{x4} = \frac{x_p^2}{z_p^3} - \frac{\mu x_o^2}{z_o^3} + \frac{\mu x_r^2}{z_r^3} - \frac{x_4^2}{z_4^3}$$

(4.45)

$$F_3 = \frac{x_p^2 + y_p^2}{z_p^3} + \frac{\mu\left(x_o^2 + y_o^2\right)}{z_o^3} - \frac{\mu\left(x_r^2 + y_r^2\right)}{z_r^3} - \frac{x_4^2 + y_4^2}{z_4^3}$$

(4.46)

$$F_4 = \frac{x_p^2 + y_p^2}{z_p^3} - \frac{\mu\left(x_o^2 + y_o^2\right)}{z_o^3} + \frac{\mu\left(x_r^2 + y_r^2\right)}{z_r^3} - \frac{x_4^2 + y_4^2}{z_4^3}.$$

Both astigmatism and coma coefficients can be made to vanish provided that $\mu = 1$, $M_{Lat} = 1$, and non-planar reconstruction beams (i.e., convergent or divergent) are used.

4.4.4 Distortion Aberration Coefficient

The final third order aberration coefficient is distortion and is a function of the cube of the ratio of the lateral to longitudinal source point coordinates. The distortion coefficients are given by [1]:

$$D_{x3} = \frac{x_p^3 + x_p y_p^2}{z_p^3} + \frac{\mu\left(x_o^3 + x_o y_o^2\right)}{z_o^3} - \frac{\mu\left(x_r^3 + x_r y_r^2\right)}{z_r^3} - \frac{x_4^3 + x_4 y_4^2}{z_4^3}$$

(4.47)

$$D_{x4} = \frac{x_p^3 + x_p y_p^2}{z_p^3} - \frac{\mu\left(x_o^3 + x_o y_o^2\right)}{z_o^3} + \frac{\mu\left(x_r^3 + x_r y_r^2\right)}{z_r^3} - \frac{x_4^3 + x_4 y_4^2}{z_4^3}.$$

The cube relation allows the value of this coefficient to be reduced by choosing opposite algebraic signs for the lateral coordinates of the construction and reconstruction source points. In fact, making $x_p/z_p = -x_r/z_r$ with $\lambda_1 = \lambda_2$ will eliminate distortion for all values of z_o, z_r, and z_p.

4.4.5 Example of a Holographic Lens Formed with Spherical Waves and Methods to Reduce Image Aberration

As noted previously, aberrations result when either the location or wavelength of the reconstruction source deviates from the construction parameters. The magnitude of aberrations can be quite large even with small deviations from the construction conditions. The following example illustrates how holographic ray trace and aberration coefficient analysis can be used to guide the design of holographic lenses to reduce image degradation.

First consider a holographic lens formed with an object point source located at $(x_o = 1, z_o = -5)$ and a collimated reference beam at $+10°$ to the z-axis as shown in Figure 4.14. The hologram aperture has a diameter of 1 unit, and the construction and reconstruction wavelengths are the same, therefore $\mu = 1$. The hologram is reconstructed with a conjugate planar beam (p) at $-11°$ to the z-axis. The spot diagram from the ray trace in Figure 4.15 shows that at the paraxial image plane located at $z = -5$, the image spot extends by $\Delta x = 0.020$ and $\Delta y = 0.009$ about the paraxial image point. Evaluation of the aberration coefficients shows moderate levels of coma ($C_x = -7.22 \times 10^{-4}$) and significant levels of astigmatism ($A_x = -0.0014$).

Now consider the holographic lens formed with the point source on the z-axis and an off-axis planar reference wave as shown in Figure 4.16. All other values for the hologram construction are the same as in Figure 4.14, and the corresponding spot diagram at the paraxial image plane is shown in Figure 4.17. In this case, the extent of the spot diagram is $\Delta x = 0.0058$, $\Delta y = 0.004$ and the aberration coefficients are $C_x = -7.22 \times 10^{-4}$ and $A_x = 6.525 \times 10^{-5}$, showing a significant decrease in astigmatism. This result illustrates the advantage of keeping an object point source on or close to the optical axis of the hologram to minimize aberrations. Notice as well that the position of the reconstructed image point shifts by roughly the same amount in the negative x-direction (i.e., from 1 to approximately 0.9125 for the off-axis object point case and from 0 to $x = -0.0855$ for the on-axis case).

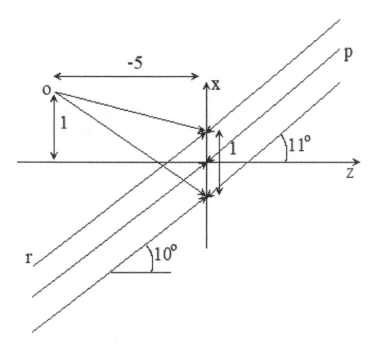

FIGURE 4.14 Construction of a holographic lens with an off-axis point source and planar reference wave.

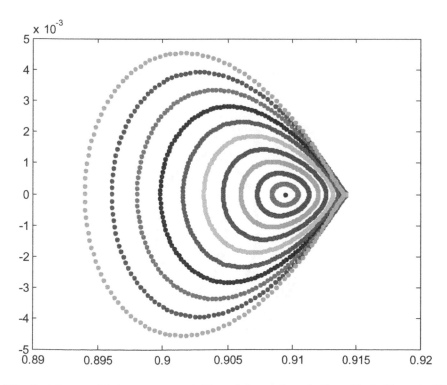

FIGURE 4.15 Spot diagram of the image reconstructed from a hologram formed with an off-axis object point (Figure 4.14). Reconstruction is with a conjugate planar reference wave at 11° to the z-axis. The horizontal axis is x and the vertical axis is y.

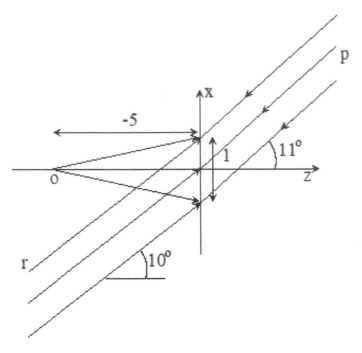

FIGURE 4.16 Construction/reconstruction geometry for holographic lens with an on-axis object point source and off-axis planar reference wave. A conjugate plane wave is used for the reconstruction.

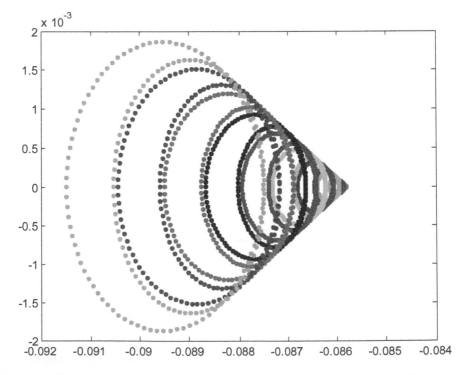

FIGURE 4.17 Spot diagram for the hologram in Figure 4.16 reconstructed with the planar reference wave at 11°. Reconstruction is with a conjugate planar reference wave at 11° to the z-axis. The horizontal axis is x and the vertical axis is y.

4.5 Dispersion Compensation

When a holographic lens is recorded and reconstructed at different wavelengths, the image is displaced and chromatic aberrations form. The effects of wavelength mismatch can be reduced by recording the focal point along the optical axis in the same manner as described in Section 4.4.5 to minimize degradation due to changes in the reconstruction angle. However, this method has limits in correcting the problem. For example, consider the hologram shown in Figure 4.18 recorded with an on-axis object point source located 5 units from the hologram, a diameter of 2 units, and an off-axis reference beam incident at an angle of 25° to the optical axis. If the hologram is constructed at a wavelength of 775 nm and reconstructed with a conjugate beam at this wavelength, the image forms at ($x = 0$, $y = 0$). However, if the hologram is reconstructed with 785 nm light, the image location will shift to (0.0112,0) and the spot size increases considerably as shown in the lower diagram of Figure 4.18.

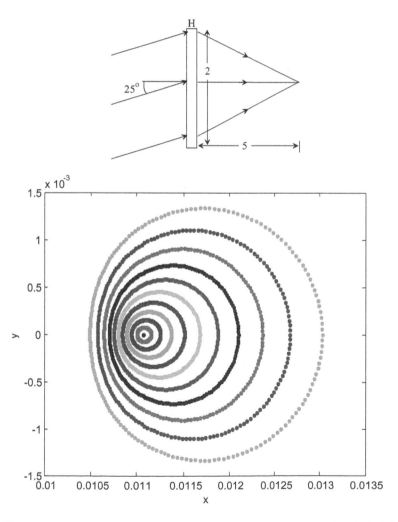

FIGURE 4.18 The upper diagram shows an illustration of the geometry and dimensions for a holographic lens that is recorded at a wavelength of 775 nm. The lower figure shows the corresponding spot diagram at $z = 5$ reconstructed with 785 nm light. The size of the spot has increased, and the location has shifted from the original (0,0) position.

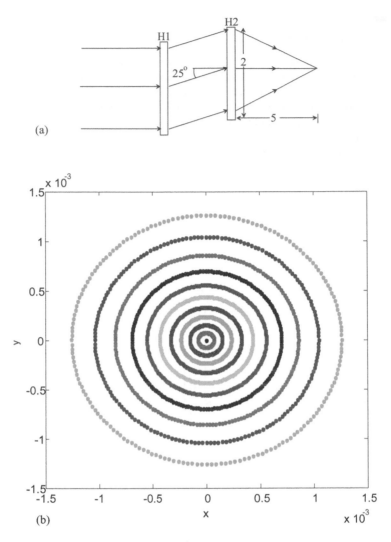

(a)

(b)

FIGURE 4.19 (a) Grating pair configuration for eliminating the shift in the reconstruction image point location with a reconstruction wavelength that differs from the construction wavelength. (b) The resulting spot diagram for the reconstructed image using the setup in a). The holograms were formed with 775 nm light and reconstructed with 785 nm light.

It is possible to eliminate the lateral shift in image position resulting from wavelength change by using the grating pair arrangement shown in Figure 4.19 [7]. In this approach, both gratings H1 and H2 are fabricated with the same wavelength (i.e., 775 nm for this example). Grating H1 is formed with two plane waves at angles equal to the incident and diffracted angle at the center of focusing hologram H2 (in this example at 0° and 25° to the z-axis). When the grating pair is reconstructed with a beam along the axis at a wavelength that differs from the construction wavelength, the beam diffracted by H1 emerges at a new angle that exactly matches the angle necessary for H2 to diffract the beam along the optical axis. As a result, the image does not shift when illuminating the hologram at wavelengths that differ from the construction wavelength as illustrated with the ray trace shown in Figure 4.19.

An experimental demonstration of this correction technique is also shown in Figure 4.20. In this experiment, the holographic lens shown in Figure 4.18 is illuminated with a laser diode and the wavelength is temperature tuned over a range of 2 nm. In this case, the focused spot shifts by 280 µm.

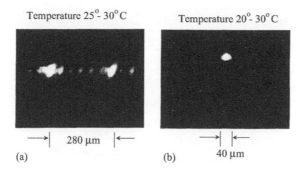

Temperature 25°- 30°C   Temperature 20°- 30°C

|→ 280 µm |←   →| |←
(a)   (b)   40 µm

FIGURE 4.20   Change in the position of focused spot formed with an F/1 holographic lens when illuminated with a laser diode. When the temperature of the laser is changed, the output wavelength also changes. Image (a) is formed with a single holographic lens while image (b) is formed with a second grating to compensate for the lateral shift. In (a) the wavelength changes by 2.1 nm and in (b) by 4.2 nm. (From Kostuk R.K. et al., *Appl. Opt.*, 28, 4939–4944, 1989. With permission from the Optical Society of America.)

However, if a matched grating H1 is used in combination with H2 as shown in Figure 4.20, the focused spot does not shift even when the wavelength is varied by 4 nm.

4.6 Analyzing Holographic Lenses with Optical Design Tools

Most optical system simulation programs such as ZEMAX and CODE V/Synopsis [9,10] have options for using holographic optical elements in the optical system design. Holographic optical elements provide different options for the design that may not be possible with refractive and reflective components. In addition, the powerful optimization routines available in these programs can be applied to improve both the hologram and the overall system performance.

In the ZEMAX software, a holographic lens is entered by specifying it as a "surface type" at a specific location in the system. There are three different ways of designating the hologram surface type: *Hologram 1*, *Hologram 2*, or as an *Optically Fabricated Hologram*. The *Hologram 1* surface type has both construction beams either converging or diverging from point sources; *Hologram 2* allows one of the construction beams to be converging and one diverging from a point source; and, the *Optically Fabricated Hologram* surface type simulates holograms formed with aspheric construction beams. A discussion of hologram fabrication with aspheric construction beams is presented in the next section.

For the *Hologram 1* and 2 surface types, eight parameters are entered to specify the hologram: x_1, y_1, z_1, x_2, y_2, z_2, λ_1, and m. x_1, y_1, z_1, x_2, y_2, z_2 are the coordinates for the point sources for, respectively, the reference and object beams, λ_1 is the construction wavelength, and m is the diffraction order. The aperture diameter and reconstruction wavelengths are specified in the system editors, and the reconstruction angles are specified in the field editors. The image evaluation plane can either be set either as a "thickness" with respect to the hologram surface or determined through a paraxial image "solve." An example of a hologram modeled in ZEMAX is shown in Figure 4.21 with $x_1 = y_1 = 0$; $z_1 = -20$; $x_2 = y_2 = 0$; $z_2 = -10^7$; aperture diameter equal 20; and a construction and reconstruction wavelength of 550 nm. The field angles are 0°, 1°, and 2°. Figure 4.21 shows the layout of the optical system and the three image locations corresponding to the three field angles. The images can be evaluated in a variety of ways including spot diagrams (Figure 4.22) and ray fan plots (Figure 4.23). As indicated in the spot diagrams and ray fan plots, a point image forms when reconstructed with the construction source location and wavelength, however, significant off-axis coma results even with small changes in reconstruction angle (1° and 2°).

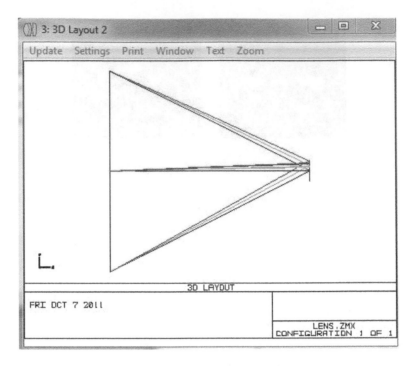

FIGURE 4.21   Zemax layout of a hologram and reconstruction at three field angles: 0°, 1°, and 2°.

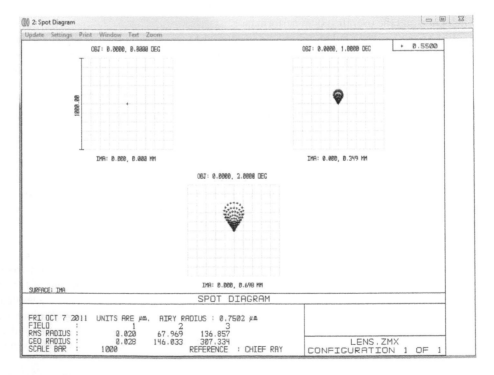

FIGURE 4.22   Spot diagram of the reconstruction rays at three field angles for the hologram example described in the text.

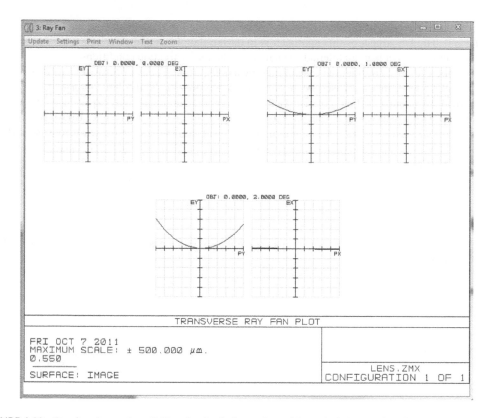

FIGURE 4.23 Ray fan plots at three field angles for the image formed from the hologram described in the text.

4.7 Hologram Formation with Aspheric Wavefronts

We have seen that holograms constructed with spherical wavefronts can form perfect images when the construction and reconstruction conditions are identical. However, it was also shown that the reconstructed image quality rapidly degrades when the reconstruction conditions deviate even by small amounts from the construction values. It is possible to trade off some of the image quality when the reconstruction and construction conditions match for better performance at non-matched reconstruction conditions. One approach to achieve this is to use non-spherical or aspheric wavefronts to record the hologram rather than spherical waves from point sources. The aspheric waves can be formed using a computer-generated hologram. The design and fabrication of computer-generated holograms will be treated in detail in Chapter 6, however, for the present discussion, we will only need to assume that it is an optical component that can form a desired aspherical wavefront.

In order to examine this approach more closely, consider the holographic process discussed previously as a way to convert the phase of an incident wavefront $\varphi_{inc}(x, y, 0)$ into a desired output wavefront $\varphi_{out}(x, y, 0)$ [11]. From this perspective, the hologram produces a phase transformation:

$$\hat{\varphi}_H(x, y) = \varphi_{out}(x, y, z = 0) - \varphi_{inc}(x, y, z = 0). \tag{4.48}$$

The hologram phase $\hat{\varphi}_H$ can also be expressed in terms of the construction conditions:

$$\hat{\varphi}_H(x, y) = \varphi_{obj}(x, y, z = 0) - \varphi_{ref}(x, y, z = 0), \tag{4.49}$$

with φ_{obj} and φ_{ref} being, respectively, the phase of the object and reference fields at the hologram plane located $z = 0$. The key aspect of this approach is to design the object and reference waves to form the desired output phase that is more tolerant to changes in the reconstruction conditions.

As a specific example, consider the case of making a holographic Fourier transform lens that converts an incident plane wave into a focused beam [12]. As shown previously, if a lens is formed with an on-axis spherical wave and an off-axis planar wave, the reconstructed image will experience considerable aberrations at reconstruction angles that slightly differ from the construction angle. However, if instead of using spherical waves to form the hologram, a wave with a quadratic phase wavefront is used, the hologram will become more tolerant to changes in the reconstruction angle [13]. The quadratic phase in this case is given by:

$$\hat{\varphi}_H\left(x, y\right) = -\frac{2\pi}{\lambda} \frac{x^2 + y^2}{2f}, \tag{4.50}$$

where f is the focal length of the Fourier transform lens. If the hologram is then reconstructed with an off-axis plane wave with phase:

$$\varphi_{inc} = \frac{2\pi}{\lambda}\left(\alpha x + \beta y\right), \tag{4.51}$$

the output phase becomes:

$$\begin{aligned}
\varphi_{out} &= \hat{\varphi}_H + \varphi_{inc} \\
&= -\frac{2\pi}{\lambda} \frac{\left(x - f\alpha\right)^2 + \left(y - f\beta\right)^2}{2f} + const.
\end{aligned} \tag{4.52}$$

Therefore, the output phase from the hologram is also a quadratic function that converges to a point $\left(\alpha f, \beta f\right)$ in the focal plane with a radius of curvature that is independent of the incident angle (α, β). A quadratic wavefront can be recorded into the hologram using the setup as shown in Figure 4.24. The results for a hologram with $f = 110$ mm and a diameter of 1 cm that is reconstructed at $0°$ (the ideal reconstruction angle), $10°$, and $20°$ are shown in Figure 4.25 [13]. Note that the image reconstructed at the ideal reconstruction angle $0°$ is somewhat degraded from the image reconstructed from a hologram formed with spherical wavefronts. However, the performance with off-axis or non-construction reconstruction angles is greatly improved.

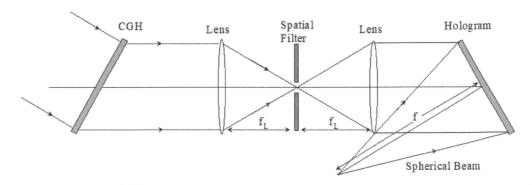

FIGURE 4.24 System for recording an aspheric wavefront diffracted from a computer-generated hologram (CGH) into an analog hologram for aberration correction with non-ideal reconstruction conditions. (From Kedmi, J. and Friesem, A.A., *Appl. Opt.*, 23, 4015–4019, 1984. With permission from the Optical Society of America.)

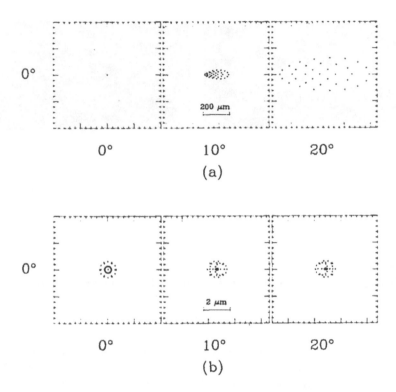

FIGURE 4.25 Image results for a Fourier transform lens with a focal length of 110 mm and a diameter of 10 mm: (a) formed spherical waves and (b) formed with a quadratic reference wave. (From Kedmi, J. and Friesem, A.A., *Appl. Opt.*, 23, 4015–4019, 1984. With permission from the Optical Society of America.)

4.8 Holographic Lenses Recorded and Reconstructed at Different Wavelengths

An important problem in holography is to design a hologram for construction at one wavelength and to reconstruct it with good performance at a different wavelength. The need for this operation results from the somewhat limited wavelength sensitivity of holographic recording materials. For planar gratings or holograms recorded with two collimated plane waves, this is not a problem since there is only one grating vector for the entire aperture, and it can be Bragg matched at different wavelengths by changing the reconstruction angle of incidence. However, as shown earlier, a holographic lens has a grating vector that varies across the aperture. Therefore, one reconstruction angle will only Bragg match one location on the aperture. Contributions from the other points on the aperture will suffer chromatic aberration effects that can be severe. This problem is equivalent to the aberration correction problem described in the previous section and can be corrected by exposing with an aspheric wavefront that compensates for degradation by pre-aberrating the holographic recording material with a computer-generated hologram that produces the desired pre-aberration.

Other correction methods have also been demonstrated that utilize point sources. In a technique described by Latta and Pole [14], the positions of the reference and object beams at a short wavelength (488 nm) are repositioned to optimize both the image quality and diffraction efficiency of the reconstruction at a longer wavelength (632.8 nm). In another approach developed by Amatai et al. [15,16], a series of holograms are recorded with point sources and used to form a final hologram with the correct pre-aberration to compensate for using a different reconstruction wavelength. Figure 4.26 shows that the chromatic aberration is significantly reduced using this approach. In this example, the spot size is reduced from approximately 1 mm–50 µm for a holographic lens with an f-number (f/#) of 2.5 recorded at 488 nm and reconstructed at 632.8 nm.

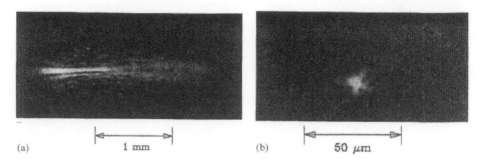

(a) |←—— 1 mm ——→| (b) |←—— 50 μm ——→|

FIGURE 4.26   (a) shows the image from an $f/\#2.5$ holographic lens recorded at 488 nm and reconstructed at 632.8 nm without wavelength compensation. (b) shows the image for a similar $f/\#$ 2.5 holographic lens using the compensation method of Amatai. (From Amitai, Y. et al., *J. Opt. Soc. Am.*, 7, 80–86, 1990. With permission from the Optical Society of America.)

4.9 Combining Image Analysis with Localized Diffraction Efficiency

This chapter has focused on the ray analysis properties of holographic images. However, in order to obtain a true representation of an image formed with a hologram, the diffraction efficiency properties must also be included. The diffraction efficiency describes the power in the image distribution, whereas the ray trace indicates phase matching at the boundary of the grating and surrounding medium. In some cases, a reconstructed image may have excellent ray analysis characteristics, but minimal power at the image due to low diffraction efficiency. The diffraction efficiency of holograms is described in detail in the next chapter. Later in Chapter 11 on holographic optical elements, the complete hologram design process that combines image analysis with the diffraction efficiency is described to provide the full picture of hologram performance.

PROBLEMS

1. Write a ray tracing program (i.e., in MATLAB or MathCad) using the algorithm outlined in Section 4.2.1 for exact ray tracing. Generate a ray intercept plot for the following conditions: (Hologram is 2 cm square. Evaluate the image at z_o)

    a. $z_o = -6$ cm, $z_r = -20$ cm, $z_p = -19.8$ cm, $\lambda_1 = \lambda_2 = 0.60$ μm

    b. $z_o = -12$ cm, $z_r = -\infty$, ($x_p = 225$ cm, $y_p = 0$ cm, $z_p = -10000$ cm), $\lambda_1 = \lambda_2 = 0.60$ μm

    c. $z_o = -7.5$ cm, $z_r = -20$ cm, $z_p = -20$ cm, $\lambda_1 = 0.60$ μm, $\lambda_2 = 0.601$ μm.

2. Determine the paraxial image positions and the longitudinal magnification for the conditions given in question 2 a, b, c.

3. Determine the spherical and coma aberration coefficients for the conditions given in questions 2 a, b, c.

4. A holographic lens is formed with the geometry shown below in Figure 4.27. The object point is located on the z-axis, the reference beam is a plane wave at 30° to the z-axis, and the reconstruction beam is a plane wave at 32° to the z-axis. The construction wavelength is 550 nm and the reconstruction wavelength is 555 nm. Determine the position of the primary paraxial image.

5. Compute the *direction cosines* for the image ray at point A on the hologram aperture in Figure 4.27 for the $q = -1$ diffraction order when the hologram is reconstructed with the conditions shown in the figure and given in problem 4. What is the angle that the diffracted ray makes with the z-axis?

6. A hologram is formed with an object point at $O(1.5,0,-10)$ and a collimated reference beam at 10° to the z-axis shown in the Figure 4.28. The hologram is reconstructed with a plane

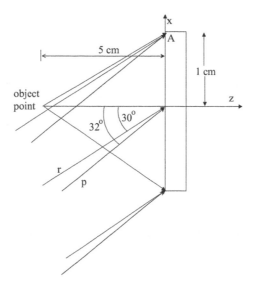

FIGURE 4.27   Figure for Problems 4.4, 4.5, and 4.8.

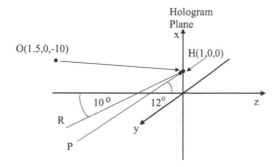

FIGURE 4.28   Figure for Problem 4.6.

wave at 12° to the z-axis with the same wavelength as used during construction. The hologram plane is located at $z = 0$. Consider a point on the hologram at $H(1,0,0)$:

a. Compute the direction cosines of the image ray that is diffracted from the point H on the hologram plane that corresponds to the +1 diffraction order.

b. Compute the intercept point of this ray with the z-axis (i.e., z coordinate of the intercept).

c. Compute the position of the secondary paraxial image (x_4, z_4).

7. Figure 4.29 below shows two configurations (Figure 4.29 a and b) for forming a holographic lens. Indicate which one is better and explain why it is better in terms of the imaging properties of holographic lenses. Note that o, r, and p indicate, respectively, the object point, reference point, and reconstruction point.

8. The spherical aberration coefficient is given by:

$$S_4 = \frac{1}{z_p^3} - \frac{\mu}{z_o^3} + \frac{\mu}{z_r^3} - \frac{1}{z_4^3}$$

For the holographic lenses shown above in Figure 4.27 r and p are point sources at $-\infty$. Derive the relation that shows that when $\lambda_1 = \lambda_2$, the spherical aberration coefficient is 0 when r and p are point sources at $-\infty$.

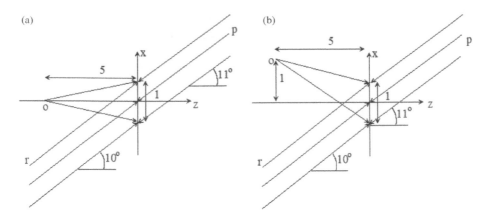

FIGURE 4.29 Figure for Problem 4.7. (a) Holographic lens formed with an on axis object point source and (b) with an off-axis object point source.

REFERENCES

1. R. W. Meir, "Magnification and third-order aberrations in holography," *J. Opt. Soc. Am.*, Vol. 55, 987–992 (1965).
2. J. N. Latta, "Computer-based analysis of holography using ray tracing," *Appl. Opt.*, Vol. 10, 2698–2710 (1971).
3. W. T. Welford, "A vector raytracing equation for hologram lenses of arbitrary shape," *Opt. Comm.*, Vol. 14, 322–323 (1975).
4. H. W. Holloway and R. A. Ferrante, "Computer analysis of holographic systems by means of vector ray tracing," *Appl. Opt.*, Vol. 20, 2081–2084 (1981).
5. J. N. Latta, "Computer-based analysis of hologram imagery and aberrations. I. Hologram types and their nonchromatic aberrations," *Appl. Opt.*, Vol. 10, 599–608 (1971).
6. W. T. Welford, *Aberrations of the Symmetrical Optical System*, Ch. 6, Academic Press, London, UK (1974).
7. D. J. De Bitetto, "White-light viewing of surface holograms by simple dispersion compensation," *Appl. Phy. Lett.*, Vol. 9, 417–418 (1966).
8. R. K. Kostuk, Y.-T. Huang, D. Hetherington, and M. Kato, "Reducing alignment and chromatic sensitivity of holographic optical interconnects with substrate-mode holograms," *Appl. Opt.*, Vol. 28, 4939–4944 (1989).
9. www.zemax.com (accessed on 15 October 2018).
10. www.synopsys.com (accessed on 15 October 2018).
11. K. A. Winick and J. R. Fienup, "Optimum holographic elements recorded with nonspherical wavefronts," *J. Opt. Soc. Am.*, Vol. 73, 208–217 (1983).
12. R. C. Fairchiled and J. R. Feinup, "Computer-originated aspheric holographic optical elements," *Opt. Eng.*, Vol. 21 133–140 (1982).
13. J. Kedmi and A. A. Friesem, "Optimal holographic Fourier-transform lens," *Appl. Opt.*, Vol. 23, 4015–4019 (1984).
14. M. R. Latta and R. V. Pole, "Design techniques for forming 488 nm holographic lenses with reconstruction at 633 nm," *Appl. Opt.*, Vol. 18, 2418–2421 (1979).
15. Y. Amitai, A. A. Friesem, and V. Weiss, "Designing holographic lenses with different recording and readout wavelengths," *J. Opt. Soc. Am.*, Vol. 7, 80–86 (1990).
16. Y. Amitai and J. W. Goodman, "Design of substrate-mode holographic interconnects with different recording and readout wavelengths," *Appl. Opt.*, Vol. 30, 2376–2381 (1991).
17. K. Kostuk, Appearing in Chapter 4, "Optical Holography," *Photonics Technology and Instrumentation*, D. L. Andrews (Eds.), John Wiley & Sons, Hoboken, NJ (2015).

5

Hologram Diffraction Efficiency

5.1 Introduction

The diffraction efficiency describes the amount of optical power diffracted by a hologram into a desired wavefront. More specifically, diffraction efficiency is defined as the power diffracted into one or more diffraction orders relative to the power incident on the hologram. According to this definition, the diffraction efficiency is:

$$\eta_i = \frac{P_i}{P_{inc}}, \tag{5.1}$$

where P_i is the optical power diffracted into the ith order and P_{inc} is the power incident on the hologram. Other definitions for the diffraction efficiency are sometimes used such as the ratio of the diffracted light relative to the optical power remaining in the transmitted zero order. However, this definition does not accurately reflect the performance as an optical element and is not used in the remaining discussion.

Many approaches have been used to analyze and compute the diffraction efficiency of holographic and other grating types. In this chapter, several methods will be reviewed including Fourier analysis of thin gratings, approximate coupled wave analysis of thick or volume gratings, and rigorous coupled wave analysis which while more complex provides the most comprehensive treatment of grating diffraction. A description of the criteria for deciding if a grating is "thin" or "thick" and the appropriate approximate model for analyzing them is also provided. The diffraction efficiency of gratings with absorption and phase modulation, transmission and reflection modes of operation, and with surface relief modulation are investigated. A method for determining the polarization properties of holographic gratings with the approximate coupled wave model is also developed.

5.2 Fourier Analysis of Thin Absorption and Sinusoidal Phase Gratings

The diffraction efficiency of a thin sinusoidal grating can be evaluated in a straightforward way by using far-field diffraction analysis as described in Chapter 2. A grating may be considered "thin" if the period (Λ) is much greater than the thickness (d). Later it will be shown that a more exact definition of a "thin" grating occurs when more than one, non-zero diffraction order occurs. In either case, when the grating is "thin," it can be treated as a transmittance function with periodic modulation of either the optical phase or absorption. The following is a description of these two types of thin holograms based on the analysis of Goodman [1].

5.2.1 Diffraction by a Thin Sinusoidal Absorption Grating

In this case, the transmittance function has sinusoidal absorption modulation in the x_1-direction within a rectangular aperture as shown in Figure 5.1 with:

$$t_M(x_1, y_1) = \left[\frac{1}{2} + \frac{a_M}{2}\cos\left(\frac{2\pi}{\Lambda}x_1\right)\right]rect\left(\frac{x_1}{a_x}\right)rect\left(\frac{y_1}{a_y}\right), \tag{5.2}$$

where a_M is the peak-to-peak variation in the absorption, Λ is the period of the absorption modulation, and a_x and a_y are the widths of the grating aperture in the x- and y-directions, respectively. To compute the far-field diffraction pattern, the Fourier transform of the transmittance function is performed as described in Chapter 2. It is possible to simplify the transform calculation by noticing that the transmittance function is the product of two terms: the modulation term in brackets and the rectangular aperture functions. According to the convolution theorem (Appendix A, Equation A.6), the Fourier transform of the product of two functions is the convolution of the Fourier transform of each function:

$$FT\{u \cdot w\} = U * W. \tag{5.3}$$

In this case, the Fourier transform of the modulation term in brackets is:

$$FT\left\{\frac{1}{2} + \frac{a_M}{2}\cos\left(\frac{2\pi}{\Lambda}x_1\right)\right\} = \frac{1}{2}\delta(v_x, v_y) + \frac{a_M}{4}\delta\left(v_x + \frac{1}{\Lambda}, v_y\right) + \frac{a_M}{4}\delta\left(v_x - \frac{1}{\Lambda}, v_y\right), \tag{5.4}$$

where $v_x = x_2/\lambda z$; $v_y = y_2/\lambda z$, and the transform of the rectangular functions are:

$$FT\left\{rect\left(\frac{x_1}{a_x}\right)rect\left(\frac{y_1}{a_y}\right)\right\} = a_x a_y \text{sinc}(a_x v_x)\text{sinc}(a_y v_y). \tag{5.5}$$

The δ functions simplify the convolution operation and the resulting far-field pattern becomes:

$$E(x_2, y_2) = \frac{a_x a_y}{j2\lambda z}e^{jkz}e^{j\frac{\pi}{\lambda z}(x_2^2 + y_2^2)}\left\{\text{sinc}\left(\frac{a_x x_2}{\lambda z}\right) + \frac{a_M}{2}\text{sinc}\left[\frac{a_x}{\lambda z}\left(x_2 + \frac{\lambda z}{\Lambda}\right)\right] + \frac{a_M}{2}\text{sinc}\left[\frac{a_x}{\lambda z}\left(x_2 - \frac{\lambda z}{\Lambda}\right)\right]\right\}$$

$$\times \text{sinc}\left(\frac{a_y y_2}{\lambda z}\right). \tag{5.6}$$

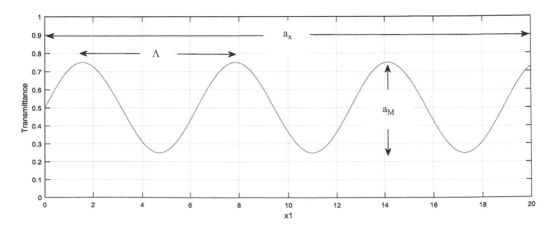

FIGURE 5.1 Transmittance function for a thin absorption grating.

The corresponding intensity distribution is found by taking the product of the field and its complex conjugate. Although there are some interference effects between the different field terms, when $\Lambda \ll a_x, a_y$, most of the power is in three main orders and the resulting intensity function can be approximated as:

$$I(x_2, y_2) = |E(x_2, y_2)|^2$$

$$\simeq \left(\frac{a_x a_y}{2\lambda z}\right)^2 \left\{ \text{sinc}^2\left(\frac{a_x x_2}{\lambda z}\right) + \frac{a_M^2}{4}\text{sinc}^2\left[\frac{a_x}{\lambda z}\left(x_2 + \frac{\lambda z}{\Lambda}\right)\right] + \frac{a_M^2}{4}\text{sinc}^2\left[\frac{a_x}{\lambda z}\left(x_2 - \frac{\lambda z}{\Lambda}\right)\right] \right\} \quad (5.7)$$

$$\times \text{sinc}^2\left(\frac{a_y y_2}{\lambda z}\right).$$

A plot of the intensity is illustrated in Figure 5.2. Note that there are three diffraction orders (-1, 0, $+1$) with each having a width of $\lambda z / a_x$ and have the diffraction pattern of the rectangular aperture of the hologram. The $+$ and $-$ diffraction orders are located at $\pm \lambda z / \Lambda$ from the axis. The corresponding angle that this distance makes with the axis is:

$$\tan \theta_1 = \frac{\lambda z / \Lambda}{z} = \frac{\lambda}{\Lambda} \simeq \sin \theta_1, \quad (5.8)$$

and is exactly the same as that predicted by the grating equation with the incident angle equal to zero.

The diffraction efficiency is equal to the ratio of the diffracted optical power in a particular order to the incident optical power. Using arguments based on the radiance theorem, it can be shown that the power incident on the grating and diffracted into the (-1, 0, $+1$) orders is proportional to:

$$\left[\frac{a_x a_y}{\lambda z}\right]^2. \quad (5.9)$$

Therefore, the peak diffraction efficiency for:

$$\eta_{tot} = \frac{1}{4}\left[1 + \frac{a_M^2}{4} + \frac{a_M^2}{4}\right] = \eta_0 + \eta_{-1} + \eta_{+1}, \quad (5.10)$$

where the value for each sinc function is taken at its maximum value. This result shows that the maximum diffraction efficiency for the $+1$ and -1 orders is 6.25% and 25% for the zero order. The remaining power illuminating the grating is absorbed.

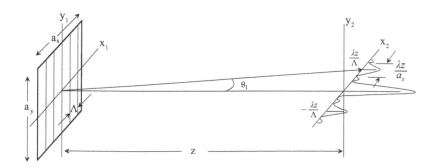

FIGURE 5.2 Diagram illustrating the diffraction pattern from an absorption grating that is illuminated with a normally incident plane wave.

5.2.2 Diffraction by a Thin Sinusoidal Phase Grating

The previous section showed that absorption gratings result in low diffraction efficiency of the non-zero diffraction orders as well as high residual loss. More efficient grating can be formed by modulating the phase of a grating transmittance function. For this analysis, the grating is again considered thin (i.e., $\Lambda \gg d$) with modulation of the refractive index and given by:

$$n(x_1) = n_o + \Delta n \sin(K_x \cdot x_1), \tag{5.11}$$

where n_o is the average refractive index, Δn is the amplitude of the index modulation, and $K_x = 2\pi/\Lambda_x$ with Λ_x the grating period in the x-direction. The corresponding phase modulation is:

$$\varphi(x_1) = \frac{2\pi\, n(x_1)\, d}{\lambda} = \frac{2\pi n_o d}{\lambda} + \frac{2\pi\,\Delta n\, d \sin(K_x \cdot x_1)}{\lambda}$$

$$= \frac{2\pi n_o d}{\lambda} + a_M \sin\left(\frac{2\pi \cdot x_1}{\Lambda_x}\right), \tag{5.12}$$

where the thickness d is a constant. It should be noted that the phase modulation can also be formed by keeping the refractive index constant and modifying the height of the material. This occurs when photoresist is exposed with an interference pattern and processed.

An illustration of this grating is shown in Figure 5.3. It consists of the phase modulation section surrounded by an aperture with dimension a_x in the x-direction and a_y in the y-direction. The corresponding transmittance function is:

$$t_p(x_1) = \exp\left\{ j\frac{2\pi n_o d}{\lambda} + ja_M \sin\left(\frac{2\pi \cdot x_1}{\Lambda_x}\right) \right\} rect\left(\frac{x_1}{a_x}\right) rect\left(\frac{y_1}{a_y}\right)$$

$$= \exp\left[j\frac{2\pi n_o d}{\lambda} \right] \exp\left\{ ja_M \sin\left(\frac{2\pi \cdot x_1}{\Lambda_x}\right) \right\} rect\left(\frac{x_1}{a_x}\right) rect\left(\frac{y_1}{a_y}\right). \tag{5.13}$$

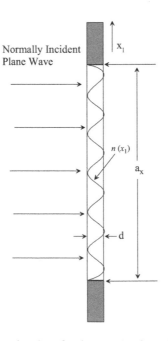

FIGURE 5.3 Diagram of the transmittance function of a phase grating formed by modulating the refractive index and keeping the thickness d constant.

Note that the constant phase term in the first exponential does not affect the resulting intensity of the diffraction pattern and can be dropped.

As in the case of the absorption grating, when illuminated by a normally incident plane wave, the far-field diffraction distribution is determined by taking the Fourier transform of the transmittance function. Here also the transmittance function consists of the product of two terms, an exponential phase and the rectangular aperture function, and the convolution theorem can be applied to simplify the transform operation. To further simplify the computation, the exponential phase function is expressed as a series [1]:

$$\exp\{ja_M \sin(2\pi x_1 / \Lambda_x)\} = \sum_{m=-\infty}^{\infty} J_m(a_M) \exp[j2\pi m x_1 / \Lambda_x], \tag{5.14}$$

where J_m is a Bessel function of the first kind of order m and is not a function of x_1. The Fourier transform of this function is:

$$FT\left\{\exp j[a_M \sin(2\pi x_1 / \Lambda_x)]\right\} = \sum_{m=-\infty}^{\infty} J_m(a_M) \delta\left(\frac{x_2}{\lambda z} - \frac{m}{\Lambda_x}\right) \delta\left(\frac{y_2}{\lambda z}\right), \tag{5.15}$$

where use of the shift theorem (Appendix A, Equation A.5) was made. As before, the Fourier transform of the rect aperture function in the second term is a sinc function. Therefore, the Fourier transform of the transmittance function, and the resulting far-field distribution is:

$$E(x_2, y_2) = \frac{a_x a_y}{j\lambda z} e^{jkz} e^{j\frac{k}{2z}(x_2^2 + y_2^2)} \left[\operatorname{sinc}\left(\frac{a_x x_2}{\lambda z}\right) \operatorname{sinc}\left(\frac{a_y y_2}{\lambda z}\right)\right]$$

$$\otimes \left[\sum_{m=-\infty}^{\infty} J_m(a_M) \delta\left(\frac{x_2}{\lambda z} - \frac{m}{\Lambda_x}\right) \delta\left(\frac{y_2}{\lambda z}\right)\right] \tag{5.16}$$

$$= \frac{a_x a_y}{j\lambda z} e^{jkz} e^{j\frac{k}{2z}(x_2^2 + y_2^2)} \sum_{m=-\infty}^{\infty} J_m(a_M) \operatorname{sinc}\left[\frac{a_x}{\lambda z}(x_2 - m\lambda z / \Lambda_x)\right] \operatorname{sinc}\left(\frac{a_y y_2}{\lambda z}\right),$$

where the symbol $\otimes$ represents the convolution operation. Assuming again that the side-lobe field components are relatively weak compared to the primary diffraction lobes, the approximate intensity distribution is:

$$I(x_2, y_2) = |E(x_2, y_2)|^2$$

$$\simeq \left(\frac{a_x a_y}{\lambda z}\right)^2 \sum_{m=-\infty}^{\infty} J_m^2(a_M) \operatorname{sinc}^2\left[\frac{a_x}{\lambda z}(x_2 - m\lambda z / \Lambda_x)\right] \operatorname{sinc}^2\left(\frac{a_y y_2}{\lambda z}\right). \tag{5.17}$$

Notice that the resulting far-field diffraction pattern consists of a set of sinc² functions weighted by the values of J_m^2 with arguments determined by the amplitude of the phase modulation of the grating (i.e., a_M). The resulting diffraction efficiency for each order is:

$$\eta_m \simeq \left(\frac{J_m(a_M)}{\lambda z}\right)^2, \tag{5.18}$$

where it is assumed that the incident power is proportional to the aperture area of the grating.

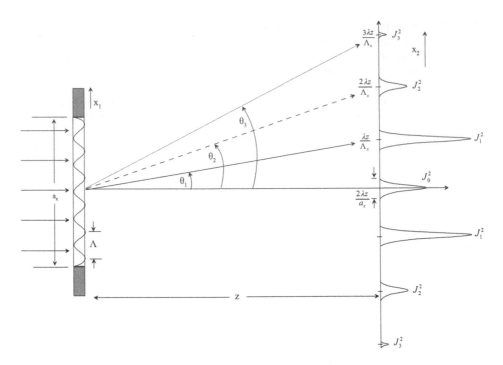

FIGURE 5.4   Diagram showing the far-field diffraction pattern from a grating with $\lambda/\Lambda_x = 1/4$ and $a_M = 1.85$.

The resulting intensity function for the phase grating shows that, unlike the absorption grating where only three diffraction orders form, an infinite number of orders may arise. This of course will not occur because of the limit on the diffraction angle of ±90° for propagating orders. Figure 5.4 shows the ±3 orders for far-field diffraction pattern from a grating with $\lambda/\Lambda_x = 1/4$. The ± 4th diffraction order is diffracted at ±90° and therefore is not shown. The argument for the Bessel functions is $a_M = 1.85$ radians and produces a maximum value for the J_1^2 function of 0.339. Therefore, a significant improvement in diffraction efficiency over absorption type gratings can be realized with phase modulation. Figure 5.5 shows a plot of the square of the Bessel functions of the first kind as

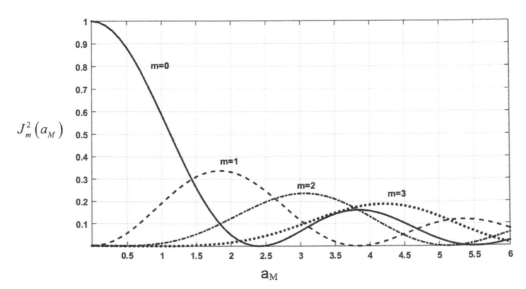

FIGURE 5.5   Plot of $J_m^2(a_M)$ for different values of m for Bessel functions of the first kind.

a function of a_M. It can be seen that with a value for a_M of ~1.25, it is possible to have significant diffracted power in the ±1st orders. In this case, the diffraction efficiency in the ±1st orders is ~0.26 and is significantly greater than the efficiency of an absorption grating (0.0625). Note that at this modulation value there is some efficiency in the $m = 2$ diffraction order (0.04) and significant power remaining in the zero order (0.38). The peak diffraction efficiency for $m = \pm 1$ orders is 0.339 and occurs with a modulation value a_M ~1.85.

5.3 Coupled Wave Analysis

Coupled wave analysis is one of the most useful and intuitive methods for modeling the diffraction efficiency of a volume or thick hologram. Considerable development of this theory has taken place, most notably by Kogelnik, Moharam and Gaylord, and Solymar and Syms [2–4]. In this section, an overview of coupled wave models is provided summarizing the results of prior work and providing working models for design and experimental applications.

Coupled wave models can be divided into two categories: rigorous and approximate methods. The rigorous models can be applied to a grating of any type (i.e., transmission, reflection, thin, or thick) because the diffraction process is evaluated as a fundamental electromagnetic problem. Approximate models impose restrictive assumptions and eliminate certain general aspects of the electromagnetic problem. Although one might feel compelled to focus on the rigorous method, the approximate models work reasonably well when the assumptions are satisfied. Furthermore, the analytical form of the approximate models allows greater insight into the diffraction problem and the ability to include them into other optical simulation programs. As a result, a useful approach to evaluating the diffraction properties of a hologram is to first use an approximate model to obtain an estimate of the diffraction efficiency and then to check accuracy of the results with the rigorous method.

The approximate coupled wave analysis (ACWA) is first examined to provide the essential ideas of coupled wave theory. This description is followed by a discussion of the criteria for deciding if the approximate coupled wave or "Kogelnik" model [2] as it is sometimes referred to is valid. The discussion of the approximate method is followed by an outline of the rigorous coupled wave analysis (RCWA) method as developed by Gaylord and Moharam [3]. Situations where RCWA differs with ACWA are discussed followed by a comparison of the diffraction efficiency predicted by the two approaches for different types of gratings.

5.3.1 Approximate Coupled Wave Analysis ("Kogelnik" Model)

5.3.1.1 Assumptions and Background Conditions

The ACWA model is perhaps the most useful of the approximate approaches. The assumptions of this model are:

1. Only the reconstruction wave and one diffracted wave have significant magnitude
2. The grating is a modulated region embedded in an optical material with the same refractive index. This implies that there are no abrupt boundaries for the grating
3. The grating has infinite extent in the *X–Y* lateral directions and has finite thickness in the longitudinal propagation direction (*Z*)
4. The wavelength λ is the free-space value and the angle θ is inside the medium.

The grating is a modulation of the permittivity or conductivity of the medium. The electromagnetic wave passing through this medium must satisfy the spatial wave or Helmholtz equation:

$$\nabla^2 E(x,z) + k^2 E(x,z) = 0. \tag{5.19}$$

where k is the magnitude of propagation vector:

$$k^2 = \left(\frac{\omega}{c}\right)^2 \varepsilon - j\omega\mu\sigma, \tag{5.20}$$

where ε is the permittivity (farad/m), σ is the conductivity (henry/m), and μ is the free-space permeability. $\omega = 2\pi v$, and v is the optical frequency (cn/λ). Both ε and σ can be modulated in the x- and z-directions to form the grating. The modulation of the material parameters can be expressed as:

$$\varepsilon(\vec{r}) = \varepsilon_a + \varepsilon_1 \cos\left(\vec{K} \cdot \vec{r}\right)$$
$$\sigma(\vec{r}) = \sigma_a + \sigma_1 \cos\left(\vec{K} \cdot \vec{r}\right) \tag{5.21}$$

ε_a and σ_a are the average permittivity and conductivity values for the medium, ε_1 and σ_1 are the modulation amplitude (0 to peak), and $\vec{r}$ is a position vector relative to the origin of a coordinate system. $\vec{K}$ is the grating vector that indicates the spatial frequency of the grating and is in a direction perpendicular to the grating planes (Figure 5.6). As discussed in Chapter 3, the grating vector is a very useful parameter for describing a grating and is defined as:

$$\vec{K} = \frac{2\pi}{\Lambda}\hat{K}$$
$$= \frac{2\pi}{\Lambda}[\sin\Phi\,\hat{x} + \cos\Phi\,\hat{z}] . \tag{5.22}$$

Combining Equations 5.20 and 5.21 yields:

$$k^2 = \beta^2 - 2j\alpha\beta + 2\kappa\beta\left[\exp\left(j\vec{K} \cdot \vec{r}\right) + \exp\left(-j\vec{K} \cdot \vec{r}\right)\right], \tag{5.23}$$

where $\beta^2 = \left(2\pi/\lambda\right)^2 \varepsilon_a$ is the average propagation constant, $\alpha = \left(\mu c\sigma_a/2\sqrt{\varepsilon_a}\right)$ is the average absorption, and the last term represents the grating modulation with κ the coupling constant.

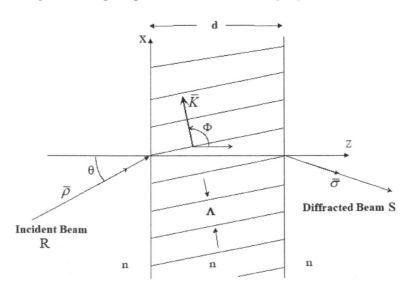

FIGURE 5.6 Schematic diagram showing the basic model for diffraction from a volume hologram.

The general form of κ when the grating has both phase and absorption modulation is:

$$\kappa = \frac{1}{4}\left[\frac{2\pi}{\lambda}\frac{\varepsilon_1}{\varepsilon_a^{1/2}} - j\mu c\frac{\sigma_1}{\varepsilon_a^{1/2}}\right]. \tag{5.24}$$

κ is one of the central parameters in the coupled wave theory. It describes the interaction between the amplitude of the incident beam (R) and the diffracted beam (S). If $\kappa = 0$, the propagation constant k^2 in the wave equation (Equation 5.19) reverts to a simple phase delay and absorption in the medium without the formation of a diffracted wave (S).

The fact that we are primarily concerned with optical materials allows making several useful simplifications to the grating model. First, an optical medium is characterized by a refractive index and absorption constant with the following general properties:

$$\frac{2\pi n}{\lambda} >> \alpha;\ \frac{2\pi n}{\lambda} >> \alpha_1;\sqrt{\varepsilon_a} = n >> n_1. \tag{5.25}$$

The above inequalities simply state that the medium is not highly absorbing, the absorption modulation, α_1, is relatively small compared to the average phase, and that the refractive index modulation, n_1, is a small percentage of the average refractive index. These conditions are necessary for light to propagate through the medium and retain appreciable power. If these conditions are applied to the expression for the coupling constant (Equation 5.24), it can be simplified to:

$$\kappa \cong \frac{\pi n_1}{\lambda} - j\frac{\alpha_1}{2}. \tag{5.26}$$

The next step in setting up the coupled wave analysis is to assume that the total field within the grating consists of just two components:

$$E(x, y, z) = R(z)\exp(-j\vec{\rho}\cdot\vec{r}) + S(z)\exp(-j\vec{\sigma}\cdot\vec{r}), \tag{5.27}$$

where $R(z)$ and $S(z)$ are, respectively, the amplitudes of the reconstruction and diffracted fields, and $\vec{\rho}$ and $\vec{\sigma}$ are, respectively, the propagation constants of the reconstruction and diffracted beams. Other diffraction orders do not have significant power and are not considered in the analysis. It is also assumed that the amplitude of the reconstruction beam $R(z)$ and the diffracted beam $S(z)$ only exchange power in the z- or thickness direction. These assumptions are valid in many experimental situations.

5.3.1.2 The Bragg Condition

A fundamental underlying assumption of coupled wave theory is the relationship between the propagation vectors and the grating vector:

$$\vec{\sigma} = \vec{\rho} - \vec{K}. \tag{5.28}$$

This is the Bragg condition or "K-vector closure" condition described earlier in Chapter 3 and indicates the phase matching between the propagation vectors and the grating. Implicit in Equation 5.27 is that only one diffraction order is involved. When additional diffracted waves are included in the analysis, the Bragg condition becomes:

$$\vec{\sigma} = \vec{\rho} - m\vec{K}, \tag{5.29}$$

where m is the number of diffracted orders. This general form for the Bragg condition will later be used in the rigorous coupled wave analysis.

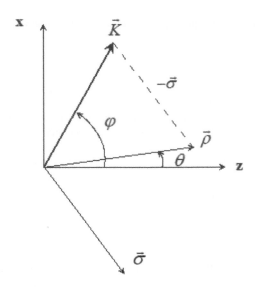

FIGURE 5.7 Diagram showing the vector relation between the propagation and grating vectors. (From Kogelnik, H., *Bell Syst. Tech. J.*, 48, 2909–2947, 1969.)

The Bragg condition (Figure 5.7) has several useful forms that provide insight to understanding the properties of volume holograms. Using the expressions for the propagation and grating vectors:

$$\vec{\rho} = \beta\left[\sin\theta\hat{x} + \cos\theta\hat{z}\right]$$
$$= \rho_x\hat{x} + \rho_z\hat{z} \tag{5.30}$$

$$\vec{K} = \frac{2\pi}{\lambda}\left[\sin\varphi\hat{x} + \cos\varphi\hat{z}\right]$$
$$= K_x\hat{x} + K_z\hat{z}, \tag{5.31}$$

and applying them to the Bragg condition yields:

$$\vec{\sigma} = \beta\left[\left(\sin\theta - \frac{K}{\beta}\sin\varphi\right)\hat{x} + \left(\cos\theta - \frac{K}{\beta}\cos\varphi\right)\hat{z}\right]. \tag{5.32}$$

At the Bragg condition:

$$\left|\vec{\rho}\right| = \left|\vec{\sigma}\right| = \beta. \tag{5.33}$$

Therefore:

$$\beta^2 = \vec{\sigma}\cdot\vec{\sigma} = \sigma_x\cdot\sigma_x + \sigma_z\cdot\sigma_z. \tag{5.34}$$

Combining these expressions yield:

$$\beta^2 = \beta^2\left[\left(\sin\theta - \frac{K}{\beta}\sin\varphi\right)^2 + \left(\cos\theta - \frac{K}{\beta}\cos\varphi\right)^2\right]$$
$$1 = 1 + \left(\frac{K}{\beta}\right)^2 - 2\frac{K}{\beta}\left(\sin\theta\sin\varphi + \cos\theta\cos\varphi\right). \tag{5.35}$$

This results in a scalar form for the Bragg relation:

$$\cos(\varphi - \theta) = \frac{K}{2\beta} = \frac{\lambda}{2n\Lambda}. \tag{5.36}$$

This condition is only satisfied when $\theta = \theta_o$, with θ_o the angle that $\vec{\rho}$ makes with the z-axis during the construction of the grating.

As discussed in Chapter 3, a very useful graphic for interpreting the Bragg condition is the Bragg circle (Figure 5.8). The radius of the Bragg circle is the length of the propagation vectors used during the formation of the hologram (β). However, when the reconstruction vector deviates from the Bragg condition ($\vec{\rho}'$ in Figure 5.8), both ends of the K-vector no longer touch the circle and the diffracted beam propagation vector ($\vec{\sigma}'$) must be longer or shorter than β. This implies that the diffracted beam propagation vector undergoes a change in wavelength since $\beta = 2\pi n/\lambda$ which is not possible in a linear optical system. However, the change in propagation vector length can also be interpreted as a deviation or detuning from the Bragg condition. As will be shown later in this chapter, when detuning increases the power in the diffracted beam decreases. The amount of detuning can be quantified through a parameter ϑ given by:

$$\vartheta \equiv \left(\beta^2 - \sigma^2 \right)$$

$$= K \left[\cos(\varphi - \theta) - \frac{K}{4\pi n} \lambda \right]. \tag{5.37}$$

When the angular and wavelength deviation ($\Delta\theta; \Delta\lambda$) from the Bragg condition ($\theta_0; \lambda_o$) are small such that the reconstruction angle and wavelength can be expressed as $\theta = \theta_o + \Delta\theta$ and $\lambda = \lambda_o + \Delta\lambda$ with $\Delta\theta/\theta_o$; $\Delta\lambda/\lambda \ll 1$, then ϑ can be approximated as:

$$\vartheta \approx \frac{2\pi}{\Lambda} \left[\Delta\theta \sin(\varphi - \theta_o) - \frac{\Delta\lambda}{2n\Lambda} \right]. \tag{5.38}$$

This expression shows that the largest angular selectivity occurs when $\sin(\varphi - \theta_o) = 1$, and the sensitivity or selectivity of the grating is greatest for small values of Λ. The smallest grating period occurs with two counter-propagating construction beams forming an unslanted reflection grating as shown in Figure 5.9. The grating period in this case is $\Lambda = \lambda/2n$. This relation can be used to establish a limit on angular and wavelength selectivity for a volume grating.

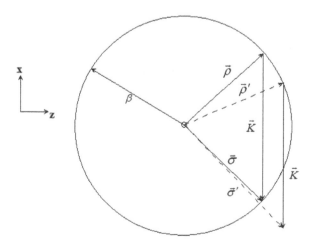

FIGURE 5.8 Bragg circle diagram showing the relation between the propagation vectors and the grating vector during construction and reconstruction of the grating.

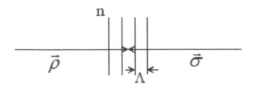

FIGURE 5.9 Construction geometry for an unslanted reflection grating.

5.3.1.3 Dispersion Properties of a Volume Grating

Grating dispersion is a measure of the change in diffraction angle θ with wavelength λ. There are some differences in the dispersion relation for volume and thin gratings. For a volume grating, the dispersion relation can be derived from the Bragg relation by differentiating Equation 5.36 with respect to λ:

$$-\sin(\varphi-\theta)\left(-\frac{d\theta}{d\lambda}\right)=\frac{K}{4\pi n} \tag{5.39}$$

$$\therefore \frac{d\theta}{d\lambda}=\frac{1}{2n\Lambda\sin(\varphi-\theta)}, \tag{5.40}$$

where θ refers to the incident angle of the reconstruction vector ρ.

For comparison, the dispersion relation for a thin grating is derived by differentiating the grating equation:

$$\sin\theta_i-\sin\theta_d=-\frac{\lambda}{\Lambda}$$

$$\frac{d\theta}{d\lambda}=\frac{1}{\cos\theta_d\Lambda}. \tag{5.41}$$

Note that the expressions are similar in that they both depend on $1/\Lambda$. However, there is a subtle difference in the meaning of the two relations. For the thin grating, the relation indicates the change in the direction of phase matching as the incident wavelength or angle varies. Whereas, for a volume grating, $d\theta/d\lambda$ indicates the change in the incident angle to obtain maximum diffraction efficiency (i.e., the Bragg match condition) when reconstructed with a different wavelength. The direction of the new diffracted beam for the volume grating is found from the Bragg relation $\vec{\sigma}=\vec{\rho}-\vec{K}$ using the new wavelength in the expressions for $\vec{\rho}$ and $\vec{\sigma}$.

5.3.1.4 Solving the Coupled Wave Equations

The coupled wave equations are solved by substituting the assumed field (Equation 5.27) into the Helmholtz Equation (5.19) and collecting terms containing the field amplitudes R and S and their derivatives. Noting that:

$$\nabla^2 E=\frac{\partial^2}{\partial x^2}+\frac{\partial^2}{\partial y^2}+\frac{\partial^2}{\partial z^2};$$

$$\vec{\sigma}=\sigma_x\hat{x}+\sigma_y\hat{y}+\sigma_z\hat{z};$$

$$\vec{\rho}=\rho_x\hat{x}+\rho_y\hat{y}+\rho_z\hat{z} \tag{5.42}$$

$$\vec{r}=x\hat{x}+y\hat{y}+z\hat{z}$$

$$\beta^2=\sigma^2=\sigma_x^2+\sigma_y^2+\sigma_z^2$$

$$\beta^2=\rho^2=\rho_x^2+\rho_y^2+\rho_z^2,$$

substituting into the Helmholtz equation, and collecting terms yields the second order differential equations:

$$R'' - 2jR'\rho_z - 2j\alpha\beta R + 2\kappa\beta S = 0$$

$$S'' - 2jS'\sigma_z - 2j\alpha\beta S + \left(\beta^2 - \sigma^2\right)S + 2\kappa\beta R = 0. \tag{5.43}$$

These equations are coupled in the sense that each equation contains both the amplitude of the reconstruction beam R and the diffracted beam amplitude S. The coupling arises from the Bragg relation. It should be noted as well that in the expansion of:

$$2\kappa\beta\left\{\mathrm{Re}^{-j\left[\vec{\rho}-\vec{K}\right]\cdot\vec{r}} + Se^{-j\left[\vec{\sigma}-\vec{K}\right]\cdot\vec{r}} + \mathrm{Re}^{-j\left[\vec{\rho}+\vec{K}\right]\cdot\vec{r}} + Se^{-j\left[\vec{\sigma}+\vec{K}\right]\cdot\vec{r}}\right\}, \tag{5.44}$$

the argument in the first exponential term in brackets is $\vec{\sigma}$, and the argument in the fourth exponential term is $\vec{\rho}$. The other two terms yield propagation vectors for waves which were not assumed to exist and are therefore discarded.

The second order differential equations can be simplified by invoking some physical constraints. Since it was assumed that the modulated region of the grating exists within a continuous medium with no abrupt boundaries, it can also be assumed that the transfer of power between $R(z)$ and $S(z)$ is gradual. In addition, it can also be assumed that absorption occurs slowly when losses exist or not at all for a phase grating. Since the exchange of power occurs slowly, R' and S' are relatively small compared to R and S and R'' and S'' are even smaller in the sense that:

$$R'' \ll \rho_z R'$$

$$S'' \ll \sigma_z S'. \tag{5.45}$$

With these assumptions, R'' and S'' can be neglected and the second order coupled wave equations can be reduced to first order differential equations:

$$c_r R' + \alpha R = -j\kappa S$$

$$c_s S' + \left(\alpha + j\vartheta\right)S = -j\kappa R, \tag{5.46}$$

with

$$c_r = \rho_z/\beta = \cos\theta$$

$$c_s = \sigma_z/\beta = \cos\theta - \frac{K}{\beta}\cos\varphi. \tag{5.47}$$

These simplified coupled wave equations provide a great deal of insight into the exchange of power between the reconstruction and diffracted beams. Some initial observations about the expressions include:

1. Movement in the x- and y-directions does not alter the amplitudes of R and S, but only translates the beams
2. Power is exchanged in the z-direction and the strength of the power exchange is proportional to the coupling coefficient κ
3. When detuning is strong (i.e., large ϑ), the coupling between R and S is small
4. When absorption is strong, coupling is weak.

5.3.1.5 General Solution

The general solution to the approximate coupled wave equations is found by assuming a solution for the fields and then substituting into the coupled wave equations. A typical solution for a first order differential equation is an exponential:

$$R(z) = r_a e^{\gamma_1 z} + r_b e^{-\gamma_2 z}$$

$$S(z) = s_a e^{\gamma_1 z} + s_b e^{-\gamma_2 z},$$

(5.48)

where the parameters r_a, r_b, s_a, s_b are constants that depend on the boundary conditions of the hologram, and the constants γ_1 and γ_2 are parameters that depend on the physical parameters of the grating. The values for γ_1 and γ_2 are found first by substituting the assumed fields into the coupled wave equations. This produces:

$$-j\kappa s_i = (c_r \gamma_i + \alpha) r_i$$

$$-j\kappa r_i = (c_s \gamma_i + \alpha + j\vartheta) s_i,$$

(5.49)

with $i = a, b$. Multiplying the left and right sides together and eliminating the factor $r_i \cdot s_i$ results in:

$$(c_r \gamma_i + \alpha)(c_s \gamma_i + \alpha + j\vartheta) = -\kappa^2.$$

(5.50)

The resulting quadratic equation in γ_i can then be solved giving:

$$\gamma_{a,b} = -\frac{1}{2}\left(\frac{\alpha}{c_r} + \frac{\alpha}{c_s} + j\frac{\vartheta}{c_s}\right) \pm \frac{1}{2}\left[\left(\frac{\alpha}{c_r} - \frac{\alpha}{c_s} - j\frac{\vartheta}{c_s}\right)^2 - 4\frac{\kappa^2}{c_r c_s}\right]^{1/2}$$

(5.51)

with the result having the + sign providing the solution for γ_a, and the result with the − sign providing γ_b.

Having found the values for γ_a and γ_b, the boundary conditions can be applied to find R and S. However, the assumption of having only two fields and no real interface with the grating requires that the boundary conditions for transmission and reflection gratings be treated separately. A graphical illustration of the power transfer between the reconstruction and diffracted beams for these two cases is illustrated in Figure 5.10. For a transmission grating, the diffracted beam amplitude is zero at $z = 0$ and increases as it travels to the $z = d$ boundary. Both beams travel in the same direction. For the reflection grating, the diffracted beam amplitude is zero at the $z = d$ boundary and reaches a maximum at the front surface. In addition, for a reflection grating, the reconstruction and diffracted beams travel in opposite directions.

5.3.1.6 Transmission Grating Field Amplitude

The boundary conditions for a transmission hologram can be summarized with the following set of conditions:

$$R(z = 0) = 1; \; S(z = 0) = 0.$$

(5.52)

Applying these conditions to the assumed field values results in:

$$r_a + r_b = 1 \text{ at } z = 0 \text{ and}$$

$$s_a + s_b = 0 \text{ at}$$

$$z = 0.$$

(5.53)

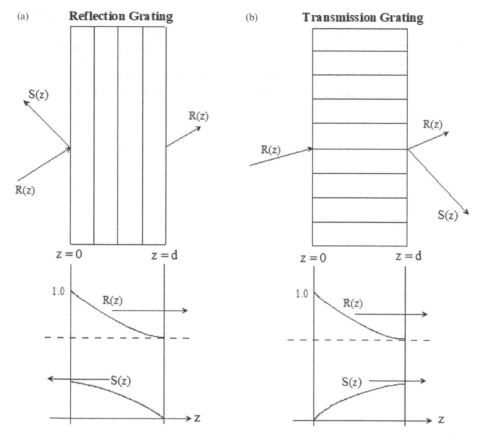

FIGURE 5.10 Graphical illustration of the reconstruction ($R(z)$) and diffracted ($S(z)$) beam amplitudes as they propagate in a volume reflection (a) and transmission (b) grating. (From Kogelnik, H., *Bell Syst. Tech. J.*, 48, 2909–2947, 1969.)

Notice that for the transmission hologram that both field values are defined on the same surface ($z = 0$). Substituting $s_a = -s_b$ into the first order coupled wave equation (Equation 5.49) results in:

$$s_a = -s_b = \frac{-j\kappa}{c_s(\gamma_a - \gamma_b)}.$$ (5.54)

At the boundary $z = d$, the amplitude of the diffracted field is:

$$S(d) = s_a e^{\gamma_1 d} + s_b e^{\gamma_2 d}.$$ (5.55)

Since s_1 and s_2 are constants and have the same values at $z = 0$ and $z = d$, the two expressions can be combined to yield:

$$S(d) = \frac{j\kappa}{c_s(\gamma_a - \gamma_b)}\left[e^{\gamma_a d} - e^{\gamma_b d}\right].$$ (5.56)

This result indicates that the amplitude of the diffracted field only depends on the coupling coefficient and the parameters γ_a and γ_b. These factors contain the absorption, detuning, and the geometrical parameters of the reconstruction process. The expression for $S(d)$ is a general result within the range of validity of the assumptions used to define the two-wave model and are applicable for non-Bragg incidence, different modulation mechanisms, and slanted grating vectors.

The boundary conditions can also be used to determine the corresponding value for the reconstruction wave at the exit face of the grating. For some experimental situations it is easier to measure the power remaining in the reconstruction beam and is therefore useful to have an analytical calculation for $R(d)$ as well [5]. To determine the power in the zero-order transmitted beam, consider the boundary condition at $z = 0$:

$$
\begin{aligned}
r_b &= 1 - r_a \\[4pt]
&= \frac{j\kappa\left(-s_b\right)}{\left(c_r\gamma_b + \alpha\right)} \\[4pt]
&= \frac{j\kappa}{\left(c_r\gamma_b + \alpha\right)}\left(\frac{j\kappa}{c_s\left(\gamma_a - \gamma_b\right)}\right) \\[4pt]
&= \frac{-\kappa^2}{\left(c_r\gamma_b + \alpha\right)c_s\left(\gamma_a - \gamma_b\right)}.
\end{aligned}
\tag{5.57}
$$

The resulting values for r_a and r_b can be substituted into the expression for $R(d)$:

$$
\begin{aligned}
R\left(z = d\right) &= r_a e^{\gamma_1 d} + r_b e^{\gamma_2 d} \\[4pt]
&= e^{\gamma_a d} + r_b\left(e^{\gamma_a d} - e^{\gamma_b d}\right),
\end{aligned}
\tag{5.58}
$$

and gives the amplitude of the transmitted beam.

5.3.1.7 Reflection Grating Field Amplitude

The boundary conditions for the reflection grating are mixed in the sense that they are specified on both the $z = 0$ and $z = d$ surfaces. This results from the fact that the reconstruction and diffracted waves travel in opposite directions. The boundary conditions for the reflection grating are:

$$
R(z = 0) = 1; \; S(z = d) = 0.
\tag{5.59}
$$

Using the assumed field terms and z-values result in:

$$
R(0) = r_a + r_b = 1 \text{ at } z = 0 \text{ and}
$$

$$
S(d) = s_a e^{\gamma_a d} + s_b e^{\gamma_b d} = 0 \text{ at } z = d.
\tag{5.60}
$$

For the reflection hologram, it is desired to find $S(z = 0) = s_a + s_b$ and requires finding the values for s_a and s_b. This can be accomplished by using the expression for $S(d)$ from Equation 5.60 and forming two equations: one by adding $s_a e^{\gamma_b d}$ and the other by subtracting $s_b e^{\gamma_a d}$ from both sides yielding:

$$
\left(s_a + s_b\right)e^{\gamma_b d} = s_a\left[e^{\gamma_b d} - e^{\gamma_a d}\right] - \left(s_a + s_b\right)e^{\gamma_a d} = s_b\left[e^{\gamma_b d} - e^{\gamma_a d}\right].
\tag{5.61}
$$

Taking the second expression for the coupled wave equations (Equations 5.49), substituting in for s_a, s_b, γ_a, and γ_b, and then summing for $i = a, b$ yields:

$$
\left(s_a + s_b\right)\left(\alpha + j\vartheta\right) + c_s\left(\gamma_a s_a + \gamma_b s_b\right) = -j\kappa\left(r_a + r_b\right) = -j\kappa.
\tag{5.62}
$$

Finally, using the fact that $r_a + r_b = 1$ and solving for $(s_a + s_b)$ gives the desired result:

$$S(0) = s_a + s_b = \frac{-j\kappa}{\left[\alpha + j\vartheta + c_s \left(\dfrac{\gamma_1 e^{\gamma_b d} - \gamma_2 e^{\gamma_a d}}{e^{\gamma_b d} - e^{\gamma_a d}} \right) \right]}. \tag{5.63}$$

5.3.1.8 Diffraction Efficiency

The diffraction efficiency of a grating is defined as the power in the diffracted beam divided by the power in the incident beam. Since power exchange only occurs in the z-direction, the projection of power in this direction must be used in the definition to maintain power conservation. Including this term, the expression for the diffraction efficiency becomes:

$$\eta = \frac{|c_s|}{c_r} SS^*. \tag{5.64}$$

The absolute value is used to allow the same expression to be used for reflection gratings where $c_s < 0$.

5.3.1.9 Properties of Specific Grating Types

5.3.1.9.1 Transmission Gratings

As discussed earlier, the diffracted beam for a transmission hologram passes through the thickness and emerges from the grating at $z = d$. Many distinct forms of transmission gratings are possible including those with phase and absorption modulation, unslanted grating vectors, and average absorption. In order to examine these properties, Equation 5.56 is developed into a more intuitive form.

Substituting the expressions for γ_1 and γ_2 into Equation 5.56 yields:

$$S(d) = -j \left(\frac{c_r}{c_s} \right)^{1/2} \exp\left(-\frac{\alpha d}{c_r} \right) e^{\xi} \frac{\sin\left(v^2 - \xi^2\right)^{1/2}}{\left(1 - \xi^2 / v^2\right)^{1/2}}, \tag{5.65}$$

with

$$v = \frac{\kappa d}{\left(c_r c_s\right)^{1/2}} \tag{5.66}$$

$$\xi = \frac{1}{2} d \left(\frac{\alpha}{c_r} - \frac{\alpha}{c_s} - j \frac{\vartheta}{c_s} \right).$$

Note that in this expression the absorption α is the amplitude absorption per unit length. The term v is the grating strength and contains the modulation and grating thickness. The term ξ contains those factors that decrease the strength of the diffracted field including the material absorption and the detuning parameter.

5.3.1.9.1.1 Transmission Phase Gratings
This is perhaps the most important type of volume transmission hologram since it can have the highest diffraction efficiency. In this case, $\alpha = \alpha_1 = 0$ and $\kappa = \frac{\pi n_1}{\lambda}$. The propagation constant becomes:

$$k^2 = \beta^2 + 2\beta\kappa \left[e^{j\vec{K} \cdot \vec{r}} + e^{-j\vec{K} \cdot \vec{r}} \right]. \tag{5.67}$$

The amplitude of the diffracted field after propagating through the grating with thickness d is:

$$S(d) = -j\left(\frac{c_r}{c_s}\right)^{1/2} e^{-j\xi} \frac{\sin\left(v^2 + \xi^2\right)^{1/2}}{\left(1 + \xi^2/v^2\right)^{1/2}}. \tag{5.68}$$

Note the sign change resulting from the j^2 factor in ξ^2. The variables v and ξ become:

$$v = \frac{\pi n_1 d}{\lambda\left(c_r c_s\right)^{1/2}}$$

$$\xi = \frac{\vartheta d}{2c_s}. \tag{5.69}$$

Substituting the diffracted field amplitude into the diffraction efficiency relation yields:

$$\eta = \frac{|c_s|}{c_r} S(d) S^*(d) = \frac{\sin^2\left(v^2 + \xi^2\right)^{1/2}}{\left(1 + \xi^2/v^2\right)}. \tag{5.70}$$

From this relation it can be seen that in order to maximize the diffraction efficiency, $(v^2 + \xi^2)^{1/2}$ should be an odd multiple of $\pi/2$, and the ratio ξ^2/v^2 should be as small as possible. The grating strength parameter v is directly proportional to the index modulation n_1 and thickness d and inversely proportional to λ, $\cos(\theta_r)$, and $\cos(\theta_s)$. The highest diffraction efficiency occurs when $\xi = 0$, implying that the Bragg condition is satisfied or equivalently that $\vartheta = 0$. Figure 5.11 shows the selectivity of a grating with $v = \pi/2$ as a function of the parameter ξ. Since ξ is directly proportional to the detuning parameter ϑ, the bandwidth of this curve indicates the spectral and angular bandwidth of the grating as expressed in Equation 5.38.

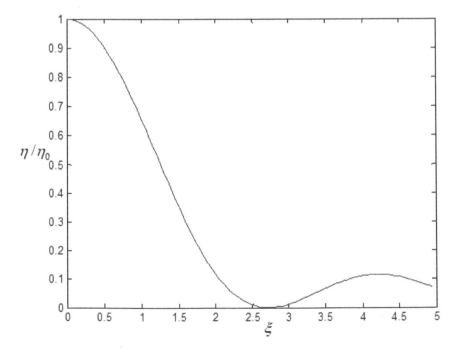

FIGURE 5.11   The normalized diffraction efficiency for a transmission hologram as a function of ξ for $v = \pi/2$.

In the special case when the grating vector $\vec{K}$ is along the x-direction (i.e., $c_r = c_s = \cos\theta_o$; the unslanted transmission grating condition) and $\xi = \vartheta = 0$, simple expressions result for the diffraction order and reconstruction beam efficiencies:

$$\eta_S = \sin^2\left(\frac{\pi n_1 d}{\lambda \cos\theta_o}\right)$$

$$\eta_R = \cos^2\left(\frac{\pi n_1 d}{\lambda \cos\theta_o}\right).$$

(5.71)

A plot of these relations in Figure 5.12 show the oscillatory nature of the coupling between the reconstruction beam and the diffracted beam as a function of v. Whenever the grating strength parameter:

$$v = \frac{n_1 d}{\lambda \cos\theta_o} = m\frac{1}{2},$$

(5.72)

where m is an odd integer, the efficiency of the diffracted beam is a maximum. After the diffraction efficiency reaches a peak, further increases to v result in an "over-coupling" condition that occurs where power from the diffracted beam is coupled back into the reconstruction beam through interaction with the grating.

5.3.1.9.1.2 Transmission Phase Grating with Loss Most holographic materials have some residual absorption and do not result in a pure phase grating. This condition exists in a variety of materials such as silver halide films, photopolymers, and photorefractive crystals. The resulting absorption can be modeled by allowing:

$$\alpha_o \neq 0; \ \alpha_1 = 0; \ \text{and} \ \kappa = \frac{\pi n_1}{\lambda}.$$

(5.73)

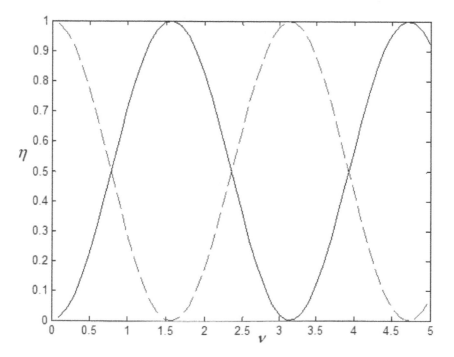

FIGURE 5.12 Signal beam diffraction efficiency (solid curve) and reconstruction beam transmittance (dashed curve) for an unslanted transmission hologram.

In this case, the propagation constant is:

$$k^2 = \beta^2 - 2j\alpha\beta + 2\kappa\beta \left[e^{jK \cdot r} + e^{-jK \cdot r} \right].$$ (5.74)

Using this expression and the simplified coupling coefficient in the general equation for a transmission grating (Equation 5.65) gives the diffraction efficiency for a transmission phase grating with loss:

$$\eta = \exp\left(-\frac{2\alpha d}{c_r} \right) e^{2\xi} \frac{\sin^2 \left(v^2 - \xi^2 \right)^{1/2}}{\left(1 - \xi^2/v^2 \right)}.$$ (5.75)

In this case, the parameters v and ξ are:

$$v = \frac{\kappa d}{\left(c_r c_s \right)^{1/2}}$$

$$\xi = \frac{1}{2} d \left(\frac{\alpha}{c_r} - \frac{\alpha}{c_s} - j\frac{\vartheta}{c_s} \right),$$ (5.76)

with α being the average absorption of the field amplitude. Therefore, the diffraction efficiency has a similar form to the pure phase grating with an added factor for the average loss within the grating as the beam propagates through the thickness d.

5.3.1.9.1.3 Absorption Transmission Gratings   It is also possible to modulate the absorption to form a volume absorption grating. This situation is similar to that described earlier in Section 5.2.1 for thin absorption gratings. In this case, however, the thickness plays a role in determining the diffraction efficiency. The volume absorption grating model is useful for evaluating low scatter noise master holograms formed using unbleached photographic film processing techniques. The parameters for an absorption grating are:

$$n_1 = 0$$

$$\kappa = -j\alpha_1/2$$ (5.77)

$$\alpha \neq 0.$$

For an unslanted transmission grating with $c_r = c_s = \cos\theta_o$, the diffracted field amplitude after propagating through the grating is:

$$S(d) = e^{-\alpha d/c_r} e^{-j\xi} \frac{\sinh\left(v^2 - \xi^2 \right)^{1/2}}{\left(1 - \xi^2 / v^2 \right)^{1/2}},$$ (5.78)

with

$$v = \frac{\alpha_1 d}{2\cos\theta}$$

$$\xi = \frac{\vartheta d}{2\cos\theta} \approx \Delta\theta\beta d \sin\theta_o - \frac{1}{2} \frac{\Delta\lambda}{\lambda} Kd \tan\theta_o.$$ (5.79)

The approximation for ξ in Equation 5.79 is for small deviations of θ and λ about the Bragg condition. At the Bragg condition, the diffraction efficiency can be written as:

$$\eta = e^{-2\alpha d/\cos\theta_o} \sinh^2\left(\frac{\alpha_1 d}{2\cos\theta_o}\right). \tag{5.80}$$

Notice that the diffraction efficiency is the product of two terms: the first being a decaying exponential and the second an increasing $\sinh^2$ function. Both the average absorption and the modulation absorption depend on the projected propagation distance through the grating (i.e., $d/\cos\theta$). The maximum absorption modulation occurs when $\alpha_1 = \alpha$ (i.e., the absorption modulation amplitude is equal to the bias absorption value). Applying these two conditions to Equation 5.80 shows that the highest efficiency occurs when:

$$\frac{\alpha_1 d}{\cos\theta} = 1.1 \approx \ln 3, \tag{5.81}$$

resulting in $\eta_{max} \approx 3.7\%$. Notice that the maximum efficiency of the volume absorption grating is significantly less than the 6.25% found earlier for the thin absorption grating. This is primarily due to the need to include the absorption of the diffracted beam as it propagates through the grating thickness as expressed in the factor $e^{-2\alpha d/\cos\theta}$ in Equation 5.80.

5.3.1.9.1.4 Mixed Phase and Absorption Transmission Gratings In several situations, it is possible to form gratings that have both phase and absorption modulation or $n_1 \neq 0$ and $\alpha_1 \neq 0$. This situation occurs in partially bleached silver halide films and in some photopolymers. In this case, the coupling coefficient becomes:

$$\kappa = \frac{\pi n_1}{\lambda} - j\frac{\alpha_1}{2}. \tag{5.82}$$

For unslanted transmission gratings at the Bragg reconstruction condition, the diffracted field amplitude is:

$$S = -j\exp(-\alpha_o d/\cos\theta)\sin(\kappa d/\cos\theta), \tag{5.83}$$

and the corresponding diffraction efficiency:

$$\eta = \left[\sin^2\left(\frac{\pi n_1 d}{\lambda\cos\theta}\right) + \sinh\left(\frac{\alpha_1 d}{2\cos\theta}\right)\right]\exp(-2\alpha_o d/\cos\theta). \tag{5.84}$$

Notice that for the absorption grating component there is residual average absorption with propagation through the thickness of the grating and will reduce the overall diffraction efficiency.

5.3.1.9.2 Reflection Gratings

A reflection grating diffracts light back into the incident medium as shown in Figure 5.10. Using the reflection grating field amplitude, Equation 5.63, the general expression for the diffraction efficiency that allows for slanted gratings and off-Bragg reconstruction conditions is of the form:

$$\eta = \frac{|c_s|}{c_r}\frac{\kappa^2}{\left[\alpha + j\vartheta + c_s\left(\dfrac{\gamma_a e^{\gamma_b d} - \gamma_b e^{\gamma_a d}}{e^{\gamma_b d} - e^{\gamma_a d}}\right)\right]^2}, \tag{5.85}$$

where the denominator of the second term is complex due to the factors $j\vartheta$, γ_a, and γ_b.

This expression can be simplified by considering the specific properties of phase and absorption gratings, and these situations are evaluated in the following sections.

5.3.1.9.2.1 Reflection Phase Grating   In this case, the average absorption and absorption modulation are zero (i.e., $\alpha = \alpha_1 = 0$), and the coupling coefficient $\kappa = \pi n_1 / \lambda$. The resulting field amplitude is:

$$S(0) = \sqrt{\frac{c_r}{c_s}} \frac{1}{\left[j\frac{\xi}{v} + \left(1 - \left(\frac{\xi}{v} \right)^2 \right)^{1/2} \coth\left(v^2 - \xi^2 \right)^{1/2} \right]}$$

$$\xi = -\vartheta d / 2 c_s \tag{5.86}$$

$$v = \frac{j\pi n_1 d}{\lambda \sqrt{(c_r c_s)}}.$$

Since the diffracted beam propagates in the $-z$-direction, c_s is negative and v will be real. Using these parameters in the relation for the diffraction efficiency (Equation 5.64) gives:

$$\eta = \frac{1}{1 + \left(\frac{1 - (\xi/v)^2}{\sinh^2 \sqrt{(v^2 - \xi^2)}} \right)}. \tag{5.87}$$

When the reflection hologram is reconstructed at the Bragg condition, the diffraction efficiency becomes:

$$\eta = \frac{\sinh^2 v}{\sinh^2 v + 1}. \tag{5.88}$$

The corresponding diffraction efficiency is shown in Figure 5.13 and increases gradually toward a maximum of 100% as a function of the grating strength parameter v. If the $\sinh^2$ function in the denominator is expanded, the efficiency at the Bragg condition for a reflection hologram reduces to:

$$\eta = \tanh^2 \left(\frac{j\pi n_1 d}{\lambda \sqrt{(c_r c_s)}} \right). \tag{5.89}$$

Note that since $c_s < 0$ that the argument of $\tanh^2$ is real. A further simplification occurs if the grating planes lie along the x-direction forming an unslanted reflection grating and the hologram is reconstructed at the Bragg condition. In this case, $\cos\theta_o = c_r = -c_s$, and the imaginary number j can be eliminated from the argument resulting in:

$$\eta = \tanh^2 \left(\frac{\pi n_1 d}{\lambda \cos\theta_o} \right). \tag{5.90}$$

This relation along with the corresponding expression for unslanted transmission gratings (Equation 5.71) are useful in providing a quick estimate of the diffraction efficiency for these two basic types of volume gratings.

At non-Bragg reconstruction conditions ($\xi \neq 0$), the efficiency relation allows us to investigate the angular and wavelength selectivity of the grating through the parameter ξ which incorporates the detuning term ϑ. Figure 5.14 shows the diffraction efficiency for a phase modulated reflection grating with

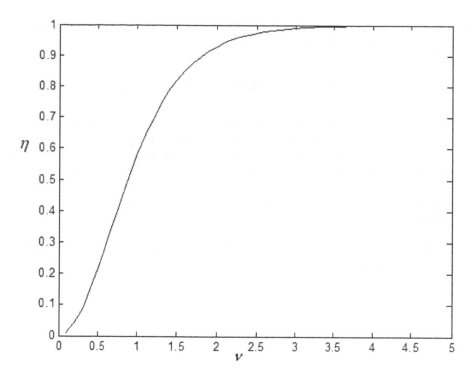

FIGURE 5.13 Diffraction efficiency for a reflection hologram reconstructed at the Bragg condition as a function of the grating strength parameter V.

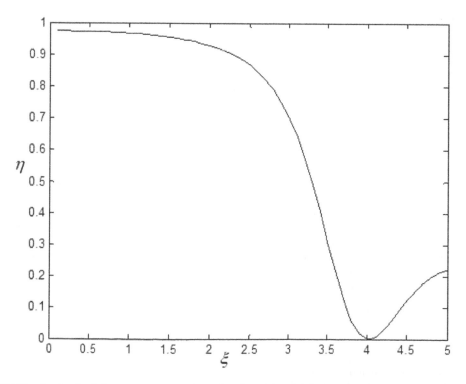

FIGURE 5.14 A reflection hologram reconstructed as a function of variation from the Bragg condition (i.e., as a function of ξ). In this case, $\xi = 0.8\pi$.

$v = 0.8\pi$ that has a maximum diffraction efficiency of 97.4% when $\xi = 0$. As shown earlier for small angle and wavelength changes about the Bragg condition, ξ can be expressed as:

$$\xi = \Delta\theta\pi d \sin\left(\theta_o - \phi\right)/\left(\Lambda c_s\right) + \Delta\lambda\pi d /\left(2\Lambda^2 c_s\right). \tag{5.91}$$

This expression shows that increasing the effective thickness of the grating (d) will make the hologram more sensitive to small deviations in wavelength or angle about the Bragg condition. Higher wavelength and angle selectivity of the volume reflection grating can also be achieved with smaller period gratings. In addition, the wavelength selectivity increases with $1/\Lambda^2$, therefore making it more sensitive than deviation of angle from about the Bragg condition. The control of angular and wavelength selectivity is particularly useful in the design of grating filters and wavefront sensors as will be shown later in the sections on applications.

5.3.1.9.2.2 Reflection Phase Grating with Loss   Just as in the case of a transmission grating, a common situation for reflection gratings is to have average absorption with phase modulation. This situation occurs with:

$$\alpha_o \neq 0; \; \alpha_1 = 0; \; and \; \kappa = \frac{\pi n_1}{\lambda}. \tag{5.92}$$

Substituting these parameters into the general expression for the diffracted amplitude of a reflection hologram (Equation 5.85) and the diffraction efficiency (Equation 5.64) gives:

$$\eta = \frac{\left|c_s\right|}{c_r} \frac{\kappa^2}{\left[\alpha_o + j\vartheta + c_s\left(\dfrac{\gamma_1 e^{\gamma_2 d} - \gamma_2 e^{\gamma_1 d}}{e^{\gamma_2 d} - e^{\gamma_1 d}}\right)\right]^2}. \tag{5.93}$$

This expression can be used to evaluate either unslanted or slanted gratings that are reconstructed on or off the Bragg condition.

Plots of the diffraction efficiency as a function of reconstruction angle for holograms formed in a material with a thickness of 10 μm, an index modulation of 0.05, average refractive index of 1.50, and a wavelength of 632.8 nm are shown in Figures 5.15 and 5.16. The hologram modeled in Figure 5.15 is unslanted (i.e., formed with one construction angle at 0° and the other at 180°). The hologram formed in Figure 5.16 is formed with propagation vectors at 0° and 150° relative to the z-axis and a grating vector, $\vec{K}$, slanted relative to the z-axis. The solid curves show the diffraction efficiency with no average absorption in the recording material while the dotted curve shows the diffraction efficiency with average amplitude absorption of 0.02/μm. The plots show several interesting features about reflection gratings. First, the unslanted reflection grating has a much broader angular bandwidth than a slanted grating formed in a recording material with the same thickness, absorption, and refractive index modulation. Also as might be expected, the angular bandwidth for the slanted grating is asymmetric. However, the drop in peak diffraction efficiency with absorption is the same for both geometries (~15%). Another characteristic feature shown in the plot for the slanted grating is the upwelling of the nulls in the diffraction efficiency plot when average absorption is present.

5.3.1.9.2.3 Reflection Absorption Grating   In this case, the average absorption, α_0, and the absorption modulation, α_1, are not equal to zero, and the coupling coefficient is $\kappa = -j\alpha_1/2$. The effective thickness of the grating is determined by the depth that the reconstruction beam can penetrate and still allow the reflected diffracted beam to emerge from the material as shown in Figure 5.17.

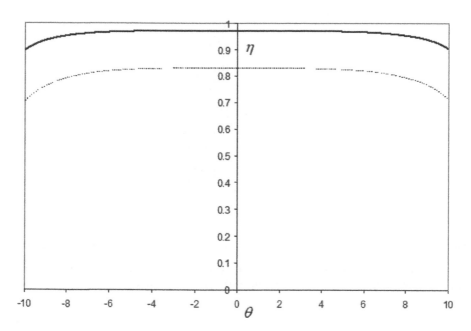

FIGURE 5.15 Light curve shows the diffraction efficiency as a function of reconstruction angle for an unslanted reflection grating with amplitude absorption of 0.02/μm, 10 μm thick, and a refractive index modulation of 0.05. The darker curve is the same grating without absorption.

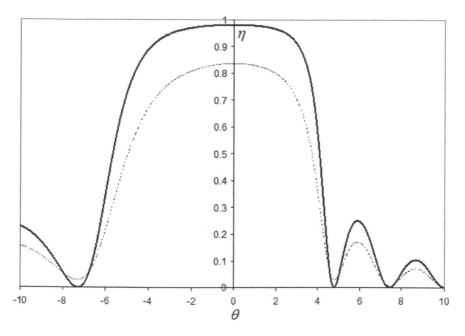

FIGURE 5.16 Diffraction efficiency as a function of reconstruction angle for a 10 μm thick slanted reflection grating with construction angles at 0° and 150°, average refractive index of 1.50, and refractive index modulation of 0.05. The dark plot indicates the efficiency when the average absorption is zero and the lighter curve is when the amplitude absorption is 0.02/μm.

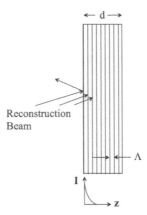

FIGURE 5.17 Reconstruction beam penetration and diffracted beam from an absorption reflection hologram.

Therefore, an optimum average absorption exists that maximizes the diffraction efficiency. For an unslanted reflection grating, the expression for the diffraction efficiency is:

$$\eta = \frac{1}{\left| \xi/v + \left[(\xi/v)^2 - 1 \right]^{1/2} \coth\left[v\left(\frac{\xi^2}{v^2} - 1 \right)^{1/2} \right] \right|^2},$$
(5.94)

with

$$v = \frac{\alpha_1 d}{2\cos\theta_o}$$

$$\xi = \frac{\alpha_o d}{\cos\theta_o} - j\chi_o$$
(5.95)

$$\chi_o = \Delta\theta\,\beta\,d\sin\theta_o + \frac{\Delta\lambda K d}{2\lambda}.$$

The expression for χ_o allows for small deviations in angle and wavelength about the Bragg wavelength and angle. When the Bragg condition is satisfied, $\chi_o = 0$ and α_1 is a maximum (i.e., $\alpha_1 = \alpha_o$), $\xi/v = 2$, and since $\coth(x) \to 1$ as $x \to \infty$, the diffraction efficiency for the unslanted reflection absorption grating approaches a maximum value of:

$$\eta = \frac{1}{\left[2 + \sqrt{3}\coth(v\sqrt{3}) \right]^2} \to \frac{1}{\left(2 + \sqrt{3} \right)^2} = 0.0718.$$
(5.96)

Therefore, the maximum diffraction efficiency for an unslanted reflection absorption grating is greater than an absorption transmission grating and the thin transmission absorption grating discussed earlier.

5.3.1.9.2.4 Mixed Phase and Absorption Reflection Gratings   Just as in the case of transmission gratings, reflection holograms can have mixed phase and absorption modulation. In this situation, $n_1 \neq 0$; $\alpha_1 \neq 0$ and also requires that the average absorption in the recording material is not zero (i.e., $\alpha_o \neq 0$). The coupling coefficient for a mixed reflection hologram is:

$$\kappa = \frac{\pi n_1}{\lambda} - j\frac{\alpha_1}{2},$$
(5.97)

and the corresponding diffraction efficiency for an unslanted reflection grating is:

$$
\eta = \left| \frac{\kappa}{\left[\alpha_o + \left(\kappa^2 + \alpha_o^2 \right)^{1/2} \coth\left\{ \dfrac{d}{\cos\theta_o} \sqrt{\left(\kappa^2 + \alpha_o^2 \right)} \right\} \right]} \right|^2 ,
\tag{5.98}
$$

where d is the hologram thickness and $\cos\theta_o = c_r = |c_s|$. Note that $d/\cos\theta_o$ is the path length of the reconstruction beam as well as the diffracted beam in the hologram. The diffraction efficiency saturates as d and κ increase and $\coth h(x) \to 1$. The maximum diffraction efficiency also increases as α_o decreases.

5.3.1.10 Polarization Aspects of Volume Holograms Using ACWA

An important aspect of Bragg holograms is their polarization properties. In general, as the hologram becomes more selective in angle and wavelength, it will also become more sensitive to the polarization state of the reconstruction beam. This can be a problem in some cases, but may also serve as a useful design parameter for specialized optical elements. To a certain extent, the polarization properties of volume gratings can be controlled in the hologram design process. In Kogelnik's classic paper [2], he describes a simple approach to explain the differences in diffraction efficiency for linearly polarized light that is either in the plane of incidence or perpendicular to this plane. Some have argued that this is a simplistic approach and is only true for these two states of polarization [3, p. 77], but it still provides an insight into the basic polarization sensitivity of volume holograms and is quite useful for hologram design. For a more complete description of the polarization characteristics of holograms, rigorous coupled wave theory can be used.

The assumptions for Kogelnik's treatment of polarization with the approximate coupled wave model include:

1. Only two orthogonal polarization states are considered: in the plane of incidence and perpendicular to this plane
2. No coupling is assumed between the polarization states
3. Only linear polarization states exist.

For this analysis, the approximate coupled wave model is modified to specify the in-plane (x–z) and out-of-plane (y–z) linear polarization states of the reconstruction field as shown in Figure 5.18.

In this case, a more general form for the Helmholtz equation is used:

$$
\nabla^2 \vec{E} - \nabla\left(\nabla \cdot \vec{E} \right) + k^2 \vec{E} = 0,
\tag{5.99}
$$

with the vector form for the allowed field given by:

$$
\vec{E} = \vec{R}(z) e^{-j\vec{\rho}\cdot\vec{r}} + \vec{S}(z) e^{-j\vec{\sigma}\cdot\vec{r}},
\tag{5.100}
$$

where the amplitudes are $\vec{A}(z) = A_x(z)\hat{x} + A_y(z)\hat{y} + A_z(z)\hat{z}$ with $\vec{A}(z) = \vec{R}(z), \vec{S}(z)$. Since the fields are transverse electro-magnetic waves, $\vec{R} \cdot \vec{\rho} = \vec{S} \cdot \vec{\sigma} = 0$. Using these relations in the general Helmholtz equation and neglecting second order derivatives results in a set of first order coupled wave equations:

$$
\begin{aligned}
-2j\rho_z\vec{R}' + j\vec{\rho}R_z' - 2j\alpha\beta\vec{R} + 2\kappa\beta\vec{S} &= 0 \\
-2j\sigma_z\vec{S}' + j\vec{\sigma}S_z' + \left(\beta^2 - \sigma^2 - 2j\alpha\beta \right)\vec{S} + 2\kappa\beta\vec{R} &= 0.
\end{aligned}
\tag{5.101}
$$

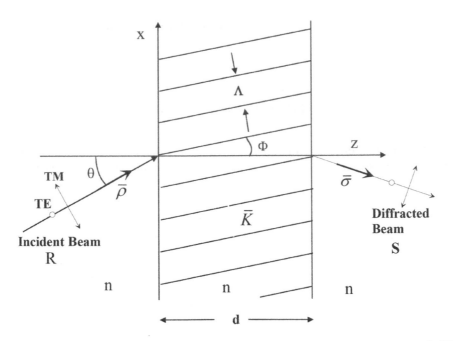

FIGURE 5.18 Hologram reconstruction geometry showing transverse electric (TE) and transverse magnetic (TM) reconstruction fields.

Applying the restriction that only orthogonal linear polarization states exist and that no polarization conversion between these states occurs within the grating, it is possible to write:

$$\vec{R}(z) = R(z)\hat{r}$$
$$\vec{S}(z) = S(z)\hat{s} \, ,$$

(5.102)

where $\hat{r}$ and $\hat{s}$ are unit polarization vectors that are not a function of z. The unit vectors also satisfy the condition that $\hat{r} \cdot \hat{r} = 1$ and $\hat{s} \cdot \hat{s} = 1$.

Using the assumed polarization properties, an additional vector expression can be specified:

$$\hat{r} \cdot \hat{\rho} = 0$$
$$\hat{s} \cdot \hat{\sigma} = 0 \, ,$$

(5.103)

which is another statement of the transverse electro-magnetic nature of the fields. Taking the dot product of $\hat{r}$ with the first coupled wave equation (Equation 5.101) and $\hat{s}$ with the second equation yields:

$$0 = -2j\rho_z R' - 2j\alpha\beta R + 2\kappa\beta S \left(\hat{s} \cdot \hat{r}\right)$$
$$0 = -2j\sigma_z S' + \left(\beta^2 - \sigma^2 - 2j\alpha\beta\right)S + 2\kappa\beta R (\hat{s} \cdot \hat{r}).$$

(5.104)

Using definitions described earlier:

$$c_r = \frac{\rho_z}{\beta}, c_s = \frac{\sigma_z}{\beta}$$

$$\vartheta = \frac{\beta^2 - \sigma^2}{2\beta},$$

(5.105)

the coupled wave equations become:

$$-j\kappa(\hat{s}\cdot\hat{r})S = c_r R' + \alpha R$$
$$-j\kappa(\hat{s}\cdot\hat{r})R = c_s S' + (\alpha + j\vartheta)S \ . \tag{5.106}$$

When the incident field is transverse electric (TE) polarized, $\hat{s}\cdot\hat{r}=1$ and κ is the same coupling coefficient found earlier (Equation 5.26). However, for transverse magnetic (TM) polarized light incident on the grating, the coupling coefficient becomes $\kappa_\parallel = \kappa(\hat{s}\cdot\hat{r}) = -\kappa\cos 2(\theta_o - \varphi)$. The coupled wave equations for the TM polarized state can be evaluated simply by replacing κ with $\kappa_\parallel$.

There are two cases where a volume grating will diffract light in one state of polarization, but not the orthogonal state. In the first case, consider an unslanted transmission grating ($\varphi = 90°$) with a Bragg angle of $\theta_o = 45°$ in the medium of the hologram with refractive index n. In this case, $\kappa_\parallel = 0$ and the diffracted beam is completely TE polarized and can be used to realize a holographic polarizer. However, one issue with this approach is that a 45° beam inside the medium will typically exceed the critical angle of the optical material for the hologram (i.e., materials with refractive indices greater than 1.414).

An alternative approach for making a holographic polarizer can be found by examining the argument of the diffraction efficiency relation for an unslanted transmission grating:

$$\eta = \sin^2[C\hat{s}\cdot\hat{r}], \tag{5.107}$$

with $C = \dfrac{\pi \Delta n d}{\lambda \cos\theta_o}$. In this case, the diffraction efficiency with incident TE and TM polarized light is:

$$\eta_{TE} = \sin^2(C)$$
$$\eta_{TM} = \sin^2[C\cdot\cos(2\theta_o)]. \tag{5.108}$$

Therefore, if $C = \pi \rightarrow \eta_{TE} = 0$. At the same time, the diffraction efficiency of the TM polarization state can be maximized when $C\cdot\cos(2\theta_o) = \frac{\pi}{2}$ or $\cos 2\theta_o = \frac{1}{2}$ and corresponds to a Bragg angle of $\theta_o = 30°$ inside the medium. This result can be achieved without exceeding the critical TIR angle for most useful optical materials.

5.3.2 Criteria for Using "Thin" and "Thick" Grating Models

When using the rigorous coupled wave theory, it is not necessary to distinguish between the "thin" and "thick" hologram regime, however, it is important to have some justification for using the approximate coupled wave model. Otherwise, the analysis will lead to incorrect results. Earlier it was argued that in a thick grating the field interacts many times as it propagates through the grating. Therefore, the diffracted light has a chance to strengthen and lead to large coupling effects. For several interactions to occur, the three physical lengths of interest in the grating: λ, d, and Λ should have the correct relationship. Two relative factors of importance are the number of grating periods within the grating thickness (d/Λ) and the relative size of the wavelength and grating period ($\lambda/n\Lambda$). Kogelnik combined these terms into a metric or quality factor to determine if the grating is "thick" [2] and is given by:

$$Q = \frac{2\pi\lambda d}{n\Lambda^2}. \tag{5.109}$$

In most cases, when $Q \geq \sim 10$ the approximate coupled wave theory will give reasonably good agreement with experimental results. However, in some cases even when this condition is true there can be significant discrepancies with experimental results [6]. In particular, multiple diffraction orders with significant power can emerge from the grating during the reconstruction process even when Q exceeds 10.

Therefore, other criteria such as the occurrence of only a single diffraction order is a better test of whether the grating is acting as a "thick" grating.

Moharam and Gaylord evaluated the conditions for "thin" and "thick" gratings and are explained in detail in references [6] and [7]. In their analysis, they compare the diffraction efficiency properties of unslanted phase transmission gratings. As shown earlier, the function that describes the diffraction efficiency of a "thin" grating (Section 5.2.2) has multiple diffraction orders and is of the form:

$$\eta_i = J_i^2(2v)$$
$$v = \frac{\pi \Delta n d}{\lambda \cos \theta_o}, \tag{5.110}$$

where i indicates a specific diffraction order. In contrast, for the case of thick or volume gratings, the approximate coupled wave theory assumes that there is only the reconstruction beam and one diffraction order. For an unslanted transmission grating, the resulting diffraction efficiency for the single diffraction order is:

$$\eta = \sin^2 v. \tag{5.111}$$

Moharam and Gaylord compared the error between the results predicted by the two approximate diffraction efficiency theories to those determined by an exact or rigorous coupled wave model using the relations:

$$\left| \eta_i - J_i^2(2v) \right| \le \varepsilon$$
$$\left| \eta_{i=1} - \sin^2 v \right| \le \varepsilon, \tag{5.112}$$

where ε is the absolute error in diffraction efficiency, and η_i and $\eta_{i=1}$ are the diffraction efficiencies predicted by the rigorous model. If an arbitrary error limit of 1% is used to establish the boundary for when the "thin" or "thick" grating criteria applies, then when:

$$Qv / \cos \theta_o \le 1, \tag{5.113}$$

the grating is in the "thin grating limit," and when:

$$\frac{Q}{2(\cos \theta_0)v} = \frac{\lambda^2}{n \Delta n \Lambda^2} \ge 10, \tag{5.114}$$

the grating is in the "thick" or "volume" grating region. In both, result θ_0 is the angle of the reconstruction beam at the Bragg condition. Note that for the thick or volume grating the results do not directly depend on the thickness (d) of the grating.

5.3.3 Rigorous Coupled Wave Analysis

RCWA is based on a complete electromagnetic model of the diffraction process from a grating. Unlike the approximate methods, no assumptions about the number of diffraction orders or the thickness of the grating are made and both transmission and reflection orders are allowed at the same time. In this section, the main features of RCWA are described to illustrate the approach and to compare the predicted diffraction efficiency characteristics to those based on approximate coupled wave models. An extensive discussion of RCWA is provided in the review paper by Gaylord and Moharam [3] as well as in references [8] and [9].

5.3.3.1 Properties of the Electric Field within the Grating

The grating configuration for RCWA is illustrated in Figure 5.19. The grating consists of a periodic modulation in the permittivity between $z = 0$ and $z = d$ similar to that used for ACWA. However, in this case, a different dielectric material can surround the grating forming a discontinuity at the front and back interface. This situation is more realistic than the ACWA case where the grating was assumed to be embedded in a material with the same dielectric constant. In general, the grating modulation can be described as:

$$\varepsilon_2(r) = \sum_{q=-\infty}^{\infty} \tilde{\varepsilon}_{2,q} \exp\left(jq\vec{K} \cdot \vec{r} \right), \tag{5.115}$$

where $\tilde{\varepsilon}_{2,q}$ is a complex permittivity constant and q is an integer. The summation describes an arbitrary modulation profile, however, for this discussion and comparison with the results for ACWA, a sinusoidal phase modulation is assumed with:

$$\varepsilon_2(r) = \varepsilon_{02} + \bar{\varepsilon}_2 \cos\left(\vec{K} \cdot \vec{r} \right), \tag{5.116}$$

where ε_{02} is the average permittivity in the grating region, and $\bar{\varepsilon}_2$ is the amplitude of the permittivity modulation.

The electric field within the grating region is not specifically known so it can be described in many ways. The only requirements are that the field satisfies the wave (or Helmholtz) equation:

$$\nabla^2 E(r) + k^2 \varepsilon(r) E(r) = 0, \tag{5.117}$$

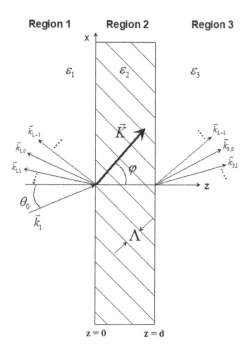

FIGURE 5.19 Grating model for the RCWA. $\vec{K}$ is the grating vector, $\vec{k}_1$ is the incident propagation vector, $\vec{k}_{1,i}$, $\vec{k}_{3,i}$ are the diffracted propagation vectors in regions 1 and 3, φ is the angle that the grating vector $\vec{K}$ makes with the z axis, and Λ is the grating period. (From Gaylord, T.K. and Moharam, M.G., *Proc. IEEE*, 73, 894–937, 1985.)

and the electromagnetic boundary conditions at $z = 0$ and $z = d$. An expansion of the electric field (TE polarized) within the grating region based on the Bragg or Floquet condition can be written as:

$$
\begin{aligned}
E_2(x,z) &= \sum_{i=-\infty}^{+\infty} S_{2i}(z) \exp\left\{-j\left[(k_{2x} - iK\sin\varphi)x + (k_{2z} - iK\cos\varphi)z\right]\right\} \\
&= \sum_{i=-\infty}^{+\infty} S_{2i}(z) \exp\left\{-j\left(\vec{k}_2 - i\vec{K}\right)\cdot\vec{r}\right\} \\
&= \sum_{i=-\infty}^{+\infty} S_{2i}(z) \exp\left\{-j\vec{\sigma}_{2,i}\cdot\vec{r}\right\}.
\end{aligned}
\tag{5.118}
$$

The next step is to solve for the $S_{2i}(z)$ amplitude functions and relate them to the transmission and reflection diffraction orders that form outside of the grating region. In Equation 5.118, the $S_{2i}(z)$ amplitudes are only a function of z. The propagation vector $\vec{k}_2$ is the ray direction of the incident reconstruction beam after it refracts into the grating (Region 2) from Region 1. The vector $\vec{\sigma}_{2,i}$ is the propagation vector of the harmonic components with amplitude $S_{2i}(z)$ within the grating. $\vec{k}_2$ and $\vec{\sigma}_{2,i}$ are related through the Bragg condition:

$$
\vec{\sigma}_{2,i} = \vec{k}_2 - i\vec{K}.
\tag{5.119}
$$

Substituting the expression for the field E_2 and the permittivity function into the wave Equation (5.17) and collecting terms result in a set of coupled wave equations that are of the form:

$$
\frac{1}{(2\pi)^2}\frac{d^2 S_{2i}(z)}{dz^2} - \frac{j}{\pi}\left[\frac{\sqrt{\varepsilon_{02}}\cos\theta}{\lambda} - i\frac{\cos\varphi}{\Lambda}\right]\frac{dS_{2i}(z)}{dz} + \frac{2i(q_B - i)}{\Lambda^2}S_{2i}(z)
$$
$$
+ \frac{\bar{\varepsilon}_2}{\lambda^2}\left[S_{2i-1}(z) + S_{2i+1}(z)\right] = 0.
\tag{5.120}
$$

This system of differential equations has several similarities to the approximate coupled wave model discussed earlier. For instance, in the approximate model the reconstruction beam R was coupled to a signal beam S, whereas in the rigorous model the *i*th diffracted beam, $S_{2i}(z)$, is coupled to the two adjacent orders $S_{2i-1}(z)$ and $S_{2i+1}(z)$. For the rigorous model, the strength of this coupling is related to the phase modulation of the grating $\bar{\varepsilon}_{02}$ just as it is for the approximate coupled wave model through the coupling coefficient $\kappa = \pi n_1/\lambda$. There is also a Bragg condition for both models which when satisfied maximizes the coupling between different diffraction orders. For the rigorous coupled wave model, the Bragg condition is:

$$
q_B = \frac{2\Lambda n_{02}}{\lambda}\cos(\theta_2 - \varphi),
\tag{5.121}
$$

where θ_2 is the angle of the reconstruction beam in the grating (Region 2), and φ is the angle of the grating vector relative to the z-axis. When q_B is an integer (*i*), the Bragg condition is satisfied for the *i*th order. For both the approximate and rigorous models, when the Bragg condition is not satisfied the diffraction efficiency can still be computed, but it will in general be lower than at the Bragg condition. A major difference between the two models is the retention of the second order derivative in the rigorous model. This leads to greater generality of the conditions in which the resultant diffracted fields are calculated, but adds more computational complexity. For instance, the boundary conditions are treated exactly the same which allows simultaneous treatment of multiple transmission and reflection diffraction orders. In the approximate coupled wave model, the grating must be designated as a transmission or reflection grating and only one diffraction order and the reconstruction beam are considered.

5.3.3.2 Fields Outside the Grating Region

The primary goal of the analysis is to determine the power in the diffraction orders that propagate outside the grating region. In general, the total field in Region 1 of Figure 5.19 is:

$$E_1(r) = \exp(-j\vec{k}_1 \cdot \vec{r}) + \sum_{i=-\infty}^{+\infty} R_{1,i} \exp(-j\vec{k}_{1i} \cdot \vec{r}), \tag{5.122}$$

where $\vec{k}_1$ is the propagation vector of the incident reconstruction beam, $\vec{k}_{1i}$ are the propagation vectors of the diffracted reflection orders, and $R_{1,i}$ are the normalized amplitudes of the diffracted reflection orders. In a similar manner, the total normalized field in the transmission Region 3 can be written as:

$$E_3(r) = \sum_{i=-\infty}^{+\infty} T_{3,i} \exp\left[-j\vec{k}_{3,i} \cdot (\vec{r} - d\hat{z})\right], \tag{5.123}$$

where $\vec{k}_{3,i}$ are the propagation vectors, and $T_{3,i}$ are the normalized amplitudes of the transmission diffraction orders.

The reflection ($\vec{k}_{1,i}$) and transmission ($\vec{k}_{3,i}$) propagation vectors outside of the grating region must be phase matched to the propagation vectors of the field within the grating region ($\vec{\sigma}_{2,i}$). This can be done by phase matching the propagation vectors along the two grating surfaces in the x-direction:

$$\vec{k}_{1,i} \cdot \hat{x} = \vec{\sigma}_{2,i} \cdot \hat{x} = \vec{k}_{3,i} \cdot \hat{x}, \tag{5.124}$$

with the magnitude of the propagation vectors in Region 1 equal to $2\pi\sqrt{\varepsilon_1}/\lambda_0 = 2\pi n_1/\lambda_0$, and in Region 3, $2\pi\sqrt{\varepsilon_3}/\lambda_0 = 2\pi n_3/\lambda_0$. The corresponding z-components of the propagation vectors are:

$$\vec{k}_{1,i} \cdot \hat{z} = \sqrt{|\vec{k}_1|^2 - (\vec{k}_{1,i} \cdot \hat{x})^2} = \sqrt{k_1^2 - (k_2 \sin\theta - iK \sin\varphi)^2}$$

$$\vec{k}_{3,i} \cdot \hat{z} = \sqrt{|\vec{k}_3|^2 - (\vec{k}_{1,i} \cdot \hat{x})^2} = \sqrt{k_3^2 - (k_2 \sin\theta - iK \sin\varphi)^2}, \tag{5.125}$$

where the last expression in each equation results from using the Floquet condition (Equation 5.119) for the propagation vectors of the diffraction orders within the grating. Note that as the number of diffraction orders increase, the term within the radical will eventually become negative and the terms become imaginary. When this happens the diffraction order is evanescent.

5.3.3.3 Solving for the Amplitudes of the Diffraction Orders

The next step in the analysis is to determine the amplitudes of the reflection ($R_{1,i}$) and transmission ($T_{3,i}$) diffraction orders. This is accomplished by solving the set of coupled wave equations (Equation 5.120) by assuming solutions for the fields and applying the electromagnetic boundary conditions. Since there is an infinite set of coupled wave equations, a more powerful approach is needed than that used for ACWA. Magnusson, Moharam, and Gaylord were the first to successfully solve this complicated problem [8,9]. To do so, they used a linear systems state-space technique that effectively reduces the second order differential coupled wave equations to first order differential equations. The technique involves the specification of state variables defined as:

$$S_{a,i} = S_{2i}(z)$$

$$S_{b,i} = \frac{dS_{2i}(z)}{dz}. \tag{5.126}$$

Substituting the state variable representation into the coupled wave equations results in two sets of first order homogeneous differential equations:

$$\frac{dS_{a,i}(z)}{dz} = S_{b,i}(z)$$

$$\frac{dS_{b,i}(z)}{dz} = -\frac{2\pi\bar{\varepsilon}_2}{\lambda^2}S_{a,i-1}(z) + \frac{4\pi^2 i(i-q_B)}{\Lambda^2}S_{a,i}(z) - \frac{2\pi^2\bar{\varepsilon}_2}{\lambda^2}S_{a,i+1}(z) \qquad (5.127)$$

$$+ j4\pi\left(\frac{n_{02}\cos\theta}{\lambda} - \frac{i\cos\varphi}{\Lambda}\right)S_{b,i}(z).$$

The two sets of state variable equations can be written in matrix form as:

$$\begin{bmatrix} \dot{S}_{a,i} \\ \dot{S}_{b,i} \end{bmatrix} = \begin{bmatrix} 0 & I \\ B & C \end{bmatrix}\begin{bmatrix} S_{a,i} \\ S_{b,i} \end{bmatrix}$$

$$\underline{\underline{A}} = \begin{bmatrix} 0 & I \\ B & C \end{bmatrix} \qquad (5.128)$$

$$\begin{bmatrix} \dot{\underline{S}} \end{bmatrix} = \underline{\underline{A}}\begin{bmatrix} \underline{S} \end{bmatrix},$$

with

$$0 = \cdots\begin{bmatrix} & \vdots & \\ 0 & 0 & 0 \\ 0 & 0 & 0 \\ 0 & 0 & 0 \\ & \vdots & \end{bmatrix}\cdots ; I = \cdots\begin{bmatrix} & \vdots & \\ 1 & 0 & 0 \\ 0 & 1 & 0 \\ 0 & 0 & 1 \\ & \vdots & \end{bmatrix}\cdots$$

$$(5.129)$$

$$B = \cdots\begin{bmatrix} & \vdots & \\ b_{-1} & a & 0 \\ a & b_0 & a \\ 0 & a & b_{+1} \\ & \vdots & \end{bmatrix}\cdots \quad C = \cdots\begin{bmatrix} & \vdots & \\ c_{-1} & 0 & 0 \\ 0 & c_0 & 0 \\ 0 & 0 & c_{+1} \\ & \vdots & \end{bmatrix}\cdots,$$

and with the coefficients defined as:

$$a = -2\pi^2\bar{\varepsilon}_2 / \lambda^2$$

$$b_i = 4\pi^2 i(i - q_B) / \Lambda^2 \qquad (5.130)$$

$$c_i = j4\pi\,[n_{20}\cos\theta / \lambda - i\cos\varphi / \Lambda].$$

Note that the matrix $\underline{\underline{A}}$ contains the physical parameters that describe the gratings such as the permittivity modulation $\bar{\varepsilon}_2$ and the grating period Λ. Inserting the matrices into Equation 5.128 gives the state variable representation for the coupled wave equations:

$$
\begin{bmatrix}
\vdots \\
\dot{S}_{a,-1} \\
\dot{S}_{a,0} \\
\dot{S}_{a,+1} \\
\vdots \\
\dot{S}_{b,-1} \\
\dot{S}_{b,0} \\
\dot{S}_{b,+1} \\
\vdots
\end{bmatrix}
=
\begin{bmatrix}
& \vdots & & & \vdots & & & \vdots & \\
\cdots & 0 & 0 & 0 & \cdots & 1 & 0 & 0 & \cdots \\
& 0 & 0 & 0 & & 0 & 1 & 0 & \\
& 0 & 0 & 0 & & 0 & 0 & 1 & \\
& \vdots & & & & \vdots & & & \\
& b_{-1} & a & & & c_{-1} & 0 & 0 & \\
\cdots & a & b_0 & a & \cdots & 0 & c_0 & 0 & \cdots \\
& & a & b_1 & & 0 & 0 & c_1 & \\
& \vdots & & & & \vdots & & &
\end{bmatrix}
\cdot
\begin{bmatrix}
\vdots \\
S_{a,-1} \\
S_{a,0} \\
S_{a,+1} \\
\vdots \\
S_{b,-1} \\
S_{b,0} \\
S_{b,+1} \\
\vdots
\end{bmatrix}.
\tag{5.131}
$$

The solutions to the first order differential equations for the state variables are exponential functions of the form:

$$
S_{h,i}(z) = \sum_{m=-\infty}^{+\infty} C_m w_{h,im} \exp(\lambda_m z),
\tag{5.132}
$$

with the subscript $h = a, b$ corresponding to the two state variables, and C_m are constants determined from the boundary conditions. λ_m are eigenvalues and $w_{h,im}$ are elements of an eigenvector. The eigenvalues and eigenvectors are obtained from the coefficient matrix $\underline{\underline{A}}$ using linear algebra techniques [10].

Theoretically, the number of orders i is infinite, however, practical solutions require limiting the number of orders to a finite value N. This implies that each of the sub matrices ($0, I, B,$ and C) have $N \times N$ elements. Therefore, the dimension of the coefficient matrix $\underline{\underline{A}} \rightarrow 2N \times 2N$ when the number of diffracted orders in Regions 1 and 3 is N. The resulting $2N$ solutions can now be written as:

$$
S_{h,k}(z) = \sum_{m=1}^{2} \sum_{q=1}^{N} C_{m,q} w_{h,k;m,q} \exp(\lambda_{m,q} z),
\tag{5.133}
$$

where $h = a, b$ for each state variable, and q ranges over the number of diffraction orders that are retained in the analysis. The eigenvalues $\lambda_{m,q}$ corresponding to the matrix $\underline{\underline{A}}$ are determined using the linear systems operation:

$$
\left| \underline{\underline{A}} - \lambda_{m,q} \underline{I} \right| = 0.
\tag{5.134}
$$

Similarly, the eigenvectors can be found by applying Cramer's Rule to the matrix $\underline{\underline{A}}$. In practice, both the eigenvalues and eigenvectors can be found using routines incorporated into programs such as MATLAB [11]. For instance, in MATLAB the eigenvalues and eigenvectors can be found using the "*eig*" operator: $\left[\lambda_m ; w_{m,q} \right] = \text{eig}\left(\underline{\underline{A}} \right)$.

At this point, the unknown parameters are the coefficients $C_{m,q}$ and the desired amplitude values for the reflection (R_i) and transmission (T_i) diffraction orders. These are obtained by using the boundary conditions that match the fields within the grating to the diffraction orders outside the grating. This requires using the boundary conditions for the electric and magnetic fields and the phase matching conditions (Equation 5.124) at $z = 0$ and $z = d$. The magnetic fields can be determined using a simplified form from the Maxwell equations (see Chapter 2):

$$
H_x = -\frac{j}{\omega\mu} \frac{\partial E_y}{\partial z}.
\tag{5.135}
$$

The boundary conditions provide four sets of equations for: the electric and magnetic fields at $z = 0$ and the electric and magnetic fields at $z = d$. In expanded form, these are:

1. Tangential electric field boundary conditions at $z = 0$:

$$\delta_{i0} + R_i = S_i(0) \tag{5.136}$$

2. Tangential magnetic field boundary conditions at $z = 0$:

$$j\left[k_1^2 - \left(k_2 \sin\theta_2 - iK \sin\varphi\right)^2\right](R_i - \delta_{i0}) = \frac{dS_i(0)}{dz} - j\left(k_2 \cos\theta_2 - iK \cos\varphi\right)S_i(0) \tag{5.137}$$

3. Tangential electric field boundary conditions at $z = d$:

$$T_i = S_i(d)\exp\left[-j\left(k_2 \cos\theta_2 - iK \cos\varphi\right)d\right] \tag{5.138}$$

4. Tangential magnetic field boundary conditions at $z = d$:

$$-j\left[k_3^2 - \left(k_2 \sin\theta_2 - iK \sin\varphi\right)^2\right]^{1/2} T_i$$

$$= \left[\frac{dS_i(d)}{dz} - j\left(k_2 \cos\theta_2 - iK \cos\varphi\right)S_i(d)\right] \cdot \exp\left[-j\left(k_2 \cos\theta_2 - iK \cos\varphi\right)d\right]. \tag{5.139}$$

Note that δ_{i0} corresponds to the incident reconstruction beam, with $\exp(-j\vec{k}_1 \cdot \vec{r})$ as previously described in Equation 5.122. For a set of $i = N$ diffraction orders there exists 4N boundary conditions. Equations 5.136 through 5.139 are now written as a vector-matrix relation:

$$\underline{S} = \underline{\underline{B}} \cdot \underline{X}, \tag{5.140}$$

where $\underline{S}$ is a $4N \times 1$ vector containing the boundary conditions, the elements of the $4N \times 4N$ matrix $\underline{\underline{B}}$ are the coefficients of the S_i parameters, and the $4N \times 1$ vector $\underline{X}$ is:

$$\underline{X} = \begin{bmatrix} C_1 \\ \vdots \\ C_{2N} \\ R_1 \\ R_2 \\ \vdots \\ R_N \\ T_1 \\ T_2 \\ \vdots \\ T_N \end{bmatrix}. \tag{5.141}$$

As discussed earlier, the parameters of interest are the R_i and T_i amplitudes. These are found by multiplying $\underline{\underline{B}}^{-1}$ on both sides of Equation 5.140. This gives the vector $\underline{X}$ which contains the desired amplitudes.

The diffraction efficiency of the different diffraction orders are found by taking the ratio of the power of a diffracted order with respect to the power incident in the power flow or z-direction. Assuming that the incident power is 1, the diffraction efficiency of the reflection orders are:

$$\eta_{1i} = \mathrm{Re}\left\{ \frac{\vec{k}_{1i} \cdot \hat{z}}{\vec{k}_{10} \cdot \hat{z}} \right\} R_i \cdot R_i^*, \tag{5.142}$$

and

$$\eta_{3i} = \mathrm{Re}\left\{ \frac{\vec{k}_{3i} \cdot \hat{z}}{\vec{k}_{30} \cdot \hat{z}} \right\} T_i \cdot T_i^*. \tag{5.143}$$

for the transmitted diffraction orders.

5.3.4 Comparison of RCWA with ACWA and Special Grating Cases

Moharam and Gaylord compared the results of RCWA to ACWA for unslanted and slanted transmission and reflection gratings [9]. Their results showed significant differences (approaching 20%) for the +1 diffraction order for slanted gratings. Also with slanted gratings, both transmission and reflection orders with significant diffraction efficiency occur. This should be expected since in this case the grating vector has components along both the *x*- and *z*-directions and the distinction of being strictly a transmission or reflection grating becomes vague. The resulting differences between the approximate and rigorous coupled wave methods indicate that when considerable accuracy is needed, RCWA should be used to evaluate grating diffraction efficiency. In addition, for cases which do not qualify for the volume regime, RCWA should be used to determine both the diffraction efficiency and number of significant diffraction orders that occur.

In the case of unslanted reflection gratings, the RCWA approach given above breaks down due to singularities for fields propagating along the optical axis. However, Moharam and Gaylord have shown that a chain matrix analysis method provides an effective solution for this situation [12]. In the special case of an unslanted reflection grating, the rigorous chain matrix approach and ACWA are in very close agreement.

RCWA can also be used to determine the diffraction efficiency of surface relief gratings [13]. Gratings of this type can be formed in photoresist using holographic exposure methods, etching using photolithographic methods, or by embossing techniques [14,15]. An illustration of a typical surface relief grating is shown in Figure 5.20. In this case, a triangular grooved relief grating exists in Region 2 and is surrounded by a material with a dielectric permittivity of ε_1 above in Region 1 and permittivity ε_3 below in Region 3. The grating has a period Λ and a total thickness of *d*. In order to solve the problem of computing the diffraction efficiency, the grating structure in Region 2 is divided into a set of P, thin slabs. Each thin slab has a different width of permittivity ε_3 which varies with depth (z_p). The permittivity for each slab is described as:

$$\varepsilon_p(x, z_p) = \varepsilon_1 + (\varepsilon_3 - \varepsilon_1) \sum_{q=-\infty}^{\infty} \tilde{a}_{q,p} \exp\left(j \frac{2\pi}{\Lambda} qx \right), \tag{5.144}$$

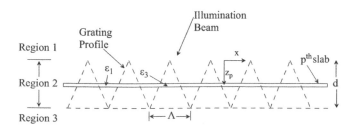

FIGURE 5.20 Illustration of the division of a surface relief grating into a set of slabs for rigorous coupled wave analysis.

where q is the harmonic index (integer) and $\tilde{a}_{q,p}$ are the normalized complex coefficients given by:

$$\tilde{a}_{q,p} = \frac{1}{\Lambda}\int_0^{\Lambda} g(x,z_p)\exp\left(-j\frac{2\pi}{\Lambda}qx\right)dx, \qquad (5.145)$$

with $g(x,z_p)$ a function that represents the grating profile. The diffracted fields and diffraction efficiency is found in a similar manner to the volume grating using the state variable approach. However, in this case, the fields emerging from each slab are matched with the adjacent slab until the slab at the full depth of the grating (i.e., $z = d$) is boundary matched and provides the transmission coefficients T_i while the reflection coefficients R_i are obtained at the first slab at $z = 0$.

PROBLEMS

1. A thin phase grating is placed in front of a thin lens as shown in Figure 5.21.

   The phase grating has a sinusoidal varying phase with a period Λ, peak-peak modulation m, and a square aperture with dimension 2w on a side. The grating is illuminated with a normally incident plane wave with unit amplitude and a wavelength λ.

   a. Write an expression for: (i) the *transmittance function* of the phase grating, and (ii) the *intensity* at the back focal plane of the lens in terms of the variables (not the numbers). (iii) What part of the expression for the intensity determines the relative strength of the diffracted orders?

   b. Assume that 2w = 1 cm; Λ = 10 μm; peak-peak modulation m = 12; and λ = 0.70 μm. (i) How many propagating diffraction orders will be formed by the grating (assume that the lens does not limit the number of propagating orders)? (ii) What are the intensities for the five diffraction orders closest to the z-axis?

2. A hologram is formed with two plane waves one at 25° and one at 0° relative to the z-axis (in air) at a wavelength of 514.5 nm. The refractive index of the recording material is 1.52 during exposure and 1.48 after processing and material thickness is 8 μm before and after processing. The hologram is reconstructed at a wavelength of 532 nm.

   a. At what angle (in air) will the diffraction efficiency be a maximum?

   b. Assuming a lossless grating, what is the minimum refractive index modulation at which the Bragg matched diffraction efficiency will be a maximum? What is the maximum value?

   c. For the grating with index modulation found in part b, the reconstruction angle remains unchanged and the wavelength is varied from 532 nm. At what wavelength does the diffraction efficiency drop to 25% of its peak value?

   d. If the material has an amplitude absorption coefficient of 0.010/μm what is the diffraction efficiency using the same refractive index modulation found in part b?

   e. What is the Q parameter for this grating? Does the grating satisfy the volume or thick grating requirement?

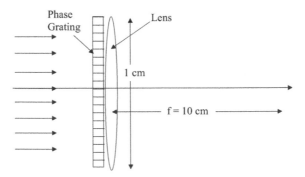

FIGURE 5.21   Configuration for Problem 1.

3. A reflection hologram is formed with 488 nm light and construction beams at 20° and 165° relative to the z-axis in the recording material. The material has a refractive index of 1.50 and is 10 μm thick.

   a. Assuming an index modulation of 0.015, determine the diffraction efficiency at the Bragg condition.

   b. For the same index modulation and assuming that the wavelength remains at 488 nm, determine the change of incident angle (inside the medium) at which the diffraction efficiency goes to zero.

4. The two-wave coupled wave equations were solved using the Helmholtz equation: $\nabla^2 E + k^2 E = 0$. The $k^2 E$ factor contains the important coupled wave terms:

$$R e^{-j[\rho \cdot r - K \cdot r]} + S e^{-j[\sigma \cdot r - K \cdot r]} + R e^{-j[\rho \cdot r + K \cdot r]} + S e^{-j[\sigma \cdot r + K \cdot r]}$$

Only two of the four factors need to be retained. Which two terms are retained? Why? Why do these factors show coupling?

5. For ACWA of volume holograms:

   a. What are the boundary conditions for R and S for a *volume transmission grating*?

   b. What are the boundary conditions for R and S of a *volume reflection grating*?

   c. Draw a sketch showing the relative amplitudes for R and S in a *transmission* grating as a function of thickness from 0 to d.

   d. Draw a sketch showing the relative amplitude for R and S in a *reflection* grating as a function of thickness from 0 to d.

6. Draw a Bragg circle for an unslanted transmission grating formed with a wavelength of 632.8 nm and with the propagation vectors at 20° to the z-axis (label the x- and z-axes in your diagram). The hologram recording material has a refractive index of 1.50 and is surrounded by air.

   a. Show both propagation vectors and their angles along with the K-vector.

   b. Assume that the hologram is now reconstructed with a beam incident at 20° to the z-axis, but with a wavelength of 514.5 nm. Draw the K-vector closure reconstruction on the *same Bragg circle diagram* used for construction. Indicate the part of the diagram that shows the amount of detuning from the Bragg condition.

7. Assume that the hologram described in Problem 6 is *lossless* and is reconstructed at the Bragg condition. Determine the index modulation that gives the *first peak* in the diffraction efficiency. The free-space wavelength is 500 nm.

8. Now assume that the reconstruction angle for the hologram in Problem 6 is shifted from the Bragg condition by 2° inside the grating. What is the wavelength shift that results in an equivalent decrease in the diffraction efficiency as the 2° angular shift? (The 2° is inside the recording medium. Assume that this is a small deviation from the Bragg condition.)

9. Assume that the reconstruction angle for the hologram for Problem 6 is shifted from the Bragg condition by 2° inside the grating and that there is no change in reconstruction wavelength. Also assume the same index modulation as found in Problem 6. Determine the diffraction efficiency of the reconstructed beam.

10. Consider the rigorous coupled wave equation:

$$\frac{1}{(2\pi)^2} \frac{d^2 S_i(z)}{dz^2} - j \frac{2}{\pi} \left[\frac{n\cos\theta}{\lambda} - \frac{i\cos\phi}{\Lambda} \right] \frac{dS_i(z)}{dz} + \frac{2i(m-i)}{\Lambda^2} S_i(z)$$

$$+ \frac{\varepsilon_1}{\lambda^2} \left[S_{i-1}(z) + S_{i+1}(z) \right] = 0$$

   1. What term in this expression represents the coupling of harmonics?

   2. What term represents the Bragg condition?

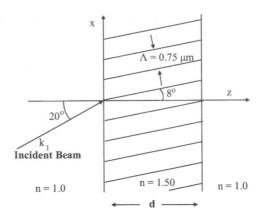

FIGURE 5.22 Grating conditions for Problem 5.11.

11. A field is incident on a volume grating as illustrated in Figure 5.22.

The grating has a period of 0.75 μm and the grating planes are at an angle of 8° relative to the z-axis as shown. The incident field has a wavelength of 632.8 nm. The field within the grating is given by:

$$E_y = \sum_{-\infty}^{+\infty} S_i(z)\exp\left[-j\left(\bar{k}_2 - i\bar{K}\right)\cdot\bar{r}\right].$$

Determine how many reflection and transmission diffraction orders are allowed to propagate (i.e., $i = \ldots,-3,-2,-1,0,+1,+2,+3,\ldots$). Use the condition for phase matching from the discussion of rigorous coupled wave analysis.

REFERENCES

1. J. W. Goodman, *Introduction to Fourier Optics*, 4th ed., Freeman & Co., New York (2017).
2. H. Kogelnik, "Coupled wave theory for thick hologram gratings," *Bell Syst. Tech. J.*, Vol. 48, 2909–2947 (1969).
3. L. Solymar and D. J. Cooke, *Volume Holography and Volume Gratings*, Academic Press, London, UK (1981).
4. T. K. Gaylord and M. G. Moharam, "Analysis and applications of optical diffraction by gratings," *Proc. IEEE*, Vol. 73, 894–937 (1985).
5. R. R. A. Syms and L. Solymar, "Planar volume phase holograms formed in bleached photographic emulsions," *Appl. Opt.*, Vol. 22, 1479–1496 (1983).
6. T. K. Gaylord and M. G. Moharam, "Thin and thick gratings: Terminology clarification," *Appl. Opt.*, Vol. 20, 3271–3273 (1981).
7. M. G. Moharam, T. K. Gaylord, and R. Magnusson, "Criteria for Raman-Nath regime diffraction by phase gratings," *Opt. Commun.*, Vol. 32, 19–23 (1980).
8. R. Magnusson and T. K. Gaylord, "Analysis of multiwave diffraction of thick gratings," *J. Opt. Soc. Am.*, Vol. 67, 1165–1170 (1977).
9. M. G. Moharam and T. K. Gaylord, "Rigorous coupled wave analysis of planar grating diffraction," *J. Opt. Soc. Am.*, Vol. 71, 811–818 (1981).
10. C. Harper, *Introduction to Mathematical Physics*, Chapter 2, Prentice Hall, Englewood Cliffs, NJ (1976).
11. https://www.mathworks.com. (accessed on 16 May 2018)
12. M. G. Moharam and T. K. Gaylord, "Coupled wave analysis of reflection gratings," *Appl. Opt.*, Vol. 20, 240–244 (1981).
13. M. G. Moharam and T. K. Gaylord, "Diffraction analysis of dielectric surface-relief gratings," *J. Opt. Soc. Am.*, Vol. 72, 1385–1392 (1982).
14. N. K. Sheridon, "Production of blazed holograms," *App. Phys. Lett.*, Vol. 12, 316–318 (1968).
15. G. Saxby, *Practical Holography*, Chapter 21, 3rd ed., IoP, Bristol, UK (2004).

6

Computer-Generated Holograms

6.1 Introduction

In a conventional analog hologram, the reference and object fields must be physically produced with an optical system. Although this approach is quite useful for many applications, there are situations where the physical realization of the optical fields is difficult to achieve and in some cases not possible to generate with traditional optical components. Computer-generated hologram (CGH) provides an alternative approach to forming a hologram. In this method, the object field is not physically generated, but is computed and used to form the hologram. This allows great flexibility in the design of "images" and new possibilities for holographic optical elements.

Computer-generated holograms have many important applications including interferometric testing of aspheric lenses and mirrors [1–3], multiple image formation for information processing [4], and optical interconnects [5,6]. It has also been demonstrated that a CGH can be "written" onto a spatial light modulator to form dynamic holograms for displays [7,8] and optical trapping and manipulation of micro and nanometer size particles and biological cells [9,10].

Several excellent reviews of CGH methods and techniques can be found in the references by Lee [11], Yaroslavskii and Merzlyakov [12], and Tricoles [13]. This chapter provides an overview of the considerations for specifying CGHs and encoding them with different representation techniques.

6.2 Preliminary Considerations for the CGH Process

6.2.1 Basic Concept

The formation of a computer-generated hologram consists of: (i) computationally specify the desired object field, (ii) computationally propagate this field to the hologram plane, (iii) compute the interference of the object field with a reference field (usually a plane wave), and (iv) physically encoding the amplitude and phase of the interference pattern into a material. This seems straightforward enough, however, the process is complicated by the fact that the computation of the object field and the CGH encoding are digital processes. This requires sampling the phase and amplitude values of the complex field at a finite and discrete set of points at the object plane. Since the resulting sampled object field is a discrete function, propagation to the hologram plane is computed using diffraction analysis with discrete Fourier transform algorithms instead of continuous Fourier transforms. The interference pattern between the computed diffracted object field and a reference field is sampled to determine the amplitude and phase at each point on the hologram aperture and then used to encode the hologram. The actual writing of the CGH is accomplished with a digital device such as a printer, lithography, or spatial light modulator. The spatial resolution of the writing device limits the sampling frequency that can be obtained at the hologram plane and must be taken into account in the encoding algorithm. The following sections will discuss the requirements for sampling, the discrete Fourier transform used to describe beam propagation and diffraction, and several encoding and CGH writing methods.

6.2.2 Sampling Continuous Functions

Sampling the continuous field functions at the correct intervals is essential to accurately represent them in digital or discrete form for encoding a CGH. If sampling is not performed at a sufficient frequency, errors or aliasing will result in the reconstructed images. The requirements for sampling a function without introducing aliasing effects are given by the Whittaker-Shannon sampling theorem [14,15]. In general, the field function must be sampled over a two-dimensional sample, however, for simplicity only a one-dimensional example will be described.

The function to be sampled, $g(x)$, is an arbitrary continuous function that is assumed to be approximately band limited as shown in Figure 6.1. The sampled function, $g_s(x)$, is found by multiplying $g(x)$ by a set of regularly spaced Dirac delta functions:

$$g_s(x) = g(x) \sum_{m=-\infty}^{m=\infty} \delta(x - m\Delta x), \tag{6.1}$$

where Δx is the interval between delta functions. The corresponding Fourier transform of $g_s(x)$ is:

$$G_s(\xi) = \frac{1}{\Delta x} \sum_{m=-\infty}^{m=\infty} G[\xi - m/\Delta x], \tag{6.2}$$

where $G(\xi)$ is the Fourier transform of $g(x)$. $G(\xi)$ has a maximum frequency value of ξ_{max}. The function $G_s(\xi)$ consists of a set of repetitions of the Fourier transform of $g(x)$ that are regularly spaced at spatial frequency intervals of $\Delta \xi = \frac{1}{\Delta x}$. When $\xi = \frac{m}{\Delta x}$, the function $G(\xi)$ has a value of $G(0)$ as shown in Figure 6.2.

The frequency representation shows that if the sampling frequency $1/\Delta x$ is too small, then the different frequency spectra of $G_s(\xi)$ will overlap. This prevents the use of filtering to recover the original object function $g(x)$. However, by sampling at the Nyquist rate:

$$\frac{1}{\Delta x} \geq 2\xi_{max}, \tag{6.3}$$

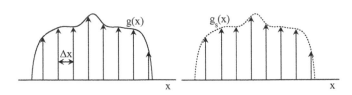

FIGURE 6.1   The plot on the left shows a band limited continuous function $g(x)$ that is sampled at intervals Δx. The plot on the right shows the resulting sampled function $g_s(x)$.

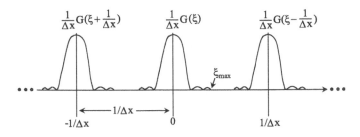

FIGURE 6.2   Illustrations of the Fourier transform of the sampled function $g_s(x)$ shown in Figure 6.1.

the ambiguity is eliminated and allows filtering to isolate one of the spectra within $G_s(\xi)$. The inverse Fourier transform can then be performed on the filtered spectrum to recover the object function. This sampling criteria is incorporated into the discrete Fourier transform operations required for the computer-generated hologram process.

6.2.3 Continuous and Discrete Fourier Transform Operations

As shown in Chapter 2, the two dimensional (2D) Fourier transform and inverse Fourier transforms are used to describe the diffraction of a field that occurs when propagating from one plane to another in an optical system. The Fourier and inverse Fourier transforms of a continuous function $h(x_1, y_1)$ can be written as:

$$H\left(f_x, f_y\right) = \Im\{h(x_1, y_1)\} = \int_{-\infty}^{\infty}\int_{-\infty}^{\infty} h(x_1, y_1)\exp\left[-j2\pi\left(f_x x_1 + f_y y_1\right)\right]dx_1 dy_1$$

$$h(x_1, y_1) = \Im^{-1}\left\{H\left(f_x, f_y\right)\right\} = \int_{-\infty}^{\infty}\int_{-\infty}^{\infty} H\left(f_x, f_y\right)\exp\left[j2\pi\left(f_x x_1 + f_y y_1\right)\right]df_x df_y,$$

(6.4)

with (x_1, y_1) the transverse coordinates at the function plane, (x_2, y_2) the coordinates at the transform plane, and the spatial frequencies:

$$f_x = x_2/\lambda z, f_y = y_2/\lambda z.$$

(6.5)

For the CGH process, however, the continuous field functions must be sampled at a finite number of points across the apertures at the object and hologram planes. Therefore, the integral form of the Fourier transform cannot be used and a discrete Fourier transform (DFT) operation is required. The DFT and inverse DFT operations are given by:

$$E_H\left(m\Delta x_2, n\Delta y_2\right) = \sum_{p=0}^{P_x-1}\sum_{q=0}^{Q_y-1} E_o\left(p\Delta x_1, q\Delta y_1\right)\exp\left[j2\pi\left(\frac{mp}{P_x} + \frac{nq}{Q_y}\right)\right]$$

(6.6)

$$E_o\left(p\Delta x_1, q\Delta y_1\right) = \frac{1}{P_x Q_y}\sum_{m=0}^{P_x-1}\sum_{n=0}^{Q_y-1} E_H\left(m\Delta x_2, n\Delta y_2\right)\exp\left[-j2\pi\left(\frac{mp}{P_x} + \frac{nq}{Q_y}\right)\right],$$

(6.7)

where P_x is the number of points in the x-direction, and Q_y the points in the y-direction. E_o is the field at the object plane, and E_H the field at the hologram plane. The sampling intervals at the hologram plane are $(\Delta x_2, \Delta y_2)$ and $(\Delta x_1, \Delta y_1)$ at the object plane.

Note that the DFT and inverse DFT operations described by Equations 6.6 and 6.7 produce only one value of the sampled field at the respective apertures. Therefore, in order to construct a representation of the discrete field function, the DFT and inverse DFT calculations must be performed at each point in the respective apertures. To make the DFT algorithm operate as efficiently and quickly as possible, a variety of fast Fourier transform (FFT) algorithms have been developed. These algorithms can decrease the number of operations (multiplications and additions) required to compute the DFT from $P_x^2 Q_y^2$ to $P_x Q_y \log_2\left(P_x Q_y\right)$. Many computing programs such as MATLAB and MathCad have 1D and 2D FFT algorithms for computing DFTs built into the software to greatly simplify the process.

6.2.4 Sampling Requirements at the Object and Hologram Planes

The sampling interval and number of points that must be encoded in the CGH depend on the spatial frequency content of the object field after propagating to the hologram plane and the size of the hologram aperture. As shown by Goodman [16], the spatial frequency content depends on whether the propagation distance from the object to the hologram plane falls within the near- or far-field diffraction regions. The two situations are illustrated in Figure 6.3. The far-field case can be realized by using the Fourier transform hologram geometry where the object is placed at the front focal plane of a lens and the hologram is encoded at the back focal plane. For the near-field or Fresnel hologram, light illuminates an object and then propagates to the hologram plane over a distance that satisfies the Fresnel diffraction requirements.

To compare the two situations, consider an object with dimensions W_{x1} by W_{y1} and a hologram with dimensions W_{x2} by W_{y2}. For the Fresnel hologram, the object is located a distance of z_{OH} from the hologram plane and by a distance of $2f$ for the Fourier transform hologram. As noted above, in order to avoid inaccuracies in the Fourier and inverse Fourier operations, the sampling interval must satisfy the Nyquist condition:

$$\frac{1}{\Delta x_2} \geq 2\xi_{x-\max}; \quad \frac{1}{\Delta y_2} \geq 2\eta_{x-\max}, \tag{6.8}$$

where $\xi_{\max}$ is the maximum spatial frequency of the Fourier transform of the object field.

In the case of the Fourier transform hologram, the field at the hologram plane is an exact Fourier transform of the object field. The maximum spatial frequency is determined by the size of the object and the focal length of the lens according to:

$$\xi_{x-\max} = \frac{W_{x1}/2}{\lambda f}; \quad \eta_{y-\max} = \frac{W_{y1}/2}{\lambda f}, \tag{6.9}$$

where λ is the wavelength, and f is the focal length of the lens. Combining this expression with the Nyquist condition gives the required sampling interval in terms of the size of the object and focal length:

$$\Delta x_2 = \frac{1}{2\xi_{x-\max}} = \frac{\lambda f}{W_{x1}}$$

$$\Delta y_2 = \frac{1}{2\eta_{y-\max}} = \frac{\lambda f}{W_{y1}}. \tag{6.10}$$

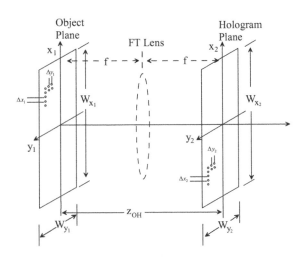

FIGURE 6.3   Fourier transform hologram (dashed diagram) and Fresnel hologram configurations.

If the size of the hologram plane is $W_{x2} \times W_{y2}$, then the number of sampling points at the hologram plane is:

$$P_x = \frac{W_{x1}}{\Delta x_2} = \frac{W_{x1}W_{x2}}{\lambda f}$$

$$Q_y = \frac{W_{y1}}{\Delta y_2} = \frac{W_{y1}W_{y2}}{\lambda f}.$$

(6.11)

Note that for this case, the sampling increment and number of sampling points are the same for both the object and hologram planes since they are related by an exact Fourier transform.

For the near-field or Fresnel hologram, the maximum spatial frequency that must be recorded at the hologram plane depends on both the size of the object and the hologram aperture. Referring to Figure 6.3, this occurs for light propagating from a point at $(-W_{x1}/2, 0)$ on the object plane to a point $(W_{x2}/2, 0)$ on the hologram plane. The corresponding maximum spatial frequency values at the Fresnel hologram plane are:

$$\xi'_{x2-max} = \frac{(W_{x2} + W_{x1})/2}{\lambda z_{OH}}; \ \eta'_{y2-max} = \frac{(W_{y2} + W_{y1})/2}{\lambda z_{OH}}.$$

(6.12)

The corresponding sampling intervals in the x- and y-directions at the hologram plane are:

$$\Delta x'_2 = \frac{1}{2\xi'_{x2-max}} = \frac{\lambda z_{OH}}{W_{x1} + W_{x2}}$$

$$\Delta y'_2 = \frac{1}{2\eta'_{y2-max}} = \frac{\lambda z_{OH}}{W_{y1} + W_{y2}},$$

(6.13)

and the total number of samples that must be taken at the hologram plane is:

$$P'_x = \frac{W_{x2}}{\Delta x'_2} = \frac{W_{x2}(W_{x1} + W_{x2})}{\lambda z_{OH}}$$

$$Q'_y = \frac{W_{y2}}{\Delta y'_2} = \frac{W_{y2}(W_{y1} + W_{y2})}{\lambda z_{OH}},$$

(6.14)

and as in the Fourier transform hologram geometry, the sampling increment required at the object plane is the same as at the hologram plane. This is due to the symmetry of the configuration when viewing the hologram plane from the object plane or vice versa. However, for the Fresnel hologram, the sampling interval is smaller and the number of sampling points is greater than the Fourier transform hologram since the spatial frequency content is greater.

6.3 CGH Encoding Methods

Having illustrated the discrete sampling requirements at the object and hologram planes, the next step is to find a suitable method to encode the sampled values at the hologram. Several approaches have been demonstrated to accomplish this either as phase and amplitude or phase-only encoding with different printing devices [11]. The holograms can also be encoded with or without a reference beam to suit a particular application. In any case, the resulting method must be able to represent the field distribution across the hologram aperture as a set of real, non-negative values.

This section describes two different CGH encoding methods: (i) binary detour phase method that encodes both amplitude and phase [17–19] and (ii) the binary phase hologram by Lee [20–22]. This is followed with an example of a binary phase Fourier transform hologram encoded with an off-axis reference beam.

6.3.1 Binary Detour Phase Encoding

The first demonstration of a CGH was performed in the mid 1960s by Brown and Lohmann using a binary detour phase method [17–19] and can be used to encode both Fresnel and Fourier transform type holograms. During this time period printing technology did not have very high spatial resolution, and this limited the ability to record complex object fields in the CGH. In addition, printing was restricted to writing binary transmittance patterns, i.e., a clear or opaque region, at each sampling point. Nonetheless, impressive examples were demonstrated by combining the capability of available printer plotting quality with photographic reduction to reconstruct images at optical wavelengths.

To illustrate the detour phase encoding process, consider the formation of a Fourier transform hologram in which the hologram is placed at the front focal plane of a lens and the image formed on the optical axis at the back focal plane (Figure 6.3). The hologram aperture is divided into a set of $P_x \times Q_y$ cells that serve as the sampling points. The location of these points are spaced distances that are consistent with the Nyquist sampling criteria given in Equation 6.8. The sampled value of the object field at a particular location on the hologram aperture is given by:

$$\tilde{c}_{m,n} = E_H\left(m\Delta x_2, n\Delta y_2\right) = \left|\tilde{c}_{m,n}\right| e^{j\varphi_{m,n}}, \tag{6.15}$$

where $\left|\tilde{c}_{m,n}\right|$ is the amplitude, and $\varphi_{m,n}$ is the phase of the object field.

The detour phase method allows both the amplitude and phase of an optical field to be encoded at each sampling point on the aperture of the hologram. The aperture of the detour plane hologram is shown in Figure 6.4. The encoding locations, $\left(m\Delta x_2, n\Delta y_2\right)$, or cells have a height of A_x and within each cell is a sub cell of width a_x. The sampled amplitude value is encoded by making the length of the sub cell, a_x, in the y-direction proportional to $\left|\tilde{c}_{m,n}\right|$. In order to encode the phase, the hologram will be reconstructed with an off-axis plane wave incident at an angle α to the optical axis as shown in Figure 6.4. The off-axis plane wave can be expressed as:

$$E_R\left(x_2, y_2\right) = \exp\left(-jk_x x_2\right), \tag{6.16}$$

with

$$k_x = \frac{2\pi \sin\alpha}{\lambda}. \tag{6.17}$$

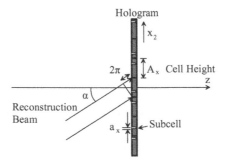

FIGURE 6.4 Diagram of a detour-phase hologram showing the cell and sub cell heights. (From Andrews, D.L.: *Photonics Technology and Instrumentation*. 2015. Copyright Wiley-VCH Verlag GmbH & Co. KGaA. Reproduced with permission.)

The height of each cell in the x-direction, A_x, is determined by letting:

$$A_x = \frac{\lambda}{\sin\alpha}. \tag{6.18}$$

The phase value $\varphi_{m,n}$ in Equation 6.15 is determined by positioning the center of a sub cell of width a_x within the cell height A_x to form a phase value between 0 and 2π. Shifting the sub cell within the cell is the "detour" operation of the encoding process. Referring again to Figure 6.4, if the sub cell is placed at the bottom of the cell, it can be used to encode a phase value of "0," at the center of the cell a value of π, and at the top of the cell a value of 2π. There are limits to this approach. For instance, the relation between the sub-cell to cell width requires:

$$a_x \ll A_x = \frac{\lambda}{\sin\alpha}. \tag{6.19}$$

If the wavelength is 0.5 µm and $\alpha = 5°$, then the cell height $A_x \sim 6$ µm. The sub-cell width in the x-direction, a_x, must be significantly smaller than A_x to provide adequate phase resolution, but must also allow sufficient light transmission. Therefore, a compromise is required to optimize the overall design.

At this point, it is useful to consider how an image will be formed with the CGH, and how the detour phase encoding of phase and amplitude is utilized. During reconstruction, the detour phase hologram is placed in the front focal plane of a lens. An image is formed at the back focal plane of the lens after light transmitted by the hologram undergoes an exact Fourier transform. The discrete Fourier transform of the sampled hologram transmittance function is:

$$E_{im}(x_i, y_i) = \sum_{m=0}^{P_x-1}\sum_{n=0}^{Q_y-1} |\tilde{c}_{mn}| e^{j\varphi_{mn}} \exp\left[\frac{2\pi}{\lambda f}(x_i m \Delta x_2 + y_i n \Delta y_2)\right]. \tag{6.20}$$

As can be seen from the above expression, the image is formed by summing the contributions to phase and amplitude from each point on the hologram. In the following paragraphs the purpose and contribution of each part of the detour phase hologram is examined to see how it contributes to the overall image formation process.

First, consider the transmittance of a sub cell (a_x) within the sample cell (A_x). The sub cell is a rectangular function that is shifted with respect to the center of the cell (x_c, y_c). As described earlier, during reconstruction, the CGH is illuminated with an off-axis plane wave. Therefore, the resulting field immediately past one cell of the hologram is given by:

$$E_H'(x_2, y_2) = \exp\left(-j\frac{2\pi \sin(\alpha)x_2}{\lambda}\right) rect\left(\frac{x_2 - x_c}{a_x}\right) rect\left(\frac{y_2 - y_c}{a_y}\right). \tag{6.21}$$

The resulting "image" from a single cell in the hologram at the back focal plane of the lens is found by taking the Fourier transform:

$$E_i^1(x_i, y_i) = \frac{a_x a_y}{\lambda f} \text{sinc}\left[\frac{a_x(x_i + f\sin\alpha)}{\lambda f}\right] \text{sinc}\left(\frac{a_y y_i}{\lambda f}\right) \exp\left\{j\frac{2\pi}{\lambda f}\left[(x_i + f\sin\alpha)x_c + y_i y_c\right]\right\}. \tag{6.22}$$

Since the width of the sub cell $a_x \ll \lambda/\sin\alpha$ and the width of the center region of the sinc function is very broad with respect to the image plane $(W_{x_i} \times W_{y_i})$ the term

$$\text{sinc}\left[\frac{a_x(x_i + f\sin\alpha)}{\lambda f}\right] \text{sinc}\left(\frac{a_y y_i}{\lambda f}\right) \approx 1.$$

Using this approximation, the simplified image of the *single* sub cell/cell reduces to:

$$E_i^1(x_i, y_i) = \frac{a_x a_y}{\lambda f} \exp\left(-j2\pi \sin(\alpha) x_c/\lambda\right) \exp\left[j\frac{2\pi}{\lambda f}(x_c x_i + y_c y_i) \right]. \tag{6.23}$$

Next, if the contributions from all sub cells/cells that comprise the hologram are considered with the assumption that all sub cells are located at the center of their respective cells the image becomes:

$$E_i^c(x_i, y_i) = \sum_{m=0}^{P_x-1}\sum_{n=0}^{Q_y-1} a_{x;mn} a_{y;mn} e^{-j2\pi m} \exp\left[j\frac{2\pi}{\lambda f}(m\Delta x_2 \cdot x_i + n\Delta y_2 \cdot y_i) \right]$$

$$= \sum_{m=0}^{P_x-1}\sum_{n=0}^{Q_y-1} a_{x;mn} a_{y;mn} \exp\left[j\frac{2\pi}{\lambda f}(m\Delta x_2 \cdot x_i + n\Delta y_2 \cdot y_i) \right], \tag{6.24}$$

where $A_x = \Delta x_2$ and $A_y = \Delta y_2$. Furthermore, the exponential $\exp\left(-j2\pi \sin(\alpha) x_c/\lambda\right)$ is replaced with $\exp(-j2\pi m)$ since $\sin\alpha/\lambda = \Delta x_2$, and the identity $e^{-j2\pi m} = 1$ is used since m is an integer. This result is essentially the discrete Fourier transform relation given earlier with the exponential term representing the basic functions for the transform. As mentioned earlier, the amplitudes of the individual Fourier coefficients can be controlled by varying a_y since no other constraints have been placed on this length.

The phase values for each coefficient can now be encoded by displacing the centers of the sub cells within the cell length (A_x) in the x-direction. As discussed earlier, the reconstruction beam angle was chosen so that different locations in this direction represent a phase value that ranges from 0 to 2π. If we now consider that the coordinates of the center of the mnth sub cell are located at:

$$x_{c;mn} = m\Delta x_2 + \delta_{x;mn}$$

$$y_{c;mn} = n\Delta y_2, \tag{6.25}$$

then the phase at this point can be written as:

$$\exp\left[-j\frac{2\pi}{A_x}(m\Delta x_2 + \delta_{x;mn}) \right] = \exp\left[-j\frac{2\pi}{A_x}(mA_x + \delta_{x;mn}) \right] = \exp\left[-j2\pi\left(m + \frac{\delta_{x;mn}}{A_x} \right) \right]$$

$$= e^{-j2\pi\frac{\delta_{x;mn}}{A_x}}. \tag{6.26}$$

The resulting image field at the back focal plane of the Fourier transform lens becomes:

$$E_i(x_i, y_i) = \sum_{m=0}^{P_x-1}\sum_{n=0}^{Q_y-1} a_{x;mn} a_{y;mn} \exp\left(-j2\pi \frac{\delta_{x,mn}}{\Delta x_2} \right) \exp\left[j\frac{2\pi}{\lambda f}\left\{ (m\Delta x_2 + \delta_{x,mn}) \cdot x_i + n\Delta y_2 \cdot y_i \right\} \right]. \tag{6.27}$$

Recognizing that if $\lambda f \gg \Delta x = A_x$ and $\delta_{x;mn} \ll \Delta x$, then the $\delta_{x;mn}$ parameter in the second exponential is negligible and simplifies the image field to:

$$E_i(x_i, y_i) = \sum_{m=0}^{P_x-1}\sum_{n=0}^{Q_y-1} a_{x;mn} a_{y;mn} \exp\left(-j2\pi \frac{\delta_{x,mn}}{\Delta x_2} \right) \exp\left[j\frac{2\pi}{\lambda f}(m\Delta x_2 \cdot x_i + n\Delta y_2 \cdot y_i) \right]. \tag{6.28}$$

This result allows the amplitude and phase at each mn sample point in the hologram to be represented as:

$$|c_{mn}| \propto a_{y;mn};$$

$$\exp(-j2\pi\delta_{x;mn}/\Delta x_2) = \exp(-j\varphi_{mn}). \tag{6.29}$$

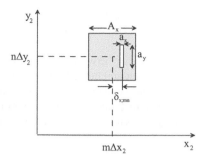

FIGURE 6.5 Diagram showing the *mn*th cell and sub cell in the hologram aperture. (From Lohmann, A.W. and Paris, D.P., *Appl. Opt.*, 6, 1739–1748, 1967. With permission from the Optical Society of America.)

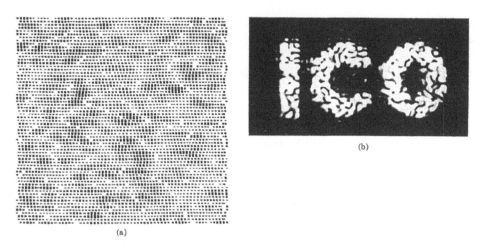

FIGURE 6.6 Illustration of a Fourier transform CGH formed using the binary detour phase method. (a) Shows the cells and sub cells at different locations across the hologram aperture and (b) shows the reconstructed image of the letters "ICO." (From Lohmann, A.W. and Paris, D.P., *Appl. Opt.*, 6, 1739–1748, 1967. With permission from the Optical Society of America.)

Where the width of the sub cell in the *y*-direction controls the amplitude, and the shift of the sub cell determines the phase at each sample point in the hologram. These parameters are illustrated in Figure 6.5, and an illustration of a binary detour-phase pattern and the corresponding image is shown in Figure 6.6 [19].

Since the detour phase method modifies the intensity of the incident beam, the diffraction efficiency of the resulting holograms is rather low. However, recent work by Turunen [20] and Sinzinger [21] suggested that combining the detour phase encoding method with other holographic techniques can overcome this limitation. In this approach, they combine binary phase diffractive optical techniques [22] with detour phase encoding to realize diffraction efficiencies of 75%–95%.

6.3.2 Binary Interferogram Computer-Generated Holograms

While successful, the binary detour-phase encoding method has some limitations. For one thing, the sample cells must be large enough to allow sufficient light through the shifted aperture and limits the phase resolution that can be achieved at the sample point. Another problem is that when recording fields with rapidly varying phase, the shifted apertures will overlap at the edge of the cell and lead to errors in the reconstruction. One way to overcome these problems was developed by Lee [23–25] who recognized that a binary encoded CGH can also be formed by sampling the computed interference pattern between an object and a reference wave. The interference pattern is similar to that of an optically recorded absorption hologram. However, in the case of the CGH, the pattern is sampled and converted into a "binary" transmittance function.

To illustrate the process, consider the interference between an arbitrary object field $O(x,y)\exp[j\varphi_o(x,y)]$ and a tilted planar reference wave with constant amplitude $R \cdot \exp[-j2\pi\sin\theta_x x/\lambda]$. In this case, R and O are, respectively, the amplitudes of the reference and object waves, φ_o is the phase of the object wave, and θ_x is the angle of the reference wave relative to a normal to the hologram surface. The resulting interference pattern and transmittance function is the same as that found earlier for an analog hologram and is given by:

$$t_A(x,y) = \left| R \cdot \exp[-j2\pi\sin\theta_x x/\lambda] + O(x,y)\exp[j\varphi_o(x,y)] \right|^2$$
$$= R^2 + O^2(x,y) + 2RO(x,y) \cdot \cos\{2\pi\sin\theta_x x/\lambda + \varphi_o(x,y)\}. \tag{6.30}$$

The fringes indicate the location of constant phase values and are found by setting the modulation term equal to a fixed phase angle:

$$2\pi\sin\theta_x x/\lambda + \varphi_o(x,y) = 2\pi m, \tag{6.31}$$

where m is an integer. The corresponding center of the fringes are located at:

$$\cos\left[2\pi\sin\theta_x x/\lambda + \varphi_o(x,y)\right] = 1. \tag{6.32}$$

A binary function describing the fringe pattern can be obtained by performing a threshold operation on the cosine modulation term. The threshold operation sets the positive values of the modulation term to "1" and negative values to "0," resulting in a binary function of the form:

$$g(x,y) = \begin{cases} 1 & for \cos\left(2\pi\sin\theta_x x/\lambda + \varphi_o(x,y)\right) \\ 0 & otherwise \end{cases}. \tag{6.33}$$

This expression can then be used to encode the phase of an object field. However, Lee also showed that if the threshold is replaced with a function $h(x,y)$ such that:

$$\cos\left[2\pi\sin\theta_x x/\lambda + \varphi_o(x,y)\right] = \cos[\pi h(x,y)], \tag{6.34}$$

then both the phase and amplitude can be encoded [22]. The width of the fringes in this case is given by:

$$2\pi\sin\theta_x x/\lambda + \varphi_o(x,y) = 2\pi m \pm \pi h(x,y), \tag{6.35}$$

where the solution with the − sign in Equation 6.35 determines the leading edge of the binary transition (1) and the + sign the trailing edge (2) as shown in Figure 6.7.

The corresponding Fourier series expression for the binary encoding method is:

$$g(x,y) = \sum_{n=-\infty}^{+\infty} \left\{ \frac{\sin\left[\pi n h(x,y)\right]}{\pi n} \right\} \cdot \exp\left\{ jn\left[2\pi\sin\theta_x x/\lambda + \varphi_o(x,y)\right]\right\}. \tag{6.36}$$

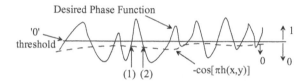

FIGURE 6.7 Representation of a phase function and sampling of the function with a threshold of 0 and with a function $h(x,y)$. Points (1) and (2) indicate the sampling locations with the threshold function $h(x,y)$. (From Lee, W.H., *Appl. Opt.*, 18, 3661–3669, 1979. With permission from the Optical Society of America.)

If both $h(x,y)$ and $\varphi_o(x, y)$ are constants, the series will become a set of rectangular functions. However, when $h(x,y)$ is not a constant, it can be used to encode information about the amplitude of the object field. This technique combined with the phase encoding performed by sampling the interference fringe contours allows both amplitude and phase information of the object field to be recorded.

As an example of a phase-only binary interferogram, consider the interference of an on-axis point source and an off-axis plane wave as shown in Figure 6.8. As demonstrated in Chapter 4, this recording effectively forms a lens. The spherical wave from the point source can be approximated with a quadratic phase function resulting in a contour given by:

$$2\pi m = \pi \frac{x^2 + y^2}{\lambda f} - \frac{2\pi}{\lambda} x \sin\theta, \tag{6.37}$$

where f is the distance from the point source to the recording plane, θ is the angle that the plane wave makes with the z-axis, and λ is the wavelength. Figure 6.8 below shows the constant phase contours when $f = 10$ units, the aperture $= 3$ units, $\theta = 3°$, and a wavelength $\lambda = 0.045$ units. The center of the contours is shifted in the x-direction due to the off-axis reference wave. The resulting pattern is then either plotted directly and converted into a transmittance function at a scale corresponding to the desired reconstruction wavelength or plotted at a larger scale and photographically reduced to the length scale corresponding to the reconstruction wavelength.

6.3.3 Example of a Binary Fourier Transform Hologram

In this section, an example of setting up a binary, phase-only Fourier transform CGH using Lee's method will be illustrated. This example is similar to the formation of an analog hologram with physical optical

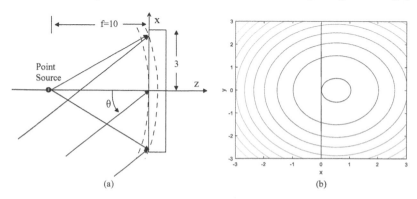

FIGURE 6.8 Off-axis lens with a point source at f and a planar reference wave at an angle θ (a) and resulting contours of constant phase (b).

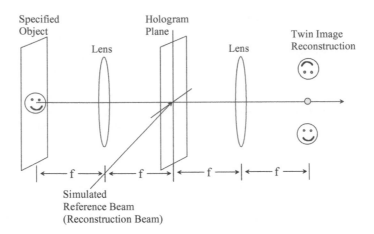

FIGURE 6.9   Construction and reconstruction geometry for a binary Fourier transform CGH.

beams and shows some of the practical considerations that must be taken into account. The recording and reconstruction geometry are illustrated in Figure 6.9. For this case, the construction algorithm is implemented in MATLAB, however, any convenient programming language can be used. (The example was provided courtesy of Dr. Shanalyn Kemme, a former University of Arizona student with permission.)

Binary Fourier Transform CGH Algorithm in MATLAB:

1. Set preliminaries such as the array size $M \times N = 256 \times 256$, λ, Δx, Δy values in the object plane; the corresponding spatial frequency in the *y*-direction in the output plane is $\Delta v = 1/(N\Delta y)$

2. Define the object as a set of "1" values within the $M \times N$ array. Set the rest of the values to "0"; (Figure 6.10a)

$$o(m,n)\Big|_{-128,-128}^{128,128} = 1; \quad object$$
$$= 0; \quad no \ object$$

(6.38)

3. Apply a random phase diffuser across the object plane: This increases the spatial extent of the object field in the Fourier/Hologram plane and reduces intensity spiking;

$$o = o. \times \exp(i \times 2 \times pi \times rand(M,N);)$$

(6.39)

4. Propagate the object field to the hologram plane by taking the 2D FT. Since the object consists of a lot of zeros it is best to first perform a fftshift() operation to randomly distribute the zeros before taking the 2D FFT (Figure 6.10b):

$$o = fftshift(o);$$
$$O = fft2(o);$$
$$O = fftshift(O).$$

(6.40)

5. The phase of the object field can then be extracted:

$$phi = angle(O).$$

(6.41)

6. At this point, a reference beam can be defined and combined with the object beam. If it desired not to have the reference and object beams overlap in the reconstruction region, a "frequency" $v_R = 3 \times d/2$ can be selected where d is the maximum number of 1's that occur in the object in the direction where the beams will overlap

7. The phase values of the object and reference beams are now added and the intensity interference (INTER) across the hologram plane computed using the interference equation (Figure 6.10c):

```
for n = 1:N;
        INTER(1:M, n) = 2 + 2*cos (2*pi*(2*vR)*n*Δv + phi(1:M,n));   (6.42)
end
```

8. Since the CGH will ultimately become a binary transparency (opaque or clear areas), it is necessary to threshold the intensity function. First the intensity function (INTER) is normalized to the maximum value. Next, the normalized intensity values that are less than half of the gray

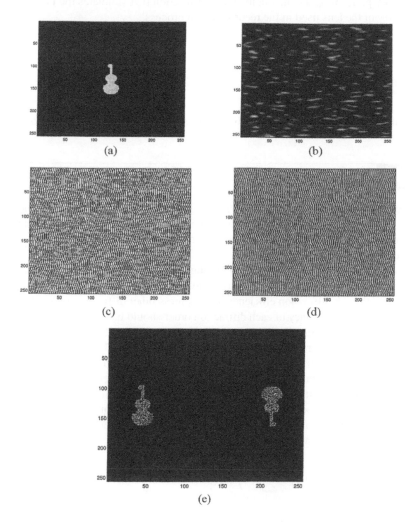

FIGURE 6.10 Steps for constructing and reconstructing a binary, phase-only Fourier transform CGH. (a) Object pattern; (b) object field after performing the discrete Fourier transform; (c) overlap of object and reference beams; (d) interference pattern after thresholding; and (e) reconstructed images.

scale range (i.e., 32 for a 64 bit gray scale) are set to 0, and those above the half gray scale value are set to 1 (Figure 6.10d):

```
normINTER = INTER/(max(max(INTER)))*64;
for m = 1:M;
  for n = 1:N:
    if normINTER(m,n) < 32;
        normINTER(m,n) = 0                              (6.43)
      else
        normINTER(m,n) = 64;
      end;
  end
end
```

9. The resulting transparency can now be printed with the correct pixel size for the reconstruction wavelength. In the case above, $(\Delta x, \Delta y) = \lambda/16$. This may be hard to realize at optical wavelengths without a very fine printer and photographic reduction. It is also possible to verify the algorithm by performing a computational reconstruction that simulates the Fourier transforming property of the lens used in the reconstruction process.

```
IMAGE = normINTER;
IMAGE = fftshift(normINTER);                            (6.44)
image = fft2(IMAGE);
image = fftshift(image).
```

The resulting image is shown in Figure 6.10e with the center intensity spike blocked for clarity.

6.4 Dammann Gratings

Another type of computer-generated hologram is the Dammann grating named after the inventor [26]. In this type of grating, the phase structure within a period is varied to produce the desired diffraction pattern. This provides a degree of freedom in the design that can be used to control the intensity of each diffraction order.

One of the first applications for the Dammann grating was to produce multiple, real, on-axis images of an object for use in early forms of projection lithography [26]. The design goal is to produce a fixed set of diffraction orders with high diffraction efficiency and suppress other diffraction orders from propagating. In addition, the diffraction efficiency of each diffraction order should be equal. The grating is specified as a phase-only transmittance function that has values of either $+1$ or -1 that change sign at a transition point located along the grating aperture. The transmittance values of $+1$ or -1 correspond to phase values for the transmitted field of either 0 or π. For a one-dimensional grating, the transition points are $x_1, x_2, \ldots x_N$ as shown in Figure 6.11. In this Figure, the pattern is symmetric about the center of the grating period which is normalized to a value of 1.0. It should be noted that in general the grating can be two-dimensional and the transition points need not be symmetrically located about the midpoint of the grating period [27].

For positive values of x, the amplitude transmittance function for one period of the grating with M transition points can be written as:

$$t(x) = \sum_{m=0}^{M} (-1)^m \text{rect} \left[\frac{x - (x_{m+1} + x_m)/2}{x_{m+1} - x_m} \right],$$ (6.45)

with $0 \le x < 0.5$. Note that the $(-1)^m$ factor is $+1$ for even values of m and -1 for odd values of m, and that the rect function has the usual meaning of having a value of "1" when the argument is less than $\frac{1}{2}$ and "0" otherwise.

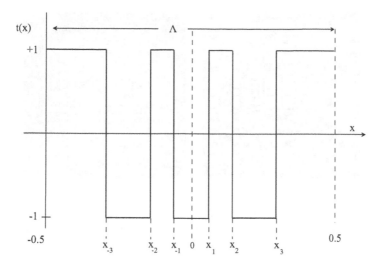

FIGURE 6.11 Single dimension surface profile for a typical Dammann grating. (From Dammann, H. and Gortler, K., *Opt. Comm.*, 3, 312–315, 1971; Mait, J.N., *J. Opt. Soc. Am. A*, 7, 1514–1528, 1990.)

If the grating is extended to infinity, the function can be expressed as a Fourier series expansion:

$$t(x) = \sum_{q=-\infty}^{+\infty} E_q \exp(-j2\pi qx), \tag{6.46}$$

where E_q are the Fourier amplitude coefficients given by:

$$E_q = 2 \int_{x=0}^{0.5} t(x)\cos(2\pi qx)\,dx. \tag{6.47}$$

For the far-field diffraction condition, the Fourier amplitude coefficients give the values of the amplitudes of the diffraction orders. In this case, the coefficients are:

$$E_q = \frac{1}{q\pi} \sum_{m=0}^{M} (-1)^m \times \left[\sin(2\pi qx_{m+1}) - \sin(2\pi qx_m) \right]$$

$$E_o = 2 \sum_{m=0}^{M} (-1)^m (x_{m+1} - x_m). \tag{6.48}$$

The total diffraction efficiency diffracted into the $2M + 1$ central diffraction orders of the grating is given by:

$$\eta_{total} = |E_o|^2 + 2 \sum_{q-1}^{M} |E_q|^2. \tag{6.49}$$

If the number of orders was infinite, then $\eta_{total} = 1$.

As can be seen, $2M + 1$ diffraction orders require M transition points in the transmittance function of the grating with:

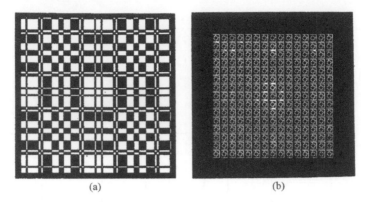

FIGURE 6.12   (a) A 2-dimensional binary Dammann grating and (b) the resulting 15 × 15 multiple images. (From Dammann, H. and Gortler, K., *Opt. Comm.*, 3, 312–315, 1971. With permission from Optical Communications.)

$$0 \le x_m \le x_{m+1} \le 1/2.$$

They can be found by using an iterative method to obtain a set of diffraction order amplitudes such that:

$$E_q = \alpha_q E_o \text{ with } 1 \le q \le M \text{ and } \alpha_q = \pm 1. \tag{6.50}$$

A variety of iterative numerical methods such as the Newton-Raphson technique, simulated annealing, and the method of steepest decent have been developed to solve this problem. These techniques are described in detail in excellent papers by Fienup [29], Mait [27], and Zhou [30] as well as the effects of fabrication errors on the diffraction efficiency and uniformity of the diffraction orders. Figure 6.12 from [28] shows the 2D transmittance function that forms a 15×15 spot array and a corresponding image array.

6.5 Dynamic CGHs Formed with a Spatial Light Modulator

The development of spatial light modulator (SLM) arrays has led to the realization of dynamic computer-generated holography. A SLM consists of a two-dimensional array of pixels that can be individually addressed to modify either the phase or amplitude of reflected or transmitted light. A variety of spatial light modulators have been developed including magneto-optic, multiple quantum well, liquid crystal, and micro-electro mechanical (MEMS). While a complete description of SLM technologies and operation is beyond the scope of this book, a brief outline of the operation of liquid crystal and MEMS devices which have been found very useful for CGH operations is provided. More in-depth reviews of SLM technologies can be found in the literature [31,32].

One of the more widely available types of liquid crystal (LC) SLM is the LC on silicon device [32–34] that is illustrated in Figure 6.13. As shown, there is an array of complementary metal oxide semiconductor (CMOS) transistors on one surface, an electrode on the other surface, and the liquid crystal material in the space between the two surfaces. When a voltage is applied to one of the pixels, the LC molecules are re-aligned and change the polarization state of the transmitted optical beam. If a polarizer (analyzer) is placed in the path of the reflected output beam, the optical intensity at the pixel location can be altered. Repeating this operation at each pixel site results in an intensity modulation of the optical field across the aperture of the SLM. This provides a pattern that can be used for an amplitude modulated CGH. This particular example is representative of an electrically addressable SLM.

Another highly successful form of SLM is based on MEMS fabrication methods developed for display and optical communications applications [35,36]. A basic form of MEMS device consists of an array of micro mirrors that can be tilted or deformed with the application of voltage to individual pixels [37]. An illustration of the tilted mirror and deformable mirror configurations are illustrated in Figure 6.14.

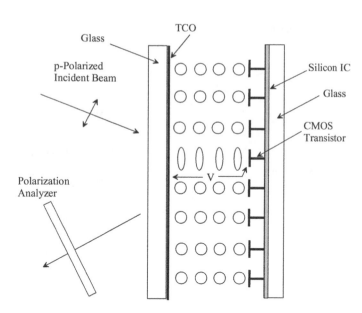

FIGURE 6.13   An illustration of an electrically addressable SLM. (From Zhou, C. and Liu, L., *Appl. Opt.*, 34, 5961–5969, 1995; Goodman, J.W., *Introduction to Fourier Optics*, 4th ed., W.H. Freeman, New York, pp. 287–308, 2017; Haist, T. and Osten, W., *J. Micro/Nanolithography MEMS MOEMS*, Vol. 14, 041309, 2015; Bleha, W.P. and Lei, L.A., *Proc. SPIE*, Vol. 8736, 2013; Lazarev, G. et al., *LCOS Spatial Light Modulators: Trends and Applications*, Wiley-VCH, Weinheim, Germany, pp. 1–29, 2012; Petersen, K.E., *Appl. Phys. Lett.*, 31, 521–523, 1977; Wu, M.C. et al., *J. Lightwave Tech.*, 24, 4433–4454, 2006.)

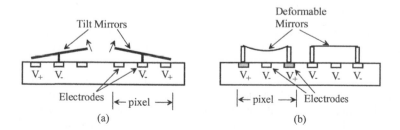

FIGURE 6.14   Illustration of a tilted mirror (a) and deformable mirror (b) MEMS type SLM. The mirror deflection in both cases is controlled by applying voltage to micro electrodes. (From Bleha, W.P. and Lei, L.A., *Proc. SPIE*, Vol. 8736, 2013; Lazarev, G. et al., *LCOS Spatial Light Modulators: Trends and Applications*, Wiley-VCH, Weinheim, Germany, pp. 1–29, 2012; Petersen, K.E., *Appl. Phys. Lett.*, 31, 521–523, 1977.)

In general, phase modulation is more efficient and has been effectively implemented using both LC on silicon and MEMS fabrication methods [38,39]. Current commercially available SLM arrays consist of 1920×1080, 4–8 µm pixels with fill factors ~90% and 8-bit or 256 grey levels. These systems correspond to high definition television (HDTV) display format requirements [37]. The fill factor is the ratio of the active region of the pixel to the total area allocated to the pixel. The non-active area is typically used for interconnections to actuate the pixel. While this performance is significant, the resolution is typically lower that what can be achieved with an etched substrate. As shown in Section 6.2.3, the sampling requirements at the hologram plane are lower for Fourier transform type holograms. Therefore, the Fourier transform geometry is frequently used for forming computer-generated holograms with SLMs. A diagram illustrating the formation of a hologram with an SLM is shown in Figure 6.15. More recent demonstrations use multiple tilted SLMs in order to overcome some of the restrictions in viewing angle due to the limited resolution of the SLM devices [38,40,41].

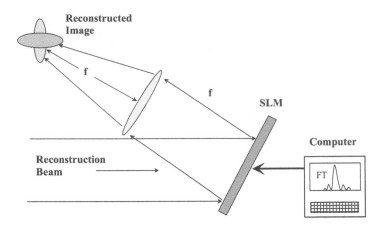

FIGURE 6.15 Illustration of the reconstruction of a reflection mode Fourier transform computer-generated hologram using a tilted SLM.

6.6 CGH Design Algorithm Optimization Methods

More generalized computer-generated holograms can be designed using different types of optimization algorithms [42,43]. In some respect, this technique is similar to the binary phase method described in Section 6.3.3. For instance, the hologram for both approaches is considered to be a transmittance phase function that forms the desired field or image at an observation plane after diffraction. For the binary phase method, the transmittance function is often simple enough so that the phase distribution can be computed analytically. However, with the generalized CGH approach, the image or optical transformation may be more complicated and requires a numerical technique to define the transmittance phase function. Several computational methods that have been found useful for this task are the direct binary search [42], simulated annealing [44,45], Gerchberg-Saxton [46], and general iterative algorithms [29]. The basic underlying approach of these methods is to use a diffraction propagation method to go back and forth between the CGH plane and the image plane to adjust changes in the phase of individual sampling points (pixels) at the CGH plane until the diffracted field at the image plane converges to the desired distribution. Each method typically uses an FFT algorithm to perform the diffraction propagation. The methods differ in how the phase of the pixels at the CGH plane is altered after each iteration. If the algorithms converge in reasonable time, these methods can be applied to form binary pattern input for SLMs for dynamic hologram formation. However, in cases that have longer computational times to reach convergence, the resulting phase distributions can readily be converted into pattern generation data for lithographically writing the CGH [42].

PROBLEMS

1. The object and hologram apertures for a CGH recording are both $1.5\,\text{cm} \times 2.5\,\text{cm}\,(x \times y)$. A wavelength of 0.5 μm will be used for the reconstruction. For a Fourier transform, CGH recorded with a 1 cm focal length lens compute: (i) the maximum spatial frequency in the x- and y-directions, (ii) the sampling interval, and (iii) the total number of sampling points. Repeat the calculation of i–iii for a Fresnel CGH recording with an object-hologram plane separation distance equal to 20 cm.

2. The *detour phase method* is used to encode a computer-generated hologram with 600 nm light. What reference beam angle should be used in order to be able to encode 2π of phase in a 25 μm pixel? What length of the sub pixel should be used to give an amplitude value for the pixel of 0.75 (a value of 1 is maximum and 0 is the minimum)? How far should the sub pixel be shifted relative to the center of the pixel to provide a phase value represented at this pixel of 10°?

3. Compute the fringe plot for the interference pattern resulting from the interference of a point source located 15 cm from the CGH recording plane ($z = -15$ cm) with a plane wave at 5° to the optical axis (z-axis). The CGH has a diameter of 5 cm, and the desired reconstruction wavelength is 0.5 μm. If this pattern is to be recorded as a binary phase CGH as discussed in Section 6.3.3, what printer resolution is needed if photo reduction of 10X can be used after printing?

4. Assume that there is one phase transition (x_1) within the half period of a Dammann grating. (i) Draw the transmittance function for the grating for a full period and (ii) compute the diffraction efficiency for the resulting diffraction orders.

REFERENCES

1. J. C. Wyant and V. P. Bennett, "Using computer generated holograms to test aspheric wavefronts," *Appl. Opt.*, 11, 2833 (1972).
2. A. Ono and J. C. Wyant, "Aspherical mirror testing using a CGH with small errors," *Appl. Opt.*, Vol. 24, 560–563 (1985).
3. T. Takahashi, K. Konno, M. Kawai, and M. Isshiki, "Computer generated holograms for testing aspheric lenses," *Appl. Opt.*, Vol. 15, 546–549 (1976).
4. B. K. Jenkins and T. C. Strand, "Computer generated holograms for space variant interconnections of optical logic systems," *Proc. SPIE*, Vol. 437, 110 (1983).
5. M. R. Feldman and C. C. Guest, "Computer generated holographic optical elements for optical interconnection of very large scale integrated circuits," *Appl. Opt.*, Vol. 26, 4373–4376 (1987).
6. B. D. Clymer and J. W. Goodman, "Optical clock distribution to silicon chips," *Opt. Eng.*, Vol. 25, 1105 (1986).
7. F. Mok et al., "Real time computer generated holography by means of liquid crystal television spatial light modulator," *Opt. Lett.*, Vol. 11, 748 (1986).
8. N. Masuda, T. Ito, T. Tanaka, A. Shiraki, and T. Sugie, "Computer generated holography using a graphics processing unit," *Opt. Exp.*, Vol. 14, 603–608 (2006).
9. D. G. Grier and Y. Roichman, "Holographic optical trapping," *Appl. Opt.*, Vol. 45, 880–887 (2006).
10. B. Sun, Y. Roichman, and D. G. Grier, "Theory of holographic optical trapping," *Opt. Exp.*, Vol. 16, 15765–15776 (2008).
11. W. H. Lee, "Computer-generated holograms: Techniques and applications," *Progress in Optics*, Vol. 16, E. Wolf ed., pp. 121–232, Amsterdam, the Netherlands (1978).
12. L. P. Yaroslavskii and N. S. Merzlyakov, *Methods of Digital Holography*, Plenum Publishing Corp, New York (1980).
13. G. Tricoles, "Computer generated holograms: An historical review," *Appl. Opt.*, Vol. 26, 4351–4360 (1987).
14. J. W. Goodman, *Introduction to Fourier Optics*, p. 29, 4th ed., W. H. Freeman New York (2017).
15. C. E. Shannon, "Communication in the presence of noise," *IRE*, Vol. 37 (1949).
16. J. W. Goodman, *Introduction to Fourier Optics*, pp. 412–418, 4th ed., W. H. Freeman, New York (2017).
17. B. R. Brown and A. W. Lohmann, "Complex spatial filtering with binary masks," *Appl. Opt.*, Vol. 5, 967–969 (1966).
18. B. R. Brown and A. W. Lohmann, "Computer generated binary holograms," *IBM J. Res. Dev.*, Vol. 13, 160–168 (1969).
19. A. W. Lohmann and D. P. Paris, "Binary Fraunhofer holograms, generated by computer," *Appl. Opt.*, Vol. 6, 1739–1748 (1967).
20. J. Turunen, J. Fagerholm, A. Vasara, and M. R. Taghizadeh, "Detour-phase kinoform interconnects: The concept and fabrication considerations," *J. Opt. Soc. Am.*, Vol. 7, 1202–1208 (1990).
21. S. Sinzinger and V. Arrizon, "High-efficiency detour-phase holograms," *Opt. Lett.*, Vol. 22, 928–930 (1997).
22. S. Sinzinger and J. Jahns, *Microoptics*, Wiley-VCH, Weinheim, Germany (1999).
23. W. H. Lee, "Sampled Fourier transform hologram generated by computer," *Appl. Opt.*, Vol. 9, 639–643 (1970).
24. W. H. Lee, "Binary synthetic holograms," *Appl. Opt.*, Vol. 13, 1677–1682 (1974).
25. W. H. Lee, "Binary computer generated holograms," *Appl. Opt.*, Vol. 18, 3661–3669 (1979).

26. H. Dammann and K. Gortler, "High-efficiency in-line multiple imaging by means of multiple phase holograms," *Opt. Comm.*, Vol. 3, 312–315 (1971).

27. J. N. Mait, "Design of binary-phase and multiphase Fourier gratings for array generation," *J. Opt. Soc. Am. A*, Vol. 7, 1514–1528 (1990).

28. H. Dammann and K. Gortler, "Coherent optical generation and inspection of two-dimensional periodic structures," *Optica Acta*, Vol. 25, 505–515 (1977).

29. J. R. Fienup, "Iterative method applied to image reconstruction and to computer-generated holograms," *Opt. Eng.*, Vol. 19, 297–305 (1980).

30. C. Zhou and L. Liu, "Numerical study of Dammann array illuminators," *Appl. Opt.*, Vol. 34, 5961–5969 (1995).

31. J. W. Goodman, *Introduction to Fourier Optics*, pp. 287–308, 4th ed., W. H. Freeman, New York (2017).

32. T. Haist and W. Osten, "Holography using pixelated spatial light modulators—Part 1: Theory," *J. Micro/ Nanolithography MEMS MOEMS*, Vol. 14, 041309 (2015).

33. W. P. Bleha, L. A. Lei, "Advances in liquid crystal on silicon (LCOS) spatial light modulator technology," *Proc. SPIE*, Vol. 8736, Display Technologies and Applications for Defense, Security, and Avionics VII, 87360A (4 June 2013).

34. G. Lazarev et al., *LCOS Spatial Light Modulators: Trends and Applications*, pp. 1–29, Wiley-VCH, Weinheim, Germany (2012).

35. K. E. Petersen, "Micromechanical light modulator array fabricated on silicon," *Appl. Phys. Lett.*, Vol. 31(8), 521–523 (1977).

36. M. C. Wu, O. Solgaard, and J. E. Ford, "Optical MEMS for lightwave communication," *J. Lightwave Tech.*, Vol. 24, 4433–4454 (2006).

37. M. E. Motamedo Ed., *MOEMS (Micro-Opto-Electro-Mechanical Systems)*, SPIE Press, Bellingham, WA (2005).

38. T. Kozacki, "Holographic display with tilted spatial light modulator," *Appl. Opt.*, Vol. 50, 3579–3588 (2011).

39. J. Extermann et al., "Spectral phase, amplitude, and spatial modulation from ultraviolet to infrared with a reflective MEMS pulse shaper," *Opt. Exp.*, Vol. 19, 7580–7586 (2011).

40. J. Hahn, H. Kim, Y. Lim, G. Park, and B. Lee, "Wide viewing angle dynamic holographic stereogram with a curved array of spatial light modulators," *Opt. Express*, Vol. 16, 12372–12386 (2008).

41. G. Finke, T. Kozacki, and M. Kujawinska, "Wide viewing angle holographic display with multi spatial light modulator array," *Proc. SPIE*, Vol. 7723 (2010).

42. M. A. Seldowitz, J. P. Allebach, and D. W. Sweeney, "Synthesis of digital holograms by direct binary search," *Appl. Opt.*, Vol. 26, 2788–2798 (1987).

43. R. Gerchberg and W. O. Saxton, "A practical algorithm for the determination of phase from image and diffraction plane pictures," *Optik*, Vol. 35, 237–246 (1972).

44. M. S. Kim, M. R. Feldman, and C. C. Guest, "Optimum encoding of binary phase-only filters with a simulated annealing algorithm," *Opt. Lett.*, Vol. 14, 545–547 (1989).

45. M. W. Farn and J. W. Goodman, "Optimal binary phase-only matched filters," *Appl. Opt.*, Vol. 27, 4431–4437 (1988).

46. R. W. Gerchberg and W. O. Saxton, "A practical algorithm for determination of phase from image and diffraction plane pictures," *Optik*, Vol. 35, 237–246 (1972).

47. K. Kostuk, Appearing in Chapter 4, "Optical Holography," *Photonics Technology and Instrumentation*, D. L. Andrews (Eds.), John Wiley & Sons, Hoboken, NJ (2015).

7

Digital Holography

7.1 Introduction

Digital or electronic holography was first demonstrated in the late 1960s [1] and has since become a major form of holography with a broad range of applications. In this technique, a digital camera is used to record the interference pattern between the object and reference waves and a numerical algorithm is used to reconstruct the image. The low resolution of early digital cameras restricted the types of objects that could be recorded and limited applications. However, recent improvements in camera design and novel recording methods has made digital holography (DH) an important technique and has been growing in importance since the early 1990s [2]. A great advantage of DH is that the holographic process is directly interfaced to a computer where a variety of computational processing methods can be applied to extract the amplitude and phase information about an object field. This is important in many applications such as microscopy and interferometry where the storage, manipulation, and evaluation, of digital data are essential to obtain the greatest information content from the recording. The prospects for digital holography continue to grow as electronic cameras evolve with smaller pixel size, larger numbers of pixels, faster frame rates, and greater dynamic range. Extensive reviews of digital holography are given in the books by Schnars and Jueptner [3] and by Kreis [4]. This chapter provides an overview of the basic DH process, construction geometries, recording techniques to overcome camera resolution issues, and highlights of applications that have benefited from this form of holography.

7.2 Digital Hologram Process

The first step of digital holography is similar to the analog hologram process described previously in that an interference pattern between an object and reference wave are formed as shown in Figure 7.1.

However, for the DH case, the interference pattern is recorded with a digital camera and then stored in computer memory. The image is then "reconstructed" by computation using a diffraction propagation algorithm and then displayed on a computer screen. Just as in analog holography, the "real" and "virtual" images can be generated using either the original or conjugate reference wave in the computational reconstruction process as shown in Figure 7.2.

Some of the most active areas of research in the field of digital holography focus on overcoming the relatively limited resolution of cameras, the development of reconstruction algorithm, and adapting DH to different applications. The remainder of this chapter highlights these activities and reviews some specialized applications in microscopy. The application of DH to interferometry is covered in more depth in Chapter 10.

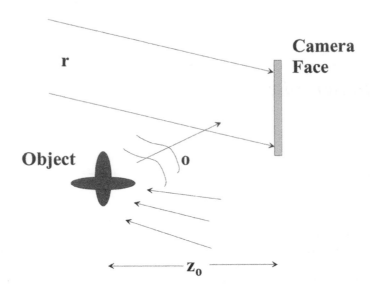

FIGURE 7.1 DH recording setup with an object located at a distance z_o from the camera face and with an off-axis planar reference wave.

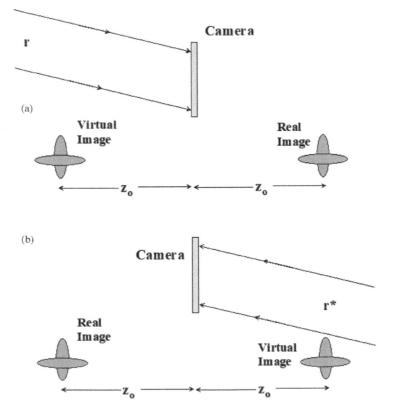

FIGURE 7.2 Numerical reconstruction of a digital hologram showing the real and virtual image reconstructed with (a) the original reference wave and (b) the phase conjugate reference wave.

7.3 DH Recording Considerations

Since the interference pattern is recorded on a digital camera, the process involves sampling a continuous function. The sampling requirements for DH are the same as those for computer-generated hologram described in Chapter 6 and are determined by the Nyquist rate. The pixel size, $\Delta\xi$, of the best digital cameras today is on the order of 2–3 μm. This resolution limit constrains the viable hologram recording geometries. The grating equation can be used to determine the maximum spatial frequency formed with a particular object and recording geometry. If the reference beam propagates along the optical axis as shown in Figure 7.3, the maximum spatial frequency formed during recording is given by:

$$f_{max} = \frac{1}{\Lambda_{min}} = \frac{\sin\theta_{max}}{\lambda}, \tag{7.1}$$

where θ_{max} is the maximum interbeam angle that is recorded, and Λ_{min} is the corresponding minimum fringe period. According to the Nyquist rate, the camera must be able to sample both the maximum and minimum intensity points of the interference fringe, therefore, the maximum spatial frequency that can be recorded is:

$$f_{max} = \frac{1}{2\Delta x_H}. \tag{7.2}$$

Therefore, if the camera has 2 μm pixels, then the maximum spatial frequency that can be recorded is 250 lp/mm. This places constraints on the object size and hologram geometry that can be used in DH. Nonetheless, a variety of recording geometries and techniques have been developed for DH configurations in different applications.

One way to increase the size of an object that can be recorded is to de-magnify the object height with a negative lens. Figure 7.4 shows an arrangement to de-magnify the extent of an object by inserting a

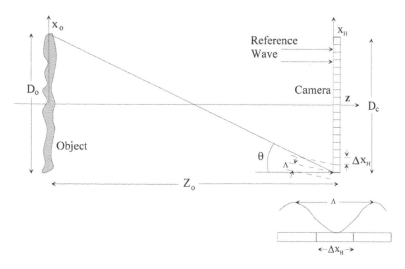

FIGURE 7.3 Digital hologram recording geometry with an object diameter (D_o), camera diameter (D_c), and separation distance Z_o. The reference beam is normally incident to the camera aperture. The camera pixel width is $\Delta\xi$, and the interference fringe period is Λ.

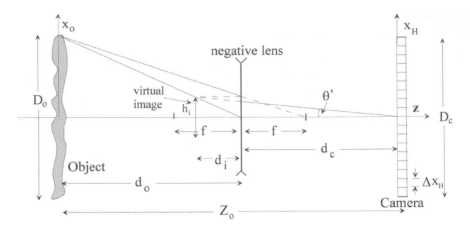

FIGURE 7.4 De-magnification of the object beam angle by using a negative lens.

negative lens with a focal length f between the object and the camera planes. The demagnification can be determined from the lens equation:

$$\frac{1}{f} = \frac{1}{d_o} - \frac{1}{d_i} \Rightarrow M = \frac{h_i}{D_o} = \frac{-f}{d_o - f}. \tag{7.3}$$

The new object beam angle θ' becomes:

$$\tan\theta' = \frac{h_i}{2(d_i + d_c)}, \tag{7.4}$$

with the distance d_c between the camera and the negative lens:

$$d_c = \frac{-D_o f}{(d_o - f)2\tan\theta'} + \frac{fd_o}{d_o - f}. \tag{7.5}$$

This technique can be used to significantly reduce the maximum spatial frequency that must be recorded while still retaining a larger object size. However, it does require the inconvenience of using an intermediate lens in the system.

7.4 Construction Geometries

Several configurations have been demonstrated for recording digital holograms that satisfy the Nyquist sampling condition. Perhaps the simplest geometry is the in-line or Gabor hologram arrangement (Figure 7.5). In this configuration, an incident beam is partially scattered by the object to produce the object beam, and the non-scattered light serves as the reference beam. However, as described previously in the discussion of Gabor type holograms, during reconstruction, light from both the real and virtual images as well as the reconstruction beam propagate along the same axis and reduces the contrast of the reconstructed images.

In order to eliminate overlapping beams during reconstruction, an off-axis tilted reference wave can be used to record the hologram as shown in Figure 7.6. In this case, the reference beam is given by:

$$E_R(x_H, y_H) = R_c \exp\left(-j\frac{2\pi}{\lambda}x_H \sin\theta_R\right), \tag{7.6}$$

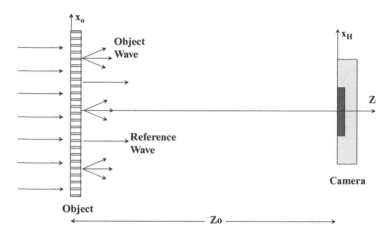

FIGURE 7.5 In-line digital hologram recording geometry.

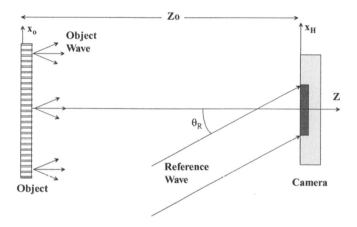

FIGURE 7.6 Off-axis digital hologram recording geometry.

where the amplitude R_c is a constant. However, the off-axis reference beam angle must be restricted to small values in order to satisfy the Nyquist sampling rate. This is often difficult to achieve experimentally using the configuration shown in Figure 7.6 without placing the object an appreciable distance from the camera plane. However, it is possible to form small reference beam angles by inserting a beam splitter and tilting the reference beam by a small angle (α) relative to the optical axis as shown in Figure 7.7. When a beam splitter is used only part of the reference and object beam power are directed to the camera with the remainder propagating out of the system. This condition increases the power required by the optical source and also affects the contrast of the resulting interference pattern. In most cases, the object beam is weaker since the irradiance of the scattered light from the object surface decreases inversely with the square of the distance to the camera surface. In this situation, it is possible to use a beam splitter with an unequal splitting ratio (i.e., 20% reflected and 80% transmitted) to allow a greater percentage of the weaker object beam to pass and form the interference pattern.

The beam splitter can also be used to form in-line holograms by not tilting the reference beam. An added benefit of using a beam splitter is that opaque objects can be used for recording in-line holograms.

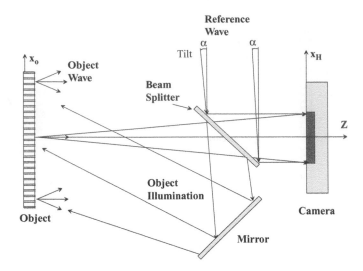

FIGURE 7.7 Use of a beam splitter to realize small reference beam angles without significantly displacing the object from the camera plane.

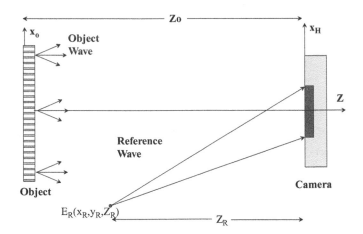

FIGURE 7.8 Digital hologram recording with a spherical reference wave located at (x_R, y_R, z_R) from the camera plane.

Another useful configuration for digital holography is to use a spherical reference beam by placing a point source at a distance Z_R from the camera plane (Figure 7.8). In this case, the reference beam has the form:

$$E_R\left(x_H, y_H, z_H\right) = R_c \exp\left(j\vec{k} \cdot \vec{\rho}\right), \qquad (7.7)$$

where

$$\left|\vec{\rho}\right| = \sqrt{\left(x_H - x_R\right)^2 + \left(y_H - y_R\right)^2 + \left(z_H - z_R\right)^2}; \left|\vec{k}\right| = 2\pi / \lambda. \qquad (7.8)$$

Note that when $Z_R = Z_o$, this arrangement is the same as the *lensless Fourier Transform hologram* [5]. The different types of reference beams are used to obtain different effects in the holographic image.

7.5 Reconstruction Methods

Several approaches have been developed to mathematically perform the diffraction propagation to reconstruct the holographic image. Two of the most commonly used methods include the Fresnel approximation and convolution integral techniques. Analog integral form of these methods have previously been discussed in Chapter 2 to describe beam propagation and diffraction from an aperture. However, in this case, the field is sampled and a discrete representation of the diffraction integral must be applied. The different types of reference beams described in the previous section modify the reconstruction conditions and allow different effects to be realized in the image field.

7.5.1 Fresnel Approximation Method

For this analysis (x_1, y_1) are the object plane coordinates, (x_2, y_2) are the camera aperture (hologram) coordinates, and (x_i, y_i) the image plane coordinates. It is assumed that the distance between planes is far enough to satisfy the Fresnel diffraction approximation or:

$$z_o^3 \gg \frac{\pi}{4\lambda} \left[(x_2 - x_1)^2 + (y_2 - y_1)^2 \right]_{max}^2,$$ (7.9)

where the coordinates are as shown in Figure 7.9. After propagating to the hologram (camera) plane, the object field has the form:

$$E(P_h) = \frac{1}{j\lambda} \iint_{A_o} E(P_o) \frac{e^{jks_{oh}}}{s_{oh}} da_o,$$ (7.10)

where $E(P_h)$ is the field at the point P_h, da_o is an increment of area on the object plane, s_{oh} is the length of a ray from an object point P_o to a point on the hologram plane P_h, and A_o is the total area of the object. The digital hologram is formed by interfering $E(P_h)$ with a reference field in the same manner as in an analog hologram that is recorded on film.

The hologram resulting from the interference pattern can be considered as a transmittance function $t(x_2, y_2)$ as previously described for analog holograms. As with the analog hologram to form an image, the transmittance function is multiplied by the reconstruction beam. The resulting field just after the hologram plane is:

$$E_H(x_2, y_2) = r(x_2, y_2) t(x_2, y_2),$$ (7.11)

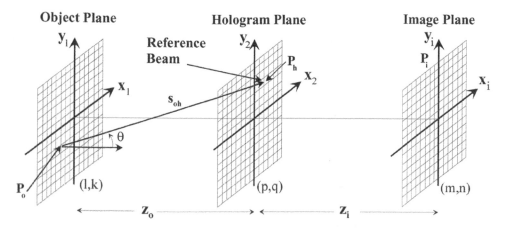

FIGURE 7.9 Coordinate system for forming and reconstructing the digital holographic image.

where $r(\xi,\eta)$ is the reconstruction field at the camera plane. The image field E_i is computed by performing another Fresnel diffraction propagation from the hologram/camera plane to the desired image plane (z_i) resulting in:

$$E_i(x_i, y_i) = \frac{\exp(jkz_i)}{j\lambda z_i} \int\limits_{-\infty}^{\infty}\int\limits_{-\infty}^{\infty} E_H(x_2, y_2)\exp\left\{ j\frac{k}{2z_i}\left[(x_i - x_2)^2 + (y_i - y_2)^2 \right] \right\} dx_2 dy_2. \qquad (7.12)$$

Expanding the quadratic factors $(x_i - x_2)^2$ and $(y_i - y_2)^2$, and using the relations for the spatial frequencies $f_x = x_i / \lambda z_o$, $f_y = y_i / \lambda z_o$, the two-dimensional Fresnel transform becomes:

$$E_i(f_x, f_y) = \frac{1}{j\lambda z_i}\exp\left[j\pi\lambda z_i\left(f_x^2 + f_y^2 \right) \right]$$

$$\times \int\limits_{-\infty}^{\infty}\int\limits_{-\infty}^{\infty} r(x_2, y_2)t(x_2, y_2)\exp\left[j\frac{\pi}{\lambda z_i}\left(x_2^2 + y_2^2 \right) \right]. \qquad (7.13)$$

$$\times \exp\left[-j2\pi\left(x_2 f_x + y_2 f_y \right) \right] dx_2 dy_2.$$

This expression is the Fourier transform of $r(x_2, y_2)t(x_2, y_2)\exp\left[jk/2z_o\left(x_x^2 + y_2^2 \right) \right]$, which is the aperture function of the hologram times a quadratic phase factor.

   For digital holography, a camera with $N \times N$ pixels does not record the continuous form of $E_h(x_2, y_2) = r(x_2, y_2)t(x_2, y_2)$, but samples it at each camera pixel. The hologram or camera aperture is digitized on a $N \times N$ grid with increments Δx_2 and Δy_2 that equal the pixel dimensions of the camera. The actual measured or sampled values are $E_h(m\Delta x_2, n\Delta y_2)$.

   The pixel dimensions on the camera $(\Delta x_2, \Delta y_2)$ and the resolution that can be achieved at the image plane $(\Delta x_i, \Delta y_i)$ are related. If the camera has $N \times N$ pixels, then the maximum extent along the x_2 axis is $\frac{N}{2}\Delta x_2$. The corresponding spatial frequency at an image plane located a distance z_i away is $f_{x_i-max} = \frac{N}{2}\frac{\Delta x_2}{\lambda z_i}$. In order to recover this frequency at the image plane, the field at the camera plane must be sampled at a rate of $\frac{1}{\Delta x_i} = N\frac{\Delta x_2}{\lambda z_i}$ (i.e., twice the highest frequency). This corresponds to an image resolution of:

$$\Delta x_i = \frac{\lambda z_i}{N\Delta x_2}; \quad \Delta y_i = \frac{\lambda z_i}{N\Delta y_2}, \qquad (7.14)$$

in the x- and y-directions at the image plane. An important implication of this result is that the image resolution is directly proportional to the image distance and reconstruction wavelength and inversely proportional to the size of the camera aperture, i.e., $N\Delta x_2 \times N\Delta y_2$. The resulting discrete Fresnel transform with a square camera aperture ($\Delta x_2 = \Delta y_2$ and $N \times N$) becomes:

$$E_i(m,n) = \frac{1}{j\lambda z_i}\exp\left[j\pi z_i\left(\frac{m^2}{N^2(\Delta x_2)^2} + \frac{n^2}{N^2(\Delta y_2)^2} \right) \right]$$

$$\times \sum_{p=0}^{N-1}\sum_{q=0}^{N-1} r(p,q)t(p,q)\exp\left[j\frac{\pi}{\lambda z_i}\left(p^2(\Delta x_2)^2 + q^2(\Delta y_2)^2 \right) \right] \qquad (7.15)$$

$$\times \exp\left[-j2\pi\left(\frac{pm}{N} + \frac{qn}{N} \right) \right].$$

In this expression, the field at each pixel on the camera plane is written as $E_h(m,n)$ where it is implied that each pixel integer is multiplied by the pixel dimension.

This expression gives the discrete field representation reconstructed from the hologram at the location (m,n) in the image plane. This relation can now be used to compute the amplitude $|E_i(m,n)|$ at each location in the image. The corresponding phase at each position is:

$$\varphi_i(m,n) = \arctan\left[\frac{\text{Im}\{E_i(m,n)\}}{\text{Re}\{E_i(m,n)\}}\right]. \tag{7.16}$$

Using these parameters, the field at each location in the image can be explicitly written as:

$$E_i(m,n) = |E_i(m,n)| e^{-j\varphi_i(m,n)}. \tag{7.17}$$

7.5.2 Convolution Method

The Fresnel reconstruction method is a useful approach, however, it is limited by the fact that the image resolution is directly proportional to the distance of the image from the hologram plane (i.e., z_i). An alternative reconstruction can be performed by recognizing that the diffraction integral:

$$E_i(P_i) = \frac{1}{j\lambda} \iint_{A_h} E_h(P_h)\frac{e^{jks_{hi}}}{s_{hi}} da_h, \tag{7.18}$$

can also be interpreted as a convolution between a kernel function $g(x,y)$ and the field at the hologram aperture $E_h(x,y)$ as:

$$E_i(x_i, y_i) = \int_{-\infty}^{\infty}\int_{-\infty}^{\infty} E_h(x_2, y_2)g(x_i - x_2, y_i - y_2)dx_2 dy_2, \tag{7.19}$$

where

$$g(x_i - x_2, y_i - y_2) = -\frac{j}{\lambda}\frac{\exp\left\{jk\left[z_i^2 + (x_i - x_2)^2 + (y_i - y_2)^2\right]^{1/2}\right\}}{\left[z_i^2 + (x_i - x_2)^2 + (y_i - y_2)^2\right]^{1/2}}. \tag{7.20}$$

The convolution theorem states that the Fourier transform of the convolution of $E_h(x_i, y_i)$ with g is the product of the Fourier transforms of these functions or:

$$\Im\{E_i(x_i, y_i)\} = \Im\{r(x_2, y_2)\cdot t(x_2, y_2)\}\cdot\Im\{g(x_2, y_2)\}. \tag{7.21}$$

This implies that an image from a digital hologram recording can be reconstructed by taking the inverse Fourier transform of $\Im\{E_i(x,y)\}$:

$$E_i(x_i, y_i) = \Im^{-1}\{\Im(r\cdot t)\cdot\Im(g)\}. \tag{7.22}$$

The kernel function *g* in digital form is:

$$g(p,q) = \frac{-j}{\lambda} \frac{\exp\left\{ jk\left[z_i^2 + \left(p - \frac{N}{2}\right)^2 \Delta x_2^2 + \left(q - \frac{N}{2}\right)^2 \Delta y_2^2 \right]^{1/2} \right\}}{\left[z_i^2 + \left(p - \frac{N}{2}\right)^2 \Delta x_2^2 + \left(q - \frac{N}{2}\right)^2 \Delta y_2^2 \right]^{1/2}}, \tag{7.23}$$

where the coordinates for this function are shifted by *N*/2 to simplify the fast fourier transform (FFT) operation. The corresponding discrete Fourier transform of *g*(*p*,*q*) becomes:

$$G(m,n) = \exp\left\{ jkz_i \left[1 - \frac{\lambda^2\left(m + \frac{N^2\Delta x_2^2}{2z_i\lambda}\right)^2}{N^2\Delta x_2^2} - \frac{\lambda^2\left(n + \frac{N^2\Delta y_2^2}{2z_i\lambda}\right)^2}{N^2\Delta y_2^2} \right] \right\}. \tag{7.24}$$

In the convolution approach, the size of the camera pixels is the same as the sampling points at the image plane:

$$\Delta x_i = \Delta x_2; \quad \Delta y_i = \Delta y_2. \tag{7.25}$$

This is an important consideration in microscopy and other applications where objects are recorded at different distances from the hologram plane since it eliminates the change in magnification that occurs with the Fresnel reconstruction method.

7.6 Digital Hologram Imaging Issues and Correction Techniques

As discussed earlier in Chapter 3, the reconstructed field from the hologram consists of four components:

$$\tilde{p}_r = t(x, y, z)\tilde{r}$$
$$= \tilde{r}\alpha\left(|\tilde{r}|^2 + |\tilde{o}|^2\right) + \alpha\tilde{r}\tilde{r}\tilde{o}* + \alpha\tilde{r}\tilde{r}*\tilde{o} \tag{3.5}$$

The first two terms result in an average bias level, while the second two terms contain the amplitude and phase of the object and conjugate object field. For an off-axis hologram, the different field components can be separated resulting in a high contrast image. However, if the spatial frequency requirements of the camera limit the recording to an on-axis recording geometry, then all four field components overlap. This obscures the image of interest and requires additional processing to improve the image quality. The zero order and twin image are difficult to remove from optical reconstructions, however, several methods can be applied to the mathematical reconstruction of digital holograms to eliminate or reduce these degrading effects.

7.6.1 Zero Order Suppression by Background Subtraction

A simple, but effective method to improve the contrast of an on-axis digital holographic image is to subtract the average background value of the field across the camera surface. If the camera has *N*×*N* pixels, and the intensity of the interference pattern across the camera aperture is $I_H(p,q)$, then the average intensity is:

$$I_{H,avg} = \frac{1}{N^2}\sum_{q=1}^{N}\sum_{p=1}^{N}I_H(p,q),$$ (7.26)

and the adjusted field becomes:

$$I_H'(p,q) = I_H(p,q) - I_{H,avg}.$$ (7.27)

This approach can potentially produce some negative values, but these can be corrected by shifting the entire array of intensity values by the most negative value.

Another way to eliminate the bias terms is to optically record r and o separately and then record the interference of these two fields. The separately recorded fields are then subtracted from the interference field before performing the reconstruction.

A different approach to background subtraction is to compute a local intensity average in the vicinity of a camera pixel, subtract the average, and then repeat this operation for each pixel across the camera aperture [3,4]. Figure 7.10 shows the basic operation.

As an example of this method, if nine pixels are used in the averaging process, the adjusted field at pixel p, q in the camera aperture is:

$$I_H'(p,q) = I_H(p,q) - \frac{1}{9}\sum_{p-1}^{p+1}\sum_{q-1}^{q+1}I_H(p,q),$$ (7.28)

where the double summation represents the sum over the eight surrounding pixels plus the target pixel (p,q). The averaging process is applied to all pixels across the camera aperture with adjustments for pixels that lie along the camera perimeter.

7.6.2 Phase Shifting Recording and Correction

While the background subtraction methods are effective at eliminating the bias levels from the reconstructed image, the r^2o^* term in Equation 7.25 still exists and obscures the desired image of an in-line hologram. In order to eliminate this term from the image a number of useful phase shifting recording methods have been developed [6–8].

One approach is to record multiple interference patterns between an object field and several reference beams with slightly different phase values. This essentially gives enough independent equations to solve for the object beam and eliminate unwanted beams from the reconstructed image.

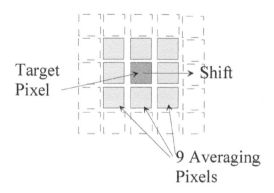

FIGURE 7.10 Localized background averaging to improve contrast of on-axis holograms.

For example, consider recording three frames of the interference pattern between r and o with phase shifts of 0, $\pi/2$, and π added to the reference beam for the three exposures. The three interference patterns at the camera plane are:

$$I_{H,0}(\xi,\eta) = |o+r|^2,$$

$$I_{H,\pi/2}(\xi,\eta) = |o+re^{j\pi/2}|^2,$$

(7.29)

$$I_{H,\pi/2}(\xi,\eta) = |o+re^{j\pi}|^2.$$

The resulting interference patterns are then combined resulting in:

$$I_{H,0}(\xi,\eta) - I_{H,\pi/2}(\xi,\eta) + j[I_{H,\pi/2}(\xi,\eta) - I_{H,\pi}(\xi,\eta)]$$

$$= ro^* + or^* - ro^* \exp\left(\frac{j\pi}{2}\right) - or^* \exp\left(\frac{-j\pi}{2}\right)$$

$$+ j\left\{ro^* \exp\left(\frac{j\pi}{2}\right) + or^* \exp\left(\frac{-j\pi}{2}\right)\right\} - j\left\{ro^* \exp(j\pi) + or^* \exp(-j\pi)\right\}$$

(7.30)

$$= ro^* + or^* - jro^* + jor^* - ro^* + or^* + jro^* + jor^*$$

$$= 2(1+j)or^*.$$

This expression can now be solved for the field of the object wave, o, in terms of the three recorded interference patterns: $I_{H,0}, I_{H,\pi/2}, I_{H,\pi}$ using the relation:

$$o(\xi,\eta) = \frac{(1-j)}{2r^*(\xi,\eta)}\left\{I_{H,0}(\xi,\eta) - I_{H,\pi/2}(\xi,\eta) + j[I_{H,\pi/2}(\xi,\eta) - I_{H,\pi}(\xi,\eta)]\right\}.$$

(7.31)

Another phase shifting approach is to separately record the intensities of the object and reference beams and then subtract $|o|^2$ and $|r|^2$ from the interference of o and r giving:

$$I'_{H,0}(\xi,\eta) = I_{H,0}(\xi,\eta) - |o(\xi,\eta)|^2 - |r(\xi,\eta)|^2.$$

(7.32)

A second interference pattern is formed between o and a phase shifted reference beam $r \cdot \exp(j\phi)$:

$$I'_{H,\phi} = |o + e^{j\phi}r|^2.$$

(7.33)

The intensities of the object and reference beams are then subtracted from the shifted interference pattern. The two modified interference patterns are then combined according to:

$$I'_H(\xi,\eta) + e^{j\phi}I'_{H,\phi}(\xi,\eta) = ro^* + r^*o + e^{2j\phi}\left[ro^* + r^*oe^{-2j\phi}\right]$$

$$= 2r^*o + ro^*\left(1 + e^{-2j\phi}\right).$$

(7.34)

If a phase shift value of $\phi = \pi/2$ is applied, the object beam can be recovered by combining terms according to the relation:

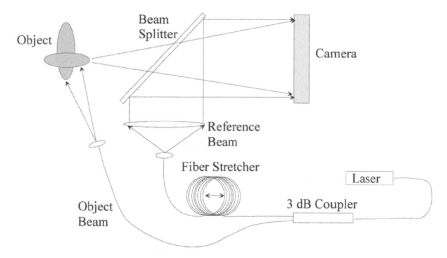

Object

Beam
Splitter

Camera

Reference
Beam

Fiber Stretcher

Laser

Object
Beam

3 dB Coupler

FIGURE 7.11   Fiber interferometer and fiber stretcher for recording phase shifted digital holograms.

$$o = \frac{1}{2r^*}\left[I'_H(\xi,\eta) + e^{j\phi}I'_{H,\phi}(\xi,\eta) \right]$$

$$= \frac{1}{2r^*}\left[I'_H(\xi,\eta) + jI'_{H,\phi}(\xi,\eta) \right]. \tag{7.35}$$

Several methods can be used to produce a phase shift including mirrors mounted on precision piezoelectric actuators or with a fiber holography setup and a thermal or mechanical fiber stretcher. Figure 7.11 shows a configuration for performing phase shifting recording using a fiber interferometer and an electrically controlled fiber stretcher. One type of fiber stretcher consists of several coils of fiber wrapped around a piezoelectric core that is expanded with the application of a voltage [9]. The coherence length of the laser must be greater than the difference in path length between the reference and object beam fiber lengths and the free-space propagation lengths.

7.7 Applications of Digital Holography

7.7.1 DH Microscopy

One of the most important applications of digital holography is in microscopy. The high magnification typically required for viewing small objects results in a very small depth of focus. This in turn necessitates either a manual or auto scanning through the object to obtain the desired information [10]. This process can be time consuming and potentially allows changes to occur in the specimen during the viewing time. For a digital hologram, the recording time is essentially the frame rate of the camera which can be considerably shorter than the time required for three-dimensional scanning. In digital holography, the complete three-dimensional information about the object is quickly stored and can then later be viewed without concern about changes to the object during the recording.

Several different configurations for DH microscopy exist. In the simplest case, a digital hologram is recorded without auxiliary optics. This is essentially the idea that Gabor had with regards to electron microscopy when he invented holography [11]. However, the resolution of the camera is still an issue and must be carefully considered in the design of the DH microscope. For instance, using the resolution limits based on the Fresnel reconstruction method:

$$\Delta x_i = \frac{\lambda z_i}{N \Delta x_2}; \quad \Delta y_i = \frac{\lambda z_i}{N \Delta y_2}, \tag{7.36}$$

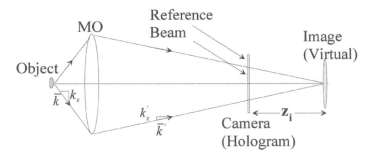

FIGURE 7.12 Typical digital holographic microscope system showing the transformation of the propagation vector and spatial frequency by using a microscope objective (MO).

along with a camera pixel size of 2 µm, $N = 3000$, $\lambda = 600\,\text{nm}$, the object would have to be 10 cm from the camera for 1 µm image resolution. Although the Fresnel approximation:

$$z_i^3 \gg \frac{\pi}{4\lambda}\left[(\Delta x)^2 + (\Delta y)^2\right]^2, \tag{7.9}$$

is satisfied in this case, it may be difficult to collect sufficient light scattered by a small object at this large distance from the camera.

Collecting light using a high numerical aperture objective solves this problem and also helps to relax the spatial frequency requirements at the camera plane [12]. As shown in Figure 7.12, if the magnification of the optical system is M, the x component of the propagation vector k_x after the lens and on the face of the camera is reduced to k_x/M. This allows using the limited spatial frequency capability of the camera to resolve small features in the object.

The DH microscope can be configured in both transmission and reflection modes as illustrated in Figures 7.13 and 7.14 [12]. Each arrangement allows the detection of the amplitude and phase of the object, and several useful techniques have been developed to increase the capabilities of these DH microscopy configurations. These include recording with multiple wavelengths [13,14] and with short coherence length sources [15]. These two approaches are related as will be shown below.

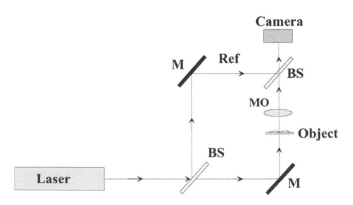

FIGURE 7.13 Transmission digital holographic microscope configuration. M—mirror, BS—beam splitter, and MO—microscope objective.

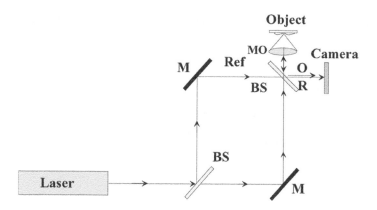

FIGURE 7.14 Reflection digital holographic microscope configuration. M—mirror, BS—beam splitter, and MO—microscope objective.

7.7.2 Multiple Wavelength DH Microscopy

The principle of multiple wavelength DH and microscopy follows from the basic relations for single wavelength holography [13,14]. Recall that the image field $E_{i,o}$ from a hologram contains the original object field according to the basic equation of holography:

$$E_i = t(x,y,z)r$$
$$= r\alpha\left(|r|^2 + |o|^2\right) + \alpha rro* + \alpha rr*o. \tag{3.6}$$

As discussed earlier, the term in this equation that represents the reconstructed object field (o) can be determined at different distances z_i from the hologram plane using the Fresnel-Huygens relation:

$$E_i(x,y,z_i) = \frac{\exp(jkz_i)}{j\lambda z_i} \int_{-\infty}^{\infty}\int_{-\infty}^{\infty} r \cdot t(\xi,\eta,0)\exp\left\{j\frac{k}{2z_i}\left[(x-\xi)^2 + (y-\eta)^2\right]\right\}d\xi d\eta.$$
$$E_{i,o}(x,y,z_i) = \alpha rr^*o \tag{7.37}$$

If holograms are recorded with different wavelengths, the propagation vectors and phase terms will change for each reconstructed field, but the object field is recovered in each reconstruction.

Now consider the case where reconstructions with different wavelengths are added forming a composite field:

$$\tilde{E}_{comp}(x,y,z_i) = \tilde{E}_{i,o}(x,y,z_i)\left\{e^{jk_1z_i} + e^{jk_2z_i} + e^{jk_3z_i} + \ldots + e^{jk_Mz_i}\right\}. \tag{7.38}$$

If the different wavelengths are selected with regular spacing of $\Delta\lambda$, then the difference in propagation constants are:

$$\Delta k = \Delta\left(\frac{2\pi}{\lambda}\right) = -\frac{2\pi\Delta\lambda}{\lambda^2}, \tag{7.39}$$

where it is assumed that $\Delta\lambda$ is small compared to the nominal wavelength λ. Inserting this value for Δk into the composite field expression yields:

$$
\begin{aligned}
\tilde{E}_{comp}\left(x,y,z_i\right) &= \tilde{E}_{i,o}\left(x,y,z_i\right)\left\{e^{jk_1 z_i}+e^{j(k_1+\Delta k)z_i}+e^{j(k_1+2\Delta k)z_i}+\ldots+e^{j(k_1+(M-1)\Delta k)z_i}\right\} \\
&= \tilde{E}_{i,o}\left(x,y,z_i\right)e^{jk_1 z_i}\left\{1+e^{j\Delta k\,z_i}+e^{j2\Delta k\,z_i}+\ldots+e^{j(M-1)\Delta k\,z_i}\right\} \\
&= \tilde{E}_{i,o}\left(x,y,z_i\right)e^{jk_1 z_i}\frac{1-e^{jM\Delta k\,z_i}}{1-e^{j\Delta k\,z_i}} \\
&= \tilde{E}_{i,o}\left(x,y,z_i\right)e^{jk_1 z_i}\frac{\sin\left(M\Delta k\,z_i\right)}{\sin\left(\Delta k\,z_i\right)}.
\end{aligned}
\tag{7.40}
$$

The corresponding intensity of the composite field with different wavelengths is:

$$
I_{comp}\left(x,y,z_i\right)=\left|\tilde{E}_{i,o}\left(x,y,z_i\right)\right|^2\frac{\sin^2\left(M\Delta k\,z_i\right)}{\sin^2\left(\Delta k\,z_i\right)}.
\tag{7.41}
$$

From this expression, it can be seen that the composite image intensity reaches a maximum when $\Delta k z_i = m\pi$ with m an integer or whenever the image distance takes on a value:

$$
z_{i,m}=\frac{m\pi}{\Delta k}=\frac{m\lambda^2}{2\Delta\lambda},
\tag{7.42}
$$

and has a depth of focus of the image is:

$$
\Delta z=\frac{2\pi}{M\Delta k}=\frac{\lambda^2}{M\Delta\lambda}.
\tag{7.43}
$$

The result shows that this method effectively samples an object at different depths $z_{i,m}$ with a depth resolution of Δz. It is important to use enough wavelengths (M) so that the depth resolution is smaller than the desired separation between images ($z_{i,m}-z_{i,m-1}$). Kim and others [13,14] have shown that this approach can be used to form 3D tomographic images of microscopic objects. The system for forming multiple wavelength holograms is similar to that shown in Figure 7.13, but with the single wavelength laser replaced with a tunable dye or swept source laser that has the desired spectral bandwidth.

7.7.3 Short Coherence Length DH Microscopy

If an extended broadband source is used to record a digital hologram, the coherence properties can be used to "gate" the recording and perform depth sectioning of the object being. The gating property is also useful in rejecting scatter that degrades the contrast of the reconstructed image. Sun and Leith made an important connection between confocal microscopy and holograms recorded with broadband extended optical sources [15,16]. This connection has significant implications for the application of digital holography to microscopy. The confocal imaging technique has significant advantages for microscopy including high resolution, background rejection, and depth discrimination. Sun and Leith specifically showed that the two processes are identical in terms of their point spread functions or impulse response and similar background rejection properties. In order to understand this relation, the coherence properties of an extended broadband source and the characteristics of a confocal microscopy are reviewed.

The spatial coherence properties of an extended source affect the resulting contrast of the interference pattern formed with this source. The Van Cittert-Zernike theorem [17] describes the relation between the source properties and the spatial coherence at a receiver located at a fixed distance from the source.

The theorem states that if an extended source of area A_s has uniform emittance, the coherence area (A_{coh}) at a distance z_s from the source is given by:

$$A_{coh} = \frac{(\lambda_s z_s)^2}{A_s} \cong \frac{\lambda_s^2}{\Omega_s}, \tag{7.44}$$

where Ω_s is the solid angle subtended by the source at the receiver. The main implication of this property for holography is that it determines the lateral extent of the region on the camera aperture where high visibility fringes will occur.

Now consider a confocal imaging system as shown in Figure 7.15 [15]. In this configuration, an image of a point source is formed at the object plane through lens (L_1) and a second matched lens (L_2) images this point image onto a pinhole. This effectively allows light from one point on the object to be detected while blocking light from other regions in the object. In order to generate an image of the entire object, the object is scanned in three dimensions and the intensity distribution is stored in computer memory and reconstructed with a mapping algorithm. If the point spread function of the first and second lenses are, respectively, g_1 and g_2, then the combined amplitude point spread function for the two lens system is $g = g_1 \times g_2$ and $|g|^2 = |g_1 \times g_2|^2$ for the intensity point spread function.

In order to understand the connection between confocal microscopy and digital holography with an extended source, consider the holographic recording system shown in Figure 7.16. Light from an extended source passes through grating $G1$ diffracting light at the same wavelength into $\pm$ diffraction orders and through a second grating to form the object and reference beams for the holographic exposure. The object beam passes through a transparency at the object plane and is imaged to the recording or camera plane through lens L_1. An equally positioned "reference plane" is imaged to the camera through the reference arm imaging system. In this arrangement, a region on the object plane is only coherent with a specific region on the reference plane (i.e., A_{coh}). One location on the object plane and a corresponding

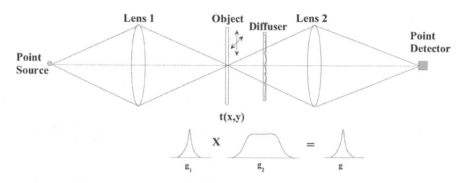

FIGURE 7.15 Confocal imaging system for generating an image of an object through a scattering medium. The point spread functions of the two lenses are g_1 and g_2.

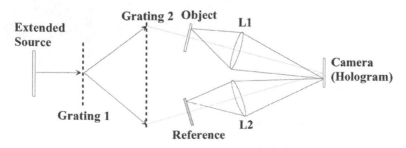

FIGURE 7.16 Dual beam interferometer for recording digital holograms with spatially incoherent light.

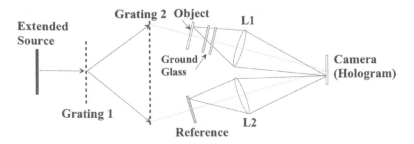

FIGURE 7.17 Experimental system to demonstrate imaging through scattering media with a spatially incoherent DH microscope to image through scattering media. (From Sun, P.-C. and Leith, E.N., *Appl. Opt.*, 33, 597–602, 1994. With permission from the Optical Society of America.)

location on the reference plane form a coherence pair. Since the object and reference planes are imaged onto the camera face with 1:1 registration, all corresponding object and reference pairs are aligned and form high contrast interference fringes at the camera. Other combinations of object and reference plane locations are not correlated and do not result in high contrast fringes that form a hologram.

Some interesting experiments have been realized with the spatially incoherent DH microscope by Leith et al. [15,18]. In particular, they have shown that objects can be reconstructed that are obscured by highly scattering media. One experiment is illustrated in Figure 7.17. In this case, the object is obscured by two highly scattering ground glass substrates inserted in the object beam path. When the object is viewed without holographic recording, it is completely obscured (Figure 7.18a). However, when an interferometric recording is made and reconstructed as a digital hologram by interfering with the reference plane, an image of the object can be recovered (Figure 7.18b). This capability has important implications for medical imaging and other forms of microscopy that require viewing objects through highly scattering media.

7.7.4 Digital Holographic Interferometry

Another very important application of digital holography is interferometry. Many of the methods that will be described in Chapter 10 in greater detail for analog holographic interferometry can readily be adapted to digital holographic interferometry.

In multiple exposure digital holographic interferometry a recording is made of each state of the object. The holograms are then reconstructed using one of the propagation methods described in Section 7.5

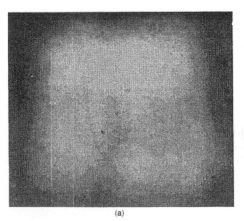
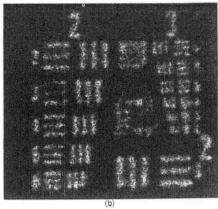

(a)   (b)

FIGURE 7.18 (a) Image of the Air Force resolution chart (AFRC) with two ground glass substrates obscuring the path using single beam imaging and (b) The holographic image of the obscured AFRC using the dual beam holographic system. (From Sun, P.-C. and Leith, E.N., *Appl. Opt.*, 33, 597–602, 1994. With permission from the Optical Society of America.)

and automatically determines the phase and amplitude information across the image plane. The phase information of the hologram can then be used to construct an interference map that represents the change in state, the surface contour, or the refractive index of the object being examined. For example, if *a* and *b* are the two states of the object (i.e., a reference condition and a perturbed condition), then using Equation 7.16, the corresponding phase values of the reconstructed holographic images are:

$$\varphi_a(x,y) = \tan^{-1}\left\{\frac{\mathrm{Im}\left[E_{i,a}(x,y)\right]}{\mathrm{Re}\left[E_{i,a}(x,y)\right]}\right\}$$

$$\varphi_b(x,y) = \tan^{-1}\left\{\frac{\mathrm{Im}\left[E_{i,b}(x,y)\right]}{\mathrm{Re}\left[E_{i,b}(x,y)\right]}\right\},$$

(7.45)

where $E_{i,a}$ and $E_{i,b}$ are the diffracted fields at the image plane corresponding to the two states of the object. These values can then be used to compute a map of the phase difference over the image plane between the object in states *a* and *b*:

$$\Delta\varphi_{a-b}(x,y) = \varphi_a(x,y) - \varphi_b(x,y); \qquad if \ \varphi_a \geq \varphi_b$$

$$= \varphi_a(x,y) - \varphi_b(x,y) + 2\pi; \quad if \ \varphi_a < \varphi_b.$$

(7.46)

Since the phase difference values cannot distinguish additive multiples of 2π phase, a phase unwrapping algorithm is usually applied to reconstruct continuous changes in the deformation or height of an object being examined [3].

PROBLEMS

1. A digital hologram is formed with the configuration shown below in Figure 7.19 and with light at a wavelength of 0.5 μm and a camera with 1.5 μm pixel size in the *x*-direction. What is the maximum reference beam angle θ_R that can be used that satisfies the sampling criteria (Nyquist rate)? (Note the reference beam angle is + with respect to the *z*-axis as shown in the figure.)

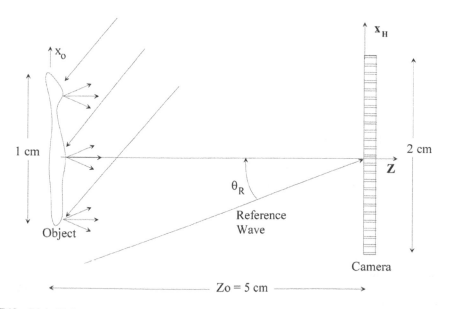

FIGURE 7.19 Digital hologram recording geometry for Problem 1.

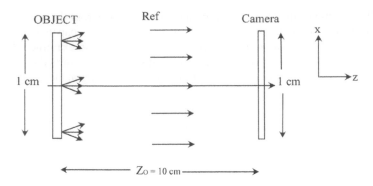

FIGURE 7.20 Digital hologram recording geometry for Problem 2.

2. Consider a digital hologram recording arrangement as shown in Figure 7.20. The wavelength used is 633 nm. (i) What is the largest camera pixel dimension that can be used to record light from the full extent of the object? (Only consider the geometry of the object and camera—not the spatial frequency content of the object.) (ii) If the field at the object is $o(\xi,\eta)$, what is the object field at the hologram plane? Leave it in integral form. Indicate the assumption that you are making.

3. A set of multiple wavelength digital holograms is used to produce a contour map of a microscopic image. The microscope uses a 0.55 NA diffraction limited objective at a wavelength of 550 nm. If a 550 nm laser is used and the wavelength step $\Delta\lambda$ is 0.20 nm, determine the number of holograms that must be recorded so that the axial depth Δz of the scanned object is 10× the lateral resolution of the microscope objective.

4. It is desired to make phase shifting digital holography system that introduces a phase shift of $\pi/2$ between holograms. The system uses 600 nm light in a single mode fiber interferometer as shown in Figure 7.11. The fiber has a refractive index of 1.50. What change in fiber length corresponds to a $\pi/2$ phase shift?

REFERENCES

1. J. W. Goodman and R. W. Lawrence, "Digital image formation from electronically detected holograms," *Appl. Phys. Lett.*, Vol. 11, 77–79 (1967).
2. W. Osten et al., "Recent advances in digital holography [Invited]," *Appl. Opt.*, Vol. 53, G44–G63 (2014).
3. U. Schnars and W. Jueptner, *Digital Holography*, Springer-Verlag, Berlin, Germany (2005).
4. T. Kreis, *Handbook of Holographic Interferometry*, Wiley-VCH, Weinheim, Germany (2005).
5. R. J. Collier, C. B. Burckhardt, and L. H. Lin, *Optical Holography*, p. 212, Academic Press, New York (1971).
6. I. Yamaguchi and T. Zhang, "Phase shifting digital holography," *Opt. Lett.*, Vol. 22, 1268–1270 (1997).
7. O. Inomato and I. Yamaguchi, "Measurements of Benard-Marangoni waves using phase-shifting digital holography," *Proc. SPIE*, Vol. 4416, 124–127 (2001).
8. I. Yamaguchi, J. Kato, S. Ohta, and J. Mizuno, "Image formation in phase shifting digital holography and applications to microscopy," *Appl. Opt.*, Vol. 40, 6177–6186 (2001).
9. T. Xi, J. Di, X. Guan, Y. Li, C. Ma, J. Zhang, and J. Zhao, "Phase-shifting infrared digital holographic microscopy based on an all fiber variable phase shifter," *Appl. Opt.*, Vol. 56, 2686–2690 (2017).
10. J. Mertz, *Introduction to Optical Microscopy*, Roberts and Company Publishers, Greenwood Village, CO (2010).
11. D. Gabor, "A new microscope principle," *Nature*, Vol. 161, 777–778 (1948).
12. E. Cuche, P. Marquet, and C. Depeursinge, "Simultaneous amplitude-contrast and quantitative phase-contrast microscopy by numerical reconstruction of Fresnel off-axis holograms," *Appl. Opt.*, Vol. 38, 6994–7001 (1999).

13. M. K. Kim, "Wavelength-scanning digital interference holography for optical section imaging," *Opt. Lett.*, Vol. 24, 1693–1695 (1999).
14. M. K. Kim, "Tomographic three-dimensional imaging of a biological specimen using wavelength-scanning digital interference holography," *Opt. Exp.*, Vol. 7, 305–310, (2000).
15. P.-C. Sun and E. N. Leith, "Broad-source image plane holography as a confocal imaging process," *Appl. Opt.*, Vol. 33, 597–602 (1994).
16. E. N. Leith, W.-C. Chen, K. D. Mills, B. D. Athey, and D. S. Dilworth, "Optical sectioning by holographic coherence imaging: A generalized analysis," *J. Opt. Soc. A*, Vol. 20, 380–387 (2003).
17. J. W. Goodman, *Statistical Optics*, pp. 207–222, John Wiley & Sons, New York (1985).
18. E. Leith, C. Chen, H. Chen, Y. Chen, D. Dilworth, J. Lopez, J. Rudd, J. Valdmanis, and G. Vossler, "Imaging through scattering media with holography," *J. Opt. Soc. A*, Vol. 9, 1148–1153 (1992).

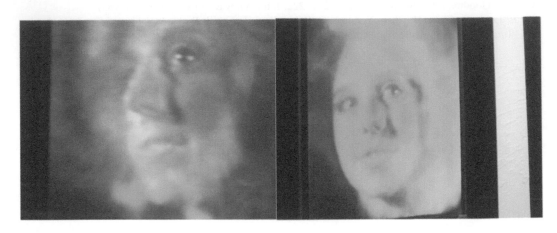

FIGURE 1.1   The figures above are photographs of a hologram taken from two different angles.

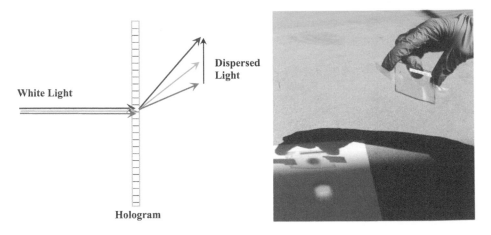

FIGURE 3.22   A schematic and experimental illustration of dispersion from a holographic grating.

FIGURE 9.4 Full color hologram of the 1911 Bay Tree Fabergé Easter Egg using an OptoClone recording system. (From Bjelkhagen, H.I. et al., *Proc. SPIE*, 9771, 2016. With permission from the SPIE.)

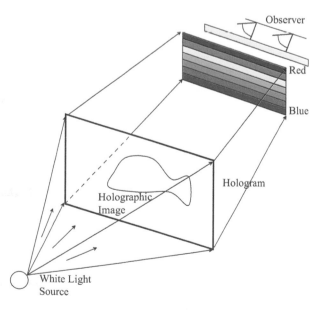

FIGURE 9.6 Reconstruction of a rainbow hologram with a white light source showing the dispersive axis in the vertical direction and observation along a horizontal axis.

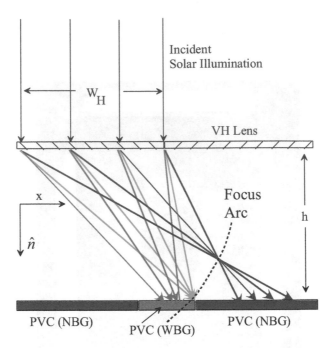

FIGURE 11.1 A volume holographic lens used in a spectrum splitting solar energy conversion system. A wide bandgap (WBG) photovoltaic cell (PVC) is surrounded by a narrow bandgap (NBG) PVC. The "green" rays indicate the light paths for illumination with the "transition" wavelength. (From Vorndran, S. et al., *SPIE Proc.*, 9937, 99370K1:8, 2016. With permission from the SPIE.)

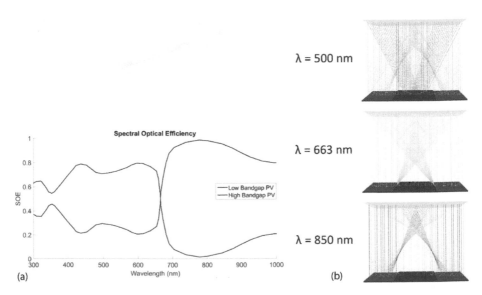

FIGURE 11.3 The spectral filtering properties (a) and raytrace (b) through the spectrum splitting volume holographic lens described in the text for a combination of silicon and indium gallium phosphide photovoltaic cells. (From Vorndran, S. et al., *Appl. Opt.*, 55, 7522–7529, 2016. With permission of the Optical Society of America.)

FIGURE 11.25 A HPC module produced by Prism Solar Technologies for rooftop and building integrated applications. (From Prism Solar Technologies. With permission.)

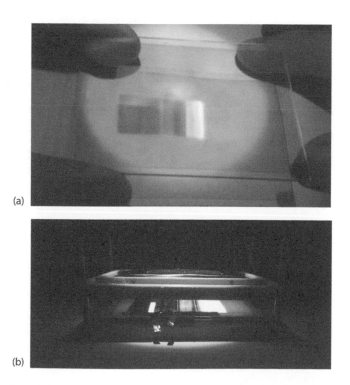

FIGURE 11.30 A volume holographic cylindrical lens fabricated in a photopolymer material showing the dispersed spectrum (a) for use in a spectrum splitting system with GaAs and silicon PV cells (b). (From Chrysler, B. et al., *SPIE Proc.*, 10368, 10368G, 2017. With permission from the SPIE.)

8

Holographic Recording Materials

8.1 Introduction

The success of a hologram design is highly dependent on the properties and capabilities of holographic recording materials. Furthermore, each holographic application generally has different requirements and places further demands on the material characteristics. In the past, the lack of an "ideal" holographic material has been an issue, however, great strides have been made using a wide variety of materials to obtain suitable results. These include silver halide films, dichromated gelatin (DCG), photopolymers, photoresists, photorefractives, and thermoplastics. A great wealth of information on these materials exists in the literature [1–5] and could be the subject of an entire book. This chapter will provide an overview of the main developments, properties, and limitations of materials that have been widely used in the past as well as some of the newer and more promising materials. In addition, several techniques for optimizing the processing of silver halide emulsions and preparing and processing DCG emulsions and phenanthrene quinone-doped poly methyl methacrylate (pq-PMMA) photopolymer is given in Appendix D to provide the reader with a starting point for experimenting with holographic materials.

Before discussing the characteristics of specific types of recording material, it is important to review the criteria for evaluating their performance. These include:

- Spectral Sensitivity
- Dynamic and Linear Response Range
- Resolution or Spatial Frequency Response
- Noise Properties and Optical Quality
- Material Change after Processing
- Multiplexing Capability
- Processing Requirements
- Stability
- Read/Write Properties

    Spectral Sensitivity: provides a measure of the energy required at a given wavelength to achieve a certain material response. The material response typically produces a change in the refractive index modulation, Δn, for a phase grating or in the absorption, $\Delta \alpha$, of the material for an amplitude grating. A photosensitizing agent is usually added to the material to increase the response at a particular wavelength. Δn and $\Delta \alpha$ can be indirectly determined by measuring the grating diffraction efficiency as discussed in Chapter 5.

    Dynamic and Linear Response Range: indicates the maximum material modulation (refractive index or absorption) that can be obtained with increasing exposure energy density (J/cm^2). Most often a linear response of the material modulation is desired since this produces holograms with high fidelity. When the maximum exposure density for linear recording is exceeded, the material response saturates leading to non-linear modulation profiles and additional diffraction orders in the reconstruction. In addition, many materials require an exposure with an irradiance (W/cm^2) value that exceeds a minimum threshold to activate the material response.

Resolution or Spatial Frequency Response: indicates the range in grating period lengths that can be formed in the material. The ability to record grating periods with different dimensions is determined by the chemical or physical mechanism that takes place in the material during the exposure and development process. This factor should be reviewed when designing a hologram especially when the grating period is very small or large to be sure that it falls within the response range of the material. If the resolution is not known, it can be determined by recording a set of two-beam plane wave holograms with different inter-beam angles to form different grating periods according to the grating equation (Equation 3.9).

Noise Properties and Optical Quality: Holographic materials include a number of different components (sensitizers, stabilizers, etc.) embedded in a host material. Differences in the optical properties between the components and host material often result in various types of volumetric and surface scattering. In addition, the method of depositing the material such as spin coating, molding, or laminating onto a substrate will affect the optical quality of the resultant hologram. This in turn affects the ability to use the resultant holograms in different optical systems.

Material Change after Processing: Changes in the holographic material's lateral and axial spatial dimensions, average absorption, and refractive index often occur with post-exposure processing. Differences between pre and post-exposure characteristics can result in aberrations and a decrease in the diffraction efficiency of the reconstructed holographic images. Most changes in the material properties can again be quantified with a set of plane wave test holograms and incorporated into the hologram design procedure.

Multiplexing Capability: Multiplexing several holograms in the same recording material is often desired for applications such as holographic data storage and multiple imaging. Multiplexed gratings can be formed either by sequentially recording several different holograms or by simultaneously recording multiple object beams in one exposure. The mechanism for grating formation such as the migration of molecules or silver halide grains must be suitable for setting up overlapping spatial frequencies.

Processing Requirements: A variety of processing techniques exist for holographic materials. Wet processing is used for a variety of emulsion-based materials and can achieve many different effects in the resulting holograms. A number of the newer photopolymers are processed with a simple white light fixing and are much more convenient for different applications.

Stability: It is generally desired that the hologram persists for long periods of time with repeated use. Many applications also require that the hologram be used in different environments. Therefore, it is important to understand the stability characteristics of the holographic material being used for a particular design.

Write/Read/Erase Properties: A number of applications such as dynamic displays and holographic data storage require that a hologram can be recorded, used for a period of time, erased, and then allow a different hologram to be recorded. Certain inorganic crystals are suitable for this purpose and have been used for some time with some success. More recently a number of photopolymers have been developed for holographic write/read/erase applications.

In the remainder of this chapter, a review of some of the more successful and commonly used materials will be given. This includes silver halide and dichromated gelatin emulsions, photorefractive crystals, photopolymers, photoresist, thermoplastics, embossed holograms, and photosensitized glass.

8.2 Emulsion-Based Materials

Silver halide and dichromated gelatin films fall into this category of recording materials. The emulsion or film consists of a gelatin that serves as a matrix for other constituents that produce the absorption or phase modulation mechanism. Typically, the gelatin is an organic material derived from animal collagen. The requirements for high performance gelatin-based emulsions were well developed for the photographic film industry [6,7]. In fact, most holography from the 1960s–1980s was performed using either photographic film emulsions directly or as a base layer for dichromated gelatin emulsions.

8.2.1 Silver Halide Emulsions

8.2.1.1 Introduction

Silver halide emulsions were initially the most widely used material for holography [8]. This was mainly due to the large variety of commercially available high resolution photographic films that existed when research on holography started in earnest during the early 1960s. Kodak, Agfa-Gevaert, and Ilford were some of the main suppliers of high quality photographic emulsions. The long history of photography with silver halide materials provided a great wealth of information about the mechanism of image formation and various chemistries to obtain different results. This background was adapted to the special requirements for hologram formation and was the subject of much investigation during the 1970s and 1980s [8–17]. However, with the advent of digital cameras, the commercial availability of photographic film suitable for holography has decreased considerably. The fabrication of high quality silver halide emulsions is an intricate process requiring many carefully controlled preparation steps for achieving good results. A detailed description of emulsion preparation can be found in the excellent book by Bjelkhagen [8]. Today only a few companies such as Geola in Slovakia still manufacture silver halide film (see for example Intergraf [18]) that is suited to holography. Nonetheless, display holograms formed in silver halide emulsions are very appealing, and it is a worthwhile effort to work in this medium. In addition, many of the techniques used for hologram formation in silver halide are relevant to other material systems.

8.2.1.2 General Silver Halide Emulsion Properties

Holographic silver halide emulsions consist of a mixture of compounds such as AgBr, AgI, AgCl, gelatin, sensitizing dyes, and stabilizers. The solution is coated on either glass plates or flexible acetate films. The silver halide grain size typically varies from 5 to 50 nm. The smaller grain size allows for smaller period holographic gratings at the expense of less sensitivity to the exposure energy. The converse is true for larger silver halide grains. Silver halide has a refractive index on the order of 2.2–2.5 while the gelatin host material has a refractive index of ~1.52–1.54. Additional anti-halation coatings are frequently applied to the substrate to reduce back-reflections from the substrate and the formation of spurious gratings.

Image formation is obtained by realizing a "latent image" in the silver halide grains within the emulsion that are later converted to a permanent image after development. A latent image can be defined as *any change in the silver halide grain that causes the probability of development of that grain to exceed 50%* [6]. A more comprehensive discussion of this process is described by Gurney-Mott [19,20], Mitchell [21], and Tani [22].

A simple model for "latent image" formation can be understood in the following manner. The silver halide grain is an n-type photoconductor with a valence and conduction band much like in a semiconductor. When a photon is absorbed an electron-hole pair is formed, and the electron can migrate throughout the grain until trapped at a defect. When the electron is trapped, the silver halide grain is reduced to a silver atom according to:

$$Br^- \xrightarrow{h\nu} Br + e^-$$

$$e^- + Ag^+ \rightarrow Ag^0.$$

(8.1)

Reducing as few as 3–4 atoms are sufficient to form a latent image site or "speck" within the silver halide grain. Each speck acts as a catalyst or development site for chemical reduction of the entire silver halide grain. After processing, the entire grain (~10^6 atoms in a grain diameter of ~50 nm) is developed (reduced to Ag^o) from the latent image site providing a large amplification of the exposure process.

When holograms are multiplexed by sequential recording in a silver halide emulsion, the latent image phenomenon will affect the resultant diffraction efficiency of each hologram according to the order of formation in the multiplexing sequence. In general, if a series of holograms are formed with equal exposures the diffraction efficiency of each successive hologram will decrease. This effect is called the "*holographic reciprocity law failure*" and results from the time delay between recordings [16]. The lifetime of an isolated silver atom on a grain can be calculated using a Boltzmann statistical lifetime estimate and the binding energy for the electron to the pre-speck (E_o = ~0.70 eV):

$$t = \tau e^{E_o/kT}, \tag{8.2}$$

where $\tau = 10^{-12} s$ is the electron collision period, k is the Boltzmann constant, and T the temperature in Kelvin. At room temperature (300 K), the lifetime is found to be about 2 seconds. Therefore, if the time between exposures (Δt) is greater than 2 seconds, then dissociation of silver atoms occurs. This situation favors the first exposure since it is reinforced by subsequent exposures. It also increases the noise of the first exposure since the photons from later exposures that reinforce the first hologram correspond to different holograms. Johnson et al. [16] found that the diffraction efficiency of multiplexed holograms in silver halide film can be equalized by first recording a set of holograms with equal exposure energy for each hologram and measuring their diffraction efficiencies, η_m. The diffraction efficiency for each hologram is then normalized to the efficiency of the first hologram η_1 which will have the highest value in the set. The exposure energy density for each hologram is then adjusted according to: $E_m = \frac{E_B}{\sqrt{\eta_{m,1}}}$ where E_B is the total bias exposure energy density for the set of holograms and $\eta_{m,1}$ is the normalized diffraction efficiency of the mth hologram. This compensation technique allows equalization of the diffraction efficiency to within ~10% for four sequentially exposed holograms [16].

8.2.1.3 Developers and Bleaches for Silver Halide Emulsions

Developers for silver halide film can be categorized as either chemical (direct) reduction or physical development. In chemical development silver halide ions (Ag^+) are reduced to silver atoms directly at the grain without halide transfer to or from that grain. In contrast for physical development, the developer enables silver to deposit at latent image sites within silver halide grains. The nature of the resulting reduced silver is different for chemical and physical development techniques [11]. As illustrated in Figure 8.1, the reduced silver for chemical development has a larger filamentary characteristic and is more compact for physical developers.

If an amplitude or absorption type hologram is being formed, the process generally consists of development followed by a fixing step to remove unexposed silver halide grains. Kodak D-19 is a good general developer for absorption hologram formation and can also be used to realize phase holograms. The composition of this developer and a holographic fixer is given in Appendix D. However, as was shown in Chapter 5 on coupled wave analysis an amplitude or absorption hologram has low diffraction efficiency (3.7% for transmission holograms and 7.2% for reflection holograms) and are therefore only

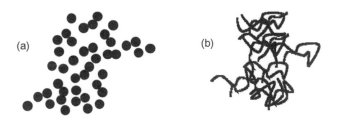

FIGURE 8.1 Silver structure resulting from (a) physical development and (b) chemical development processes.

used for special applications. The diffraction efficiency can be improved in silver halide emulsions using a bleach to rehalogenate the developed silver. Although the diffraction efficiency is generally much higher with bleaching, special attention must be paid to reduce scatter in the resulting holograms. The source of scatter in bleached silver halide holograms has been extensively studied [23,24] and is due primarily to size variations of the rehalogenated silver halide grains.

There are several techniques for bleaching silver halide materials to produce phase holograms (it should be noted that most "phase holograms" formed in silver halide have residual average absorption). These include: (1) develop-fix-bleach; (2) develop-reversal bleaching; and (3) develop followed by rehalogenating bleaching without fixing. Another technique called the silver halide sensitized gelatin process has also been developed which is a cross between more conventional silver halide and dichromated gelatin processes [25–27]. This technique will be described separately. A diagram illustrating the basic forms of silver halide processing with bleaching is shown in Figure 8.2 [1].

The basic chemistry of bleach solutions consists of: (i) an oxidizing agent; (ii) an alkali halide such as potassium bromide (KBr); and (iii) a buffer such as H_2SO_4 [1]. This prescription can generally be used for both direct (develop-fix-bleach) and develop-bleach-rehalogenate techniques. In the direct bleaching process, the emulsion is developed, undergoes a fixation step to remove unexposed silver halide grains, and then bleached as illustrated in Figure 8.2. As noted, there will be some emulsion thickness shrinkage during the fixing step that can change the highest efficiency reconstruction wavelength and angle for slanted gratings and introduce aberrations into the reconstructed image. Kodak D-19 in combination with the bleach proposed by Lehmann [24] provides high efficiency and acceptable scatter levels for transmission holograms where shrinkage is not as critical as for reflection holograms. It was found by Pennington and Harper [23] that most of the scatter in the direct bleaching process is due to reticulation of the emulsion surface and to larger grain growth and size variation during rehalogenation.

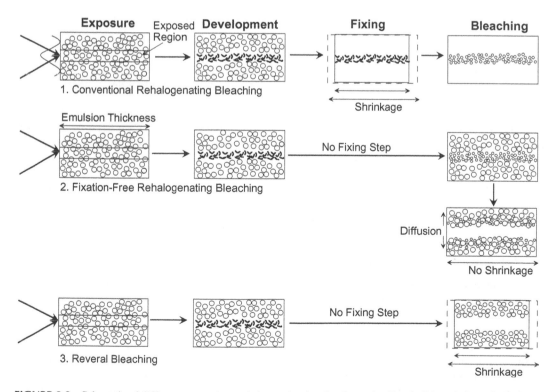

FIGURE 8.2 Schematic of different processing techniques showing the change in silver halide and silver distributions and the emulsion shrinkage. (From Bjelkhagen, H.I. (Ed.), *Holographic Recording Materials*, SPIE Milestone Series, Vol. MS 130, Engineering Press, Bellingham, WA (1996).)

In reversal bleach processes the exposed silver halide grains are removed during processing [28]. The unexposed grains experience a slight increase in size due to an effect similar to Ostwald ripening, however, grain growth is much smaller than that which occurs with direct bleaching methods. The resulting holograms have very low scatter and also undergo shrinkage. This technique is popular among those doing display holography due to the resulting low noise. They also utilize the shrinkage that occurs due to the removal of exposed silver halide grains to obtain different desired reconstruction wavelengths in the resulting holograms. The "pyrochrome process" that uses a pyrogallol developer and a potassium dichromate bleach works well as a reversal bleach system [29,30].

As shown in Figure 8.2, the fixation free bleach process is a two-step develop-bleach technique. In this process it is important to promote the diffusion of the developed silver halide grains away from the exposed fringe areas, as this process increases the index modulation and the resultant diffraction efficiency. Verification of the diffusion of silver halide grains away from the exposed fringe areas was demonstrated in [31] using phase contrast images of fringes with different spatial frequencies. Grain migration increased as a function of spatial frequency in a manner consistent with diffusion models with large gradients in the grain distributions [32]. Another advantage of the develop-bleach process is that the emulsion does not shrink appreciably and scatter can also be minimized with optimization of the bleach constituents and proportions [33].

Some developers and bleaching solutions produce a localized tanning effect or hardening of the gelatin around the silver halide grains. This contributes to the index modulation and leads to higher diffraction efficiency with some combinations of emulsions and process chemistry. A significant advance in hologram bleaching chemistry was made by Phillips et al. [14] by proposing the use of organic oxidizing agents such as para-benzo-quinone (PBQ) in the bleach. When PBQ bleach is used in combination pyrogallol or catechol tanning developers, the resulting holograms have high diffraction efficiency and low scatter. The pyrogallol developer produces oxidation products and a layer around the silver halide grain with a quinone chemical structure. Since the PBQ also has a quinone structure, it does not attack the layer around the grain during the bleaching process as do bleaches with other oxidizers such as free bromine. A typical developer-bleach combination that takes advantage of this effect is the CW-C2 catechol developer and the PBQ-2 bleach and is a fixation free process (Appendix D). Although PBQ is quite effective, it is also highly toxic and can damage the cornea of the eye with prolonged exposure. An alternative less toxic bleach based on ferric ethylene diamine tetra acetic acid (Ferric-EDTA) was developed that has similar oxidized properties [34].

When making reflection holograms special consideration should be made to ensure that processing extends to some depth within the emulsion. In addition, shrinkage is more of an issue for reflection holograms especially if it is desired to reconstruct at the original wavelength. Controlling shrinkage is particularly important for unslanted reflection gratings since it produces a significant loss in diffraction efficiency when reconstructed at the formation wavelength and angle. The fixation-free bleaches used with the CW-C2 developer listed in Appendix D work well with Agfa and Ilford emulsions to realize high efficiency reflection holograms. Good development and bleaching processes for fabricating reflection holograms also tend to work well for transmission holograms.

8.2.1.4 Silver Sensitized Holograms

Another technique for processing holograms based in silver halide emulsions is referred to as the silver halide sensitized gelatin (SHSG) process [35,36]. This method combines the sensitivity of silver halide emulsions with the high efficiency of dichromated gelatin holograms. In this method, the emulsion is exposed, developed, bleached with a tanning developer, and then fixed to remove residual silver halide grains. The bleaching agent produces Cr^- ions that promote localized tanning of the gelatin in the exposed areas. A typical chemistry for this technique is to use the phenidone, ascorbic acid, phosphate developer (PAAP) developer and modified R10 bleach using the sequence given in Appendix D. Fimia et al. [37] presented a technique for optimizing this process and was able to achieve 50% diffraction efficiency with a 50–70 nm wavelength shift from the exposing wavelength.

8.2.1.5 General Comments on Handling and Processing Silver Halide Emulsions

The properties of several commercially available silver halide materials are provided in Appendix D along with many useful processing chemistries. Commercial silver halide emulsions are available on either glass or acetate substrates. Frequently, the full glass plate area is not required for one exposure, and the plate or plastic substrate can be cut to conserve material. Glass substrates are cut by scribing the glass side of the plate with a diamond tipped tool, and then bending the glass until it cracks along the scribe mark. Note that this must be done either with a dim safe light or with no light for panchromatic film, so the process should be practiced on glass pieces prior to using the emulsion material. When making a transmission hologram, the emulsion side of the film should be placed facing the incident beams. The emulsion side of a glass plate can be identified by moistening a finger and touching the corner of the plate. The side with the emulsion will feel slightly sticky. In order to reduce back reflections from the substrate and the formation of unwanted interference patterns during the exposure, the back surface of the substrate should be index matched to an absorber. A piece of glass with one side painted with flat back paint works well as an absorber.

A typical silver halide processing sequence consists of: exposure; develop for the prescribed period; rinse in running water; fix (if using); rinse in running water; bleach (if using); rinse in running water; rinse briefly in a photoflo solution; and allow to dry. Shallow trays are preferred for the processing solutions as they allow minimum use of chemicals. The various processing chemicals should be prepared with distilled water to prevent contamination with minerals in tap water. The running water wash between processing steps, however, can be tap water. The temperature of the solutions including the running water wash should be kept at a consistent temperature usually 20°C–25°C. The plate should be processed with the emulsion side up when using trays so that the emulsion is uniformly covered with solution. The PFG emulsions are relatively soft compared to those produced by Agfa. It is useful to harden the PFG gelatin after exposure and prior to developing with a formaldehyde solution followed by processing with the JD-2 developer and bleach (see Appendix D).

Antihalation coating on the glass substrate is used on some commercial photographic plates to minimize light reflected from the back surface. This coating should be removed using a solvent such as alcohol prior to forming reflection holograms since they require light to enter the emulsion from both the front and back sides of the plate.

A major problem with bleached silver halide holograms is that they suffer from printout or darkening with exposure to ambient illumination [8]. In this process, any remaining silver halide grains continue to be reduced to silver producing a darkening of the film due to the resulting absorption. The printout effect is more pronounced when not using a fixation step in the film processing since much of the unexposed silver halide grains remain in the emulsion. Some success at minimizing this effect has been achieved with further chemical stabilization of the silver halide grains [38–40].

As a result of scattering, residual absorption, and the printout effect, silver halide films tend to have greater use in less critical holographic applications such as display artwork. However, their use in applications that require long term exposure to light is restricted as their performance can change over time. Nonetheless, some spectacular silver halide display holograms have been made during the past half century and are shown in a number of galleries around the world [41,42].

8.2.2 Dichromated Gelatin

8.2.2.1 Introduction

Dichromated gelatin or DCG is a versatile recording material that can produce holograms with very high refractive index modulation (0.08–0.10), diffraction efficiency (>95%), spatial resolution (>5000 lines/mm), and very low scatter. The recording of holograms in dichromated gelatin was first reported by T. A. Shankoff in 1968 [43] and has been used in a variety of applications that require holograms with high diffraction efficiency and optical quality. The emulsion is formed from an aqueous solution of dichromate and gelatin. The gelatin solution can be spun, molded, or dip-coated onto rigid or flexible planar as well as curved substrates providing a wide range of formats for different applications [44–46]. The hologram characteristics can be controlled by adjusting the material properties of the gelatin, the exposure conditions, and processing methods providing a great deal of design flexibility. Several groups have also been successful in developing low cost dichromated gelatin coating methods on large substrates making it suitable for window filters and solar collector applications [47].

8.2.2.2 Description of Gelatin Materials and Sensitizer

An organic gelatin derived from animal products forms the host material for DCG emulsions. The gelatin can be obtained from several manufacturers [48] and is typically specified as Type A which is derived from pig skin or Type B derived from cattle hide and bone. Both types of gelatin have been used successfully for fabricating DCG holograms provided that they satisfy properties. The most important parameters to specify for DCG gelatin are: bloom strength, viscosity, ash content, and pH. Some of these parameters are related. For instance, viscosity and gel (or bloom) strength are often used to specify the mechanical strength of the gelatin. Viscosity indicates the resistance of gelatin in liquid form when a tangential force is applied. Viscosity will affect spin coating parameters and the optical quality of the resulting layer. Acceptable viscosity values range from 40–50 mps (millipoise). Gel or bloom strength is a measure of the rigidity of the gelatin after solidification. It is measured by compressing a gelatin layer that has been deposited using a standardized method. Typical values for gel strength (measured in grams) for hologram formation range from 225 to 300 gm. Rallison suggests that a gelatin with a ratio of gel strength to viscosity ranging from 4 to 5 generally provides good results [49]. The ash or mineral content of the gelatin indicates the purity of the gelatin. For holographic film, the ash content is typically less than 1%. Both the gelatin type and the way in which it was prepared (using an acid or alkaline process) affect the pH of the gelatin. The gelatin has an isoelectric point (IEP)—pH where the molecules within the gelatin are electrically neutral. It has been shown that the viscosity reaches a minimum at the IEP [50]. For good emulsion properties of dichromate sensitized holograms, the gelatin pH should range from approximately 5.0–5.5 [51]. Typical photo-sensitizers for DCG emulsions are ammonium dichromate $(NH_4)_2Cr_2O_7$ or potassium dichromate $(K_2Cr_2O_7)$. The dichromate concentration determines the degree of absorption and the nominal reconstruction wavelength of the hologram. Typical dichromate compound to gelatin ratios (by weight) range from 1:3 to 1:10 with the 1:3 ratio commonly used for making holographic optical elements and the 1:10 ratio used to shift the reconstruction to longer wavelengths. The wavelength shift occurs due to removal of the dichromate during processing.

8.2.2.3 Mechanism for Hologram Formation in DCG

Hologram formation in DCG is the result of modulating the refractive index within the gelatin emulsion. Considerable work has been done to understand the mechanism for this process [44–46,52]. Several explanations have been proposed for grating formation with the two most popular being: (1) the formation of chromium complexes that bond with the gelatin molecules and (2) micro cracking of the gelatin after rapid dehydration that produces air voids and large localized differences in refractive index. Several studies [46] indicate that the first mechanism is primarily responsible for grating formation. Systematic experiments have shown that during exposure, the sensitizing chromate ion Cr(VI) is photochemically reduced to Cr(III) which locally hardens the gelatin by forming cross-linking bonds with gelatin molecules. The Cr(III) becomes trapped at the hardening sites forming complexes with the gelatin molecules. The exposed gelatin is then washed to remove residual Cr(VI) (which remains water soluble) and processed by rapid dehydration using solutions of isopropanol with increasing purity. The formation of regions of local hardening with Cr(III)-gelatin complexes decreases solubility and swelling. The Cr(III) molecules remain in the exposed regions of the interference fringes. Since they have a higher refractive index than the gelatin, a large localized refractive index variation occurs. The modulation in refractive index varies linearly over a wide range in exposure energy density and dichromate concentration.

8.2.2.4 Preparation of DCG Emulsions

There are a variety of methods for making DCG emulsions. These include: spin coating [51], "doctor blading" [53], molding [54], and roller coating [55]. Each method requires a degree of experimentation and standardization in order to obtain consistent results. Since the resulting emulsions are hydroscopic, the relative humidity must be monitored and the exposure parameters adjusted to the different conditions. In order to obtain consistent results, the materials should be processed, stored, and exposed in a controlled humidity and temperature environment. The company Intergraf [18] offers a DCG emulsion on glass plates (PFG-04) and is reported to give diffraction efficiencies in the range of ~75% [18]. Kodak 649F photographic plates have also been re-processed to form DCG emulsions [56], however, these plates are no longer manufactured in quantity and are difficult to obtain.

A typical procedure for spin coating DCG emulsions consists of: substrate preparation, DCG solution preparation, and emulsion deposition. A detailed procedure is given in [51]. The resulting emulsions using this procedure have a nominal thickness of 15 μm and an average refractive index [57] of 1.50. The coated plates have a shelf life of approximately 2 weeks when stored in a light tight environment with 15% relative humidity (RH) at a temperature of 25°C.

DCG emulsions can also be formed using a molding technique. Although it still requires careful environment control, it has the advantage of not requiring a spinning apparatus. A simple mold can be formed using polyvinyl tape spacers on a glass substrate using the procedure given in Appendix D. These holograms have been found to give excellent diffraction efficiency (>90%) and spectral bandwidth control.

8.2.2.5 DCG Exposure and Development Parameters

DCG emulsions sensitized with ammonium dichromate or potassium dichromate respond to wavelengths from ~350 to ~550 nm. The emulsions are most sensitive to shorter wavelengths. For example, recording made at 441 nm using 1 mJ/cm^2 will require 50 mJ/cm^2 at 514.5 nm and 100 mJ/cm^2 at 532 nm. It is also possible to sensitize the emulsion to red wavelengths using methylene blue [58,59] and has been used to make full color holograms [60,61]. DCG emulsion sensitivity also increases in warmer more humid storage and processing environments. The response sensitivity can increase by more than 15× with an increase in ambient temperature of 7°C and a RH increase of 25% [49]. This makes it necessary to characterize the DCG recording process for different laboratory conditions.

Processing of DCG emulsions consists of: pre-hardening with a fixer to strengthen the gelatin matrix; washing residual fixer and dichromate from the emulsion; dehydration of the emulsion to obtain the desired diffraction properties; and baking to remove residual water content. The amount of pre-hardening with fixer controls the amount of gelatin swelling prior to dehydration. This is an especially important step for reflection holograms for controlling the reconstruction wavelength and spectral bandwidth [60]. The temperature, pH (of the water baths), and duration in the different processing baths also can have a large impact on the results and should be noted and repeated to obtain consistent results.

Since the gelatin emulsion is hydroscopic, it must be sealed after processing to preserve the hologram. This should be performed immediately after processing and characterization. Sealing can be performed using optical cement applied directly onto the emulsion, making sure that the cement covers the edge areas, and then applying either a cover glass or non-porous plastic top layer. The hologram will last 20 years or more when properly sealed [62], making it suitable for use in exposed environments such as window filter and optics for photovoltaic systems.

8.3 Photorefractive Materials

The photorefractive process is caused by a light-induced electro-optic effect that changes the refractive index of the material [63,64]. A great attribute of photorefractive materials is that a hologram can be written, erased, and then re-written allowing the possibility for holographic data storage and dynamic displays. The first demonstration of the photorefractive effect was performed using inorganic electro-optic crystals in 1966 by Ashkin [65]. Since then, a variety of electro-optic crystals such as: ferroelectric lithium niobate (LiNbO$_3$) and lithium tantalate (LiTaO$_3$) [65,66]; the ferroelectric perovskites potassium niobate (KNbO$_3$) [67] and barium titanate (BaTiO$_3$) [68,69]; tungsten bronze crystals such as Sr$_{1-x}$Ba$_x$Nb$_2$O$_6$ (SBN) [70]; non-ferroelectric sellenites like Bi$_{12}$GeO$_{20}$ (BGO) [71]; and semiconductors GaAs [72,73] and InP [74] have been shown to be suitable for photorefractive holographic recording. In addition, a variety of photorefractive photopolymers have been developed since the early 1990s and show promise of providing larger format, updatable holographic displays at moderate cost. The remainder of this section will focus on inorganic photorefractive crystals, while photorefractive polymers will be discussed in greater depth later in the section on holographic photopolymers.

Photorefractive materials are electro-optic which allows for localized light-induced changes in the refractive index and the formation of holographic phase gratings. This process can be understood by considering an iron-doped lithium niobate crystal. Lithium niobate is electro-optic as well as photovoltaic meaning that a light-induced voltage can be formed within the material. The process of forming a holographic grating in a photorefractive material is illustrated in Figure 8.3. When the material is illuminated with two coherent beams an interference pattern is formed. In the regions of high irradiance, photons are absorbed and produce an excess of mobile charge carriers, while in the null regions relatively few carriers are produced. The areas of high and low light-induced carriers produce a charge carrier (electron) density gradient and forces electrons in high density regions to diffuse to lower density areas. The iron dopants, distributed within the crystal, trap the diffusing electrons and form a fixed charge distribution. This results in a space charge electric field (E_{SC}) according to:

$$E_{SC} = \frac{1}{\varepsilon} \int \rho(x)dx, \tag{8.3}$$

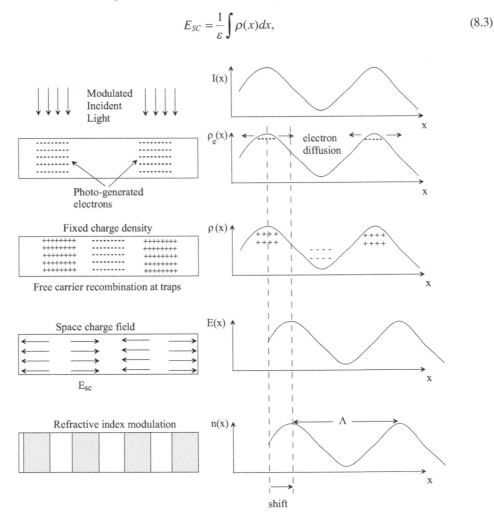

FIGURE 8.3 Steps for forming a hologram in a photorefractive material. I(x) is the spatially varying irradiance in the x direction; $\rho_e(x)$ is the electron carrier density as a function of position (x) in response to the irradiance variation; $\rho(x)$ is the fixed charge density as a function of x resulting from electron migration; E(x) is the electric field resulting from the spatially varying fixed charge density; n(x) is the refractive index modulation resulting from the spatially varying electric field and the electro optic effect; and Λ is the period of the refractive index modulation. (From Gunter, P. and Huignard, J.P., *Photorefractive Materials and Their Applications I,* Topics in Applied Physics, V. 61, Springer-Verlag, Berlin, Germany, 1988. With permission from Springer-Verlag.)

where $\rho(x)$ is the fixed charge distribution, and ε is the average permittivity of the material. The localized space charge field induces a corresponding localized variation in the refractive index through the electro-optic effect given by:

$$\Delta n = -\frac{1}{2}n^3 r E_{SC}, \tag{8.4}$$

where n is the average refractive index, and r is the electro-optic coefficient of the material. For $LiNbO_3$, $n = 2.33$ and $r = 11 \times 10^{-12}$ m/V at 633 nm. A comprehensive list of electro-optic coefficients and other parameters for photorefractive materials appears in Chapter 2 in the book by Gunter and Huignard [63].

For recording high efficiency holograms in photorefractive crystals, the crystallographic or c-axis of the crystal should be positioned parallel to the grating vector ($\vec{K}$) of the hologram. The *c-axis* indicates the preferred direction of movement of photo generated electrons and when aligned with the grating vector will maximize the electron diffusion, space charge field, and corresponding refractive index modulation. As a result, the *c-axis* of the crystal should be specified and properly arranged for the particular type of hologram that will be recorded (i.e., transmission, reflection, etc. as shown in Figure 8.4).

The hologram (refractive index distribution) can be erased by illuminating the material with a uniform, incoherent optical beam. If the hologram is illuminated with the same wavelength as used in recording, the photon energy is sufficient to transit the bandgap and will uniformly produce additional electrons across the grating region. This process erases the locally varying space charge electric field and the index modulation caused by the electro-optic effect. However, this also means that the hologram will be erased when it is illuminated with the same wavelength as used in the recording during the readout process.

In order to reduce hologram erasure during readout, the material can be fixed by processing the exposed photorefractive material to make the space charge region semi-permanent. A common way to fix $LiNbO_3$ is by heating the crystal at elevated temperatures (T>70°C–80°C) for 30 minutes to 1 hour [66]. At high temperatures, the ionic conductivity of $LiNbO_3$ is greater than at lower temperatures. When heated, the ions duplicate the space charge field pattern formed during holographic exposure. When the crystal is cooled back to room temperature the conductivity of the ions is much lower. The residual ionic grating is partially stabilized by the opposing electronic space charge field which increases the lifetime of the grating [75,76]. Lifetimes of the thermally processed $LiNbO_3$ can last from several months to several years depending on the material properties and grating strength.

Another approach to stabilizing holograms formed in photorefractive crystals is to use a combination of short and long wavelengths during the recording process, but only the long wavelength during readout. The crystal is grown with high energy gate traps, low energy writing traps, and deep traps where the energy difference with the conduction band is greater than the low energy traps as shown in Figure 8.5. During the hologram exposure step the material is exposed with short wavelength biasing (i.e., gating photons) which are absorbed by the high energy traps and help electrons transition into the conduction band. The excited electrons then decay to the energy trap nearest to the conduction band. At this point, the material is sensitized to holographic exposure at longer wavelengths which are absorbed by the low

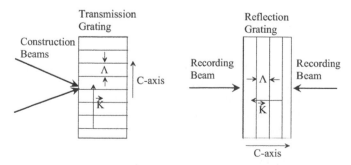

FIGURE 8.4 Orientation of the crystallographic (c-) axis of the photorefractive material with respect to the hologram formation beams. $\vec{K}$ is the grating vector; Λ is the grating period; and the c-axis is the crystallographic axis that coincides with the optical axis of the photorefractive material.

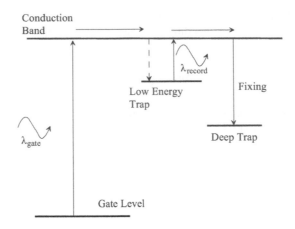

FIGURE 8.5 Energy diagram showing the two photon gating process for recording holograms in a photorefractive crystal. The material is doped with a high energy trap (gate), a low energy trap for recording, and a large energy trap for fixing the hologram. λ_{gate} is the high energy wavelength used to excite electrons from the gate energy level to the conduction band of the photorefractive material; λ_{record} is the lower energy photon used to record the hologram.

energy traps to the conduction band. The excited electrons then decay to the deep traps and fix the space charge distribution. The holograms can be erased by illumination with the gating wavelength. $LiNbO_3$ holograms can be gated with light having wavelengths near 500 nm and written with 850 nm light.

Hologram formation and decay in photorefractive materials can be described with exponential functions that in general have different time constants (τ_{form}-formation; τ_{decay}-decay) that depend on the intensity of the writing and reading beams. As discussed earlier, the formation of an index modulation Δn is achieved by absorbing photons that free electrons for transport. When the formation and decay time constants are equal, the maximum grating strength parameter occurs when $\alpha d = 2$, where α is the intensity absorption coefficient, and d is the crystal thickness. The corresponding maximum grating strength parameter (v) for an unslanted transmission hologram is:

$$v = \frac{\pi \Delta n d}{\lambda \cos \theta_o} e^{-1} = M\#,$$

(8.5)

where θ_o is the half angle between the recording beams in the material, and λ is the free-space wavelength of the recording beams. The grating strength parameter is frequently referred to as the $M\#$ since it provides a measure for the maximum number of holograms that can practically be recorded in the holographic material. It is useful for evaluating the storage capacity of the material for holographic data storage and will be discussed in more detail in Chapter 12 on applications.

8.4 Holographic Photopolymers

8.4.1 Introduction

Photopolymers show great promise as a holographic recording material. This is due to: the ability to optimize the components of the polymer for different applications; manufacture them in large formats at low cost; and that most photopolymers are dry processed. Early theoretical and experimental work on holographic photopolymers was conducted by Close et al. in 1969 [77] followed by Colburn and Haines [78] in 1971, and by Booth [79], Wopschall, and Pampalone [80] in 1972 at Dupont Laboratories. Calixto later set the foundations for realizing dry processed polymers that have higher diffraction efficiencies that are still in use today [81]. A holographic photo polymer generally consists of three components: (1) the acrylate or methacrylate vinyl monomer; (2) a sensitizer or photoinitiator; and (3) a binder that does not directly participate in the polymerization process. Other materials are also frequently added to the three basic components to improve holographic and practical handling characteristics. For instance, in pq-PMMA (see Appendix D), there is also a thermal initiated sensitizer, azobisisobutyronitrile (AIBN), that partially polymerizes the solution prior to holographic exposure that converts the liquid into a solid form [82].

8.4.2 General Photopolymer Composition

There are a large number of monomers available for use in the design of photopolymer systems. The choice of monomer is partially responsible for the hologram properties such as material shrinkage, dynamic range, and exposure sensitivity. Several monomers can also be combined in the same photopolymer to realize a desired result. Combinations of monomers are often used to promote cross-linking of polymer chains that reduce volumetric shrinking. Another class of monomer with a cationic ring-opening polymerization (CROP) has also been developed for added capability in controlling polymerization growth, termination, and reduced shrinkage in holographic photopolymers [83,84].

Polymerization (molecular chain growth) of monomer starts when the monomer reacts with a photon-activated radical molecule. The chain continues to grow in the presence of more monomer until the radical molecule is deactivated through a termination process. Photosensitizer or photo initiator molecules are used to absorb photons in a desired wavelength range and then transfer energy to a radical molecule to start the polymerization process. A variety of sensitizers can be used to absorb light in the visible (400–650 nm) wavelength range where sufficient energy is available for activating the radical molecule. In addition, a number of sensitizing molecules have been investigated for forming holograms in the near infrared (850–1100 nm). Holographic photopolymer binder molecules have a number of functions including control of the polymer viscosity to improve handling and sample preparation and in some cases are used to improve the exposure properties of the system.

The basic process of hologram exposure and grating formation in a photopolymer is illustrated in Figure 8.6 [32]. Exposure to an optical interference pattern polymerizes monomer in local regions of high illumination. This forms a gradient between regions of high and low monomer concentration. The concentration gradient initiates a diffusion process that forces additional monomer into the polymerized regions. Post-illumination with incoherent light will polymerize the remaining monomer and results in a refractive index variation that duplicates the initial exposing interference pattern.

8.4.3 Commercially Available Holographic Polymer Recording Characteristics and Physical Format

Photopolymer solutions can be coated on a variety of both flexible and rigid substrates. Some recent commercially available holographic photopolymers include those from Dupont [85,86], Bayer/Covestro [87], and Aprilis [84]. The following provides a brief overview of their characteristics.

The Dupont Omnidex polymers are coated on a Mylar polyethylene terephthalate (PET) substrate that is approximately 50 μm thick [86]. Typically, there is a thin protective cover sheet that first is removed from the polymer and Mylar which are then laminated onto a glass substrate with the polymer facing toward the glass to act as an adhesive. Typical exposure energy density for the Omnidex 352 and 706 photopolymers at 514.5 nm illumination is 30 mJ/cm^2. Both the efficiency and spectral bandwidth of the holograms can be enhanced by heating the film after exposure near 120°C for ~120 minutes. This results in a refractive index modulation of ~0.032 (for Omnidex 352) and ~0.041 (for Omnidex 706). The central reconstruction wavelength of holograms formed in the Dupont materials tends to decrease after processing (514.5 nm → 508–511 nm), indicating that the material thickness shrinks.

The Bayer/Covestro (Bayfol HX) materials are coated onto flexible polycarbonate substrates approximately 175 μm thick. Different polymer film thickness is possible with values ranging from 8 to 16 μm being reported [88,89]. The resin mixture consists of matrix precursors in addition to photosensitizers. When the resin is prepared into film layers on a substrate the matrix precursors form a rigid layer that resists lateral and axial shrinkage changes during exposure and post-exposure fixing. This approach gives minimal change in thickness and lateral dimensions of the material and does not require heating to enhance the diffraction efficiency of the holograms. The exposure energy density for the Bayfol HX film is 18 mJ/cm^2 at 632.8 nm, 25 mJ/cm^2 at 532 nm, and 30 mJ/cm^2 at 473 nm. These exposure values result in a refractive index modulation of ~0.03.

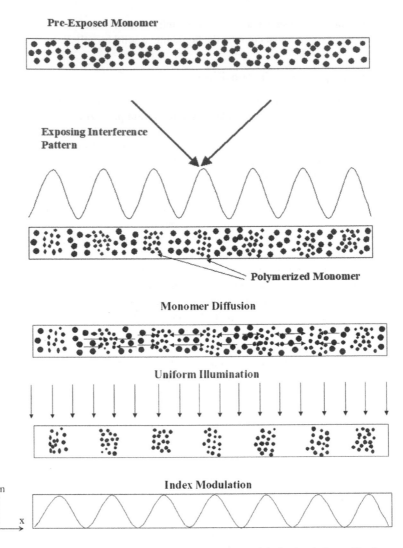

FIGURE 8.6   Schematic showing the formation of the refractive index modulation in a holographic photopolymer.

The Aprilis CROP photopolymer is typically supplied as a liquid sealed between two glass plates with spacers to set the thickness of the holograms [84]. Materials such as the ULSH-500 (100–500 μm thickness) were primarily designed for multiplexing large numbers of gratings (~1000) for holographic data storage applications [64]. Typical exposure irradiance is 4.5–5.0 mW/cm² at 514.5 nm. Refractive index modulation values of $\Delta n = 0.00275$ and diffraction efficiency of 95% were demonstrated in 100 μm thick samples. A pre-exposure of approximately (4.0–8.0 mJ/cm²) is usually necessary to overcome the threshold polymerization before holograms form with appreciable diffraction efficiency. Achieving the correct pre-exposure is necessary prior to recording a set of multiplexed data holograms. The pre-exposure also reduces distortions in the resulting holograms due to lateral and axial thickness change.

8.5 Dynamic Holographic Photopolymers

Recently, progress has been made on the development of dynamic holographic polymers. Two promising techniques include photorefractive polymers [90] and polymer-dispersed liquid crystals [91]. These materials provide a path toward write/read/erase holographic recording materials and have a variety of applications such as dynamic displays and holographic data storage. These topics are discussed in more detail in Chapters 9 and 12, respectively, on holographic applications.

8.5.1 Photorefractive Holographic Photopolymers

As described in Section 8.3, inorganic photorefractive materials utilize an electro-optic effect to form a light-induced refractive index modulation within the material. A similar effect can be realized in organic polymers, however, there are several differences with inorganic photorefractive materials. Organic materials have an amorphous rather than crystalline structure that changes the electronic band structure. In addition, the formation of the index modulation in organic photorefractive materials is only partially the result of a locally varying space charge field. The primary factor was found to be an orientational birefringence effect [92,93]. This effect results from changes in the orientation of dopants with fixed dipole moments caused by the locally varying light-induced space charge fields [see Ref. 90 p. 161].

Holographic grating formation in organic photorefractive materials results from three processes: photogeneration of charge carriers, charge transport, and formation of the locally varying refractive index modulation. In the charge carrier photogeneration process, a photon is absorbed and forms an electron hole pair or exciton. In a polymer, the exciton can rapidly decay due to recombination. To prevent rapid recombination, an electric field is applied to allow buildup of the space charge region. Another step required to form the space charge region is the transport of free carriers from regions that were illuminated to regions that were dark during the exposure. One difference between organic and inorganic photorefractive materials is that for organic polymers, transport occurs through intermolecular hopping of carriers between adjacent molecules and groups of molecules rather than through an energy band model [94]. In general, this transport mechanism is less efficient than inorganic band structure transport. However, better understanding of this process has helped to design organic materials with reasonably efficient charge transport [94,95].

The electro-optic effect in organic materials is achieved using second-order non-linearities resulting from collections of molecules in the polymer. This requires the centro-symmetry of the molecules to be broken. This is accomplished by conjugating the molecules in a manner that leaves opposite charge on either end of the molecule effectively forming a dipole. When groups of such molecules are poled with an electric field, macroscopic electro-optics properties result as well as birefringence. When this system is illuminated with an interference pattern, the locally varying light and dark regions change the orientation of the poled molecules, which in turn changes the local polarizability forming a spatially varying refractive index modulation that mimics the light pattern [64].

The magnitude of the refractive index modulation is proportional to the square of the electric field and voltage:

$$\Delta n = pE^2 = p\left(V/d\right)^2, \tag{8.6}$$

where p is a proportionality constant, E the applied electric field, V the applied voltage, and d is the thickness of the polymer sample between the electrodes. The proportionality constant p is a function of the material properties of the polymer and can be used as a figure of merit for the performance of the polymer. Values of 300–400 cm^2/V^2 have been achieved and result in index modulations of ~0.01 with moderate voltages (~5.5 kV) [96].

For hologram formation the photorefractive polymer is usually placed between two glass plates with transparent electrodes. Indium tin oxide is a common transparent electrode that allows the formation of the electric poling field across the polymer and also has high optical transmission and light to pass through the sample. The exposing beam angles relative to the sample should have a component of the resulting grating fringe planes in the direction of the electric field. This allows more efficient photogenerated charge transport through charge drift cause by the applied electric field. A typical construction geometry is shown in Figure 8.7 [64].

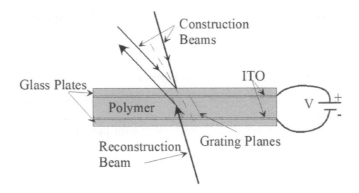

FIGURE 8.7 Diagram showing photorefractive polymer sample for hologram recording. The polymer is sandwiched between two glass plates that are coated with transparent electrodes (ITO) allowing the application of a poling and charge transport assisting electric field. ITO are indium tin oxide transparent electrodes and V is the applied voltage. (From Kippelen, B., Overview of photorefractive polymers for holographic data storage, in H. J. Coufal, D. Psaltis, and G. T. Sincerbox (Eds.), *Holographic Data Storage,* Springer-Verlag, Berlin, Germany, 2000. With permission from Springer eBook.)

8.5.2 Holographic Polymer-Dispersed Liquid Crystals

Another approach to hologram material design that provides some degree of dynamic switching is holographic polymer-dispersed liquid crystal (H-PDLC) materials. It consists of a polymer-based hologram and liquid crystal (LC) material [91]. In this approach, a hologram is recorded in the conventional manner in a solution of sensitized monomer and LC materials. The resulting hologram is an index modulation resulting from the localized displacement of monomer during exposure to the interference pattern. The grating region is filled with droplets of LC with random director vector orientations (Figure 8.8). If an electric field is now applied to the hologram, the orientation of the LC director within the droplets is aligned with the electric field vector. Since the refractive index of the LC droplets is different for directions along and perpendicular to the director vector, the refractive index can be changed. If the refractive index of the aligned LC droplets is designed to equal that of the index modulation within the fixed hologram, the grating modulation can effectively be switched off.

The H-PDLC material consists of a fast curing monomer host material, a dye photoinitiator, a coinitiator, a reactive diluent, and a liquid crystal [91]. A monomer must be chosen that provides fast gelling along with phase separation allowing the formation of submicron droplets. Several commercially available resins have been used with good success including Norland Optical Adhesive (NOA 65) [97]. The photoinitiator should have good absorption at the exposing wavelength, react with the coinitiator to produce a free radical, and undergo fast permanent bleaching. In general, dyes with high triplet quantum yield and lifetimes (on the order of micro-seconds) along with high reactivity in the triplet state provide good results.

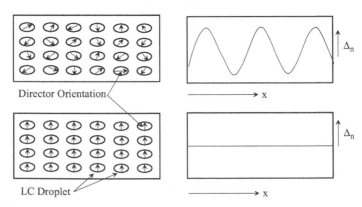

FIGURE 8.8 Holographic polymer dispersed liquid crystal showing the effect of the alignment of the liquid crystal droplets on the refractive index modulation. LC droplet are liquid crystal droplet and Δ_n is the refractive index modulation.

A dye that has been used with good success when exposing in the visible wavelength range (450–560 nm) is Rose Bengal acetate ester and has a quantum yield of ~0.70. Coinitiators serve as electron donors that respond to the triplet excited dye. Examples of coinitiators are N-phenyl glycine and triethanol amine. The reactive diluent helps to dissolve different components in the solution and make it homogeneous. The role of the reactive diluent in the curing stage of the monomer is not completely understood, but the concentration is suspected of helping to control the droplet size. A commonly used reactive diluent is N-vinyl pyrolidinone (NVP) [30]. Liquid crystals should be chosen with large dielectric anisotropy ($\Delta\varepsilon$) and optical birefringence that provide a large electro-optic effect (field-induced Δn). This allows for high resultant hologram diffraction efficiency and good switching properties. Several types of LCs that have been used with success in H-PDLC applications are smectic A LC, nematic E7 and BL, and commercially available TL compounds. The TL compounds have good environmental stability, high resistivity ($> 10^{12}\,\Omega$), high voltage holding properties, and relatively low switching field requirements (1 V/μm). However, the solubility of TL LC materials with several types of monomers is limited.

8.6 Photoresist Materials

Photoresist resins have been extensively developed for the integrated circuit fabrication industry to pattern high resolution spatial features with high quality optical properties. Sheridan performed one of the first demonstrations of fabricating holograms in photoresist in 1968 to form blazed surface relief gratings [98]. Since then, photoresist holograms have been used in a variety of applications including embossed security holograms, integrated optic grating devices, and in holographic lithography [99–102].

Photoresist materials generally consist of three components: a photoreactive element; a resin base; and a volatile solvent. Both positive and negative types of photoresist have been formulated and respond differently to an exposing light pattern. When a positive resist is exposed, the areas exposed to light are removed with processing. A negative resist responds in the opposite manner. Positive photoresist materials have primarily been used for holography although negative resists have also been used for some applications.

The most frequently used positive photoresists for holography are the Shipley AZ-1350B, J and AZ1400 series. More recently the Shipley AZ-1500 and 1800 series, and AllResist SX AR-P 3500/6 have been developed that are also useful for holography and have less toxic solvent constituents. Photoresists have refractive index values ranging from 1.60–1.75. Substrates should be chosen with refractive indices close to these values and should be indexed matched during exposure to reduce surface reflections that can result in the formation of spurious gratings.

Most photoresist materials have greater sensitivity at shorter wavelengths allowing high resolution features for lithography purposes. While there is some sensitivity at longer wavelengths (450–500 nm), their use requires significantly longer exposure times. For instance, the AR-P 3500 photoresist has a recommended exposure energy density of 40 mJ/cm^2 at the i-line (365 nm), but is 2 J/cm^2 at 488 nm [103]. Other laser wavelengths that are useful for photoresist exposure include the 457.9 nm argon ion line and the 441.56 nm line from a HeCd laser. Typically, photoresists are deposited by spin coating on a glass or other types of optical substrate. The spinner rotation rate can be varied to control the thickness for a specific type of resist. The resist viscosity can also be varied with a thinning agent to reduce the thickness. Typical photoresist grating thickness values for holography range from 0.5 to 4.0 μm. After spin coating the photoresist, it is tempered with a pre-bake treatment typically consisting of ~100°C on a hot plate for 2 minutes or at 95°C in a convection oven for 30 minutes [103]. This evaporates some of the solvent and improves adhesion to the substrate. After cooling down to room temperature, the photoresist-coated substrates can be exposed to the interference pattern and developed in a developer such as Shipley AZ 303 or 351 or AllResist AR-300 for approximately 1 minute and then rinsed in deionized water for 30 seconds.

An important consideration in using photoresist materials for holography is their non-linear response to exposure energy density. While important for integrated circuit fabrication, it leads to rectangular grating profiles and multiple diffraction orders during the reconstruction process. This effect can be controlled by biasing the photoresist into the linear response range with an incoherent pre-illumination in combination with processing using a diluted developer [104]. This process is dependent on both the photoresist and developer characteristics and requires a controlled set of experiments to optimize.

8.7 Photoconductor/Thermoplastic Materials

Thermoplastic material systems have been developed since the 1960s as a rewritable, surface relief holographic recording material. The first demonstrations of hologram recording in thermoplastic materials were performed by Urbach and Meir in 1966 using methods initially developed for photocopying [105]. The components and process for recording holograms in a thermo plastic system are illustrated in Figure 8.9. An optical surface such as glass is coated with an evaporated or sputtered layer of transparent or semitransparent conductive material such as indium tin oxide.

The glass is then coated with an organic photoconductor layer to give a thickness of 1–3 μm. Finally, the top surface of the photoconductor is coated with a thermal plastic material with a thickness determined by the spatial frequency requirements of the holographic recording. The thermoplastic material is derived from resins, and typical thickness values for holographic recording range from 0.5 to 1.0 μm. A detailed description on the preparation of thermoplastic systems for holographic recording is presented in the paper by Credelle and Spong [106].

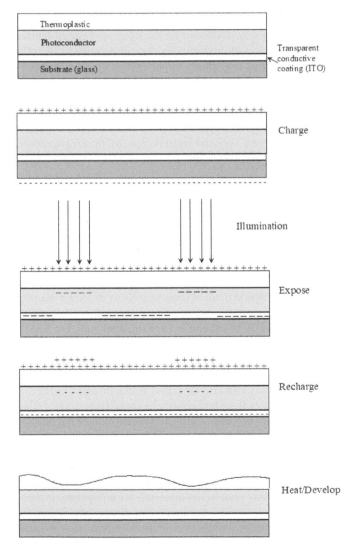

FIGURE 8.9 Thermoplastic system and process for recording holograms. (From Colburn, W.S. and Tompkins, E.N., *Appl. Opt.*, 13, 2934–2941, 1974.)

The mechanism for recording holograms in a thermoplastic material is shown in Figure 8.9. First, the top and bottom surfaces are uniformly charged with a voltage source. The surface of the thermoplastic is then exposed with a typical optical interference pattern corresponding to the holographic exposure. In regions of high illumination, the photoconductor conducts negative charge to the base of the thermoplastic material. The charge separation is maintained since the thermoplastic is an insulator. The system is charged a second time which adds more charge both to the upper surface of the thermoplastic and to the transparent conductive layer (such as indium tin oxide ITO). This produces a locally varying static electric field that mimics the intensity variation of the exposing interference pattern. The material is then heated to a temperature ranging between 50°C and 100°C which softens the thermoplastic and contracts it in the regions where the electric field is stronger producing a surface relief phase grating. The hologram can be erased by heating the thermoplastic material to temperatures slightly above the softening temperature for writing which uniformly flattens the surface. Spatial frequencies from a few hundred lines/mm to about 1500 lines/mm are typically recorded and diffraction efficiencies of 35%–40% have been reported [106,107].

Thermoplastic hologram recording systems were commercially available for many years and were used for holographic interferometry [108,109]. However, currently only one company is known to supply them (Tavex America) providing a 40 × 40 mm plate [110]. The number of write and erase cycles are limited to 3000–4000 cycles.

8.8 Embossed Holograms

While embossing is more of a processing method than a material, it provides a means of manufacturing large numbers of holograms at very low cost. Embossing is a method of replicating a master hologram [111]. The first step in the process is to form the master hologram which is usually recorded in a photoresist material to produce a surface relief type grating. Generally, the objects that are recorded do not have much depth since the hologram will be viewed under a wide variety of illumination conditions. The master hologram is formed in photoresist as a transmission type rainbow hologram as described in Chapter 9. The master hologram is then coated with an electrically conductive layer. This can be accomplished in several ways including vacuum deposition, silver spray, and electrodeless deposition. Vacuum deposition and silver spray methods deposit a uniform layer of silver on the photoresist grating surface while electrodeless deposition deposits a layer of nickel. This plated master is a negative of the desired pattern and is not very robust for the stamping operation. Therefore, a negative of this original master is formed using an electroforming technique to electrodeposit a thicker layer of metal (500 μm) onto the original plated photoresist master. This electroformed copy (working master) is a double negative or positive of the original grating profile that can be removed from the master and used to stamp copies of the desired hologram.

The working master hologram is then embossed or stamped into a plastic film (polyester) using either a single unit press for large format replicas or on a rotary press for mass produced replication. In either case, the metalized master hologram is heated and pressed into the copying material using what is referred to as either a "soft" or "hard" embossing method. The "soft" or "hard" embossing indicates whether the polyester is metalized with a reflecting coating after or before stamping. This metallization step allows the final hologram to be viewed in reflection mode.

8.9 Photosensitized Glass

A variety of glass types can be sensitized for holographic exposure to form different types of holographic optical elements. One of the most common applications is for fiber optic Bragg reflectors that are used in optical communication systems [112]. However, photo-thermo-refractive glass (PTRG) is also finding applications in a variety of laser systems, narrow bandwidth spectral filtering, and spectroscopy [113].

8.9.1 Sensitized Optical Fiber

The waveguiding properties of optical fiber are formed by increasing the refractive index of the fiber core. This is accomplished by preparing a SiO_2 pre-form that is doped with other atoms such as germanium. The pre-form is then heated and the fiber is pulled from the heated melt in a fiber drawing tower. Germanium is similar to the host material silicon in that it has four valence electrons, however, the substitution causes bonding defects or color centers in the glass. These color centers are believed to photosensitize the fiber to certain bands of near-UV radiation [114]. In particular, it was shown that the absorption of germanium-doped fiber decreases after exposure to 240 nm light suggesting a bleaching effect [115]. The change in absorption is related to a change in refractive index through the Kramers-Kronig relation [116]:

$$\Delta n(\omega) = \frac{c}{\pi} P \int_0^\infty \frac{\Delta\alpha(\omega')d\omega'}{\omega'^2 - \omega^2}, \tag{8.7}$$

where ω is the optical frequency, $\Delta\alpha$ is the change in absorption, and P is a constant equal to the principal part of the integral. A typical refractive index change of 10^{-4} can be obtained with normal levels of germanium doping (i.e., 3–6 mole% GeO_2) in the 1.3–1.6 μm wavelength range [117].

While this mechanism for generating an index change is somewhat useful, it was found that exposing the fiber to hydrogen at high pressure (hydrogen loading) can dramatically increase the sensitivity of germanium atoms and the resulting refractive index modulation. Mizrahi reported index modulation of 10^{-2} with hydrogen loading [118]. The hydrogen loading process consists of maintaining the fiber in a hydrogen rich (95% concentration) environment at a temperature of 300 K and pressure of 2000 psi for a period of approximately 9 days. This loads the fiber core with 1.6% hydrogen and allows recording light-induced gratings with high refractive index modulation [117].

Holograms can be formed in sensitized fibers by illuminating a phase mask to produce +/−1 diffraction orders that form an interference pattern. A conventional interferometer setup can also be used. In both cases, the side of the fiber is exposed to fringes that are perpendicular to the axis of the fiber. The exposing wavelength is typically 244 nm (frequency doubled 488 nm argon laser line), and the diffraction efficiency of the grating is actively monitored to determine the optimum exposure time.

8.9.2 Photo-Thermo-Refractive Glass

Another recording material used for holography is photo-thermo-refractive (PTR) glass. Photosensitizing glass was first demonstrated in 1949 by Stookey [119] and has been studied and developed for holography since the 1990s [120,121]. The PTR process consists of exposure of the sensitized glass to UV illumination followed by a thermal development treatment.

Several glass compositions have been developed that exhibit the PTR effect and have been used for recording holographic gratings. In general, these are multicomponent silicate glasses. Some specific examples include lithium aluminum silicate and sodium zinc aluminum silicate glasses doped with silver and cerium [120,121]. Recently, the compositions have been modified to include NaF and KBr in the glass with additional dopants such as Sb_2O_3 and SnO_2 to reduce scattering and improve the transmittance of the resultant holograms [122].

During exposure to UV illumination (~325 nm), the cerium atoms are ionized releasing electrons that are then trapped by positively charged silver ions and converted to a silver atom. At this point, the glass is in the equivalent of a latent image stage that occurs in other holographic materials such as silver halide emulsions after exposure and prior to development.

The hologram recorded in the PTR glass is developed by thermal treatment. The thermal development produces two effects. First, when treated at relatively low temperatures (~300°C), the reduced silver atoms diffuse within the glass and lead to the formation of nanometer size silver clusters in the high intensity regions of the holographic exposure pattern. Increasing the temperature to approximately 520°C causes a shell to form around the colloidal silver particles made of AgBr and NaBr.

Following shell growth, the crystalline phase of NaF forms around the shell during the latter part of the thermal treatment [123]. The resulting composite has a refractive index that differs with the host glass material producing a phase grating.

Typical holographic exposures of PTR glass are performed with the 325 nm emission line from a HeCd laser and exposure energy densities of 2–4 J/cm². This results in refractive index modulation values from $4 \times 10^{-4} - 10^{-3}$ [124].

PROBLEMS

1. It is desired to form a transmission hologram with 92% diffraction efficiency. The hologram is formed with recording angles of $\pm 25°$ to the surface normal (in air) at a wavelength of 514.5 nm. The Bayer/Covestro photopolymer has a refractive index modulation of 0.03 and pq-PMMA photopolymer has a refractive index modulation of 10^{-3}. What are the corresponding thickness values for each material to obtain the desired diffraction efficiency?

2. A set of four holograms are formed in a silver halide film with equal exposure energy density resulting in diffraction efficiencies of: 0.80, 0.69, 0.55, and 0.38, where 0.80 is the efficiency of the first hologram and 0.38 the efficiency of the last hologram. Assuming that the difference in diffraction efficiency results from the holographic reciprocity law failure, determine the relative exposure energy densities for each hologram to equalize the diffraction efficiency.

3. It is desired to form a fiber Bragg grating in a sensitized section of fiber to reflect light propagating in the fiber at 1535 nm. Assume that the refractive index of the glass fiber is 1.53 and the exposure wavelength is 244 nm in air. Determine the construction angles required to form the hologram.

4. A pure phase hologram is formed with recording angles of 0° and 25° to the surface normal (in air) at a wavelength of 514.5 nm. The material has an average refractive index of 1.51, refractive index modulation of 0.025, and a thickness of 15 μm. If the material shrinks by 5%, determine the shift in the Bragg reconstruction angle. Consider the Bragg reconstruction angle near 0°, and that the shrinkage only occurs in the z-direction. If reconstructed at 0° how much does the diffraction efficiency change? (See Appendix B).

5. Why would "noise" gratings be more apt to form with a silver halide film re-halogenating bleach process than with a reversal bleach process? (See Appendix B).

6. It is desired to form an unslanted transmission hologram in a PTR glass with a laser having a wavelength of 325 nm. The resulting hologram should have a Bragg reconstruction angle of 40° (in air) at a wavelength of 650 nm. Determine the construction angles at the formation wavelength.

7. A single beam Denisyuk reflection hologram is formed in a silver halide emulsion that has a residual absorption of 15% prior to processing. Compute the maximum fringe visibility that can be obtained for exposing this type of hologram in this emulsion (See Appendix B).

REFERENCES

1. H. I. Bjelkhagen, Ed., *Holographic Recording Materials*, SPIE Milestone Series, Vol. MS 130, (1996).
2. R. L. Kurtz and R. B. Owen, "Holographic recording materials—A review," *Opt. Eng.*, Vol. 14, 393–401 (1975).
3. P. Hariharan, "Holographic recording materials: Recent developments," *Opt. Eng.*, Vol. 19, 636–641 (1980).
4. R. D. Rallison and S. E. Bialkowski, "Survey of properties of volume holographic materials," *Proc. SPIE*, Vol. 1051, 68–75 (1989).
5. R. A. Lessard and G. Manivannan, "Holographic recording materials: An overview," *Proc. SPIE*, Vol. 2405, doi:10.1117/12.205350 (1995).
6. T. H. James, Ed., *The Theory of the Photographic Process*, 4th ed., Macmillian, New York (1977).

7. C. B. Neblette, *Photography: Its Materials and Processes*, 6th ed., Van Nostrand Reinhold, New York (1977).
8. H. I. Bjelkhagen, *Silver-Halide Recording Materials*, Springer-Verlag, Berlin, New York, (1993).
9. R. J. Collier, C. B. Burckhardt, and L. H. Lin, *Optical Holography*, Academic Press, New York and London, UK (1971).
10. L. Solymar and D. J. Cooke, *Volume Holography and Volume Gratings*, Academic Press, London, UK (1981).
11. R. L. van Renesse and F. A. Bouts, "Efficiency of bleaching agents for holography," *Optik*, Vol. 38, 156 (1973).
12. A. Graube, "Advances in bleaching methods for photographically recorded holograms," *Appl. Opt.*, Vol. 13, 2942 (1974).
13. L. Joly and R. Vanhorebeck, "Development effects in white-light reflection holography," *Photogr. Sci. Eng.*, Vol. 24, 108 (1980).
14. N. J. Phillips, A. A. Ward, R. Cullen, and D. Porter, "Advances in holographic bleaches," *Photogr. Sci. Eng.*, Vol. 24, 120 (1979).
15. R. R. A. Syms and L. Solymar, "Planar volume phase holograms formed in bleached photographic emulsions," *Appl. Opt.*, Vol. 22, 1479–1496 (1983).
16. K. M. Johnson, L. Hesselink, and J. W. Goodman, "Holographic reciprocity law failure," *Appl. Opt.*, Vol. 23, 218–227 (1984).
17. D. J. Cooke and A. A. Ward, "Reflection hologram processing for high efficiency in silver halide emulsions," *Appl. Opt.*, Vol. 23, 934 (1984).
18. www.integraf.com. (accessed on 2 April 2019).
19. R. W. Gurney and N. F. Mott, "The theory of the photolysis of silver bromide and the photographic latent image," *Proc. R. Soc. Ser. A*, Vol. 164, 151–167 (1938).
20. N. F. Mott, "The photographic latent image," *Photogr. J.*, Vol. 81, 62–69 (1941).
21. J. W. Mitchell, "The formation of latent image in photographic emulsion grains," *Photogr. Sci. Eng.*, Vol. 25, 170–188 (1981).
22. T. Tani, "Physics of the photographic latent image," *Phys. Today*, Vol. 42, 36–41 (1989).
23. K. S. Pennington and J. S. Harper, "Techniques for producing low noise, improved efficiency holograms," *Appl. Opt.*, Vol. 9, 1643–1650 (1970).
24. M. Lehmann, J. P. Lauer, and J. W. Goodman, "High efficiencies, low noise, and suppression of photochromic effects in bleached silver halide holography," *Appl. Opt.*, Vol. 9, 1948–1949 (1970).
25. B. J. Chang and K. Winick, "Silver halide gelatin holograms," *Proc. SPIE*, 215, 172–177 (1980).
26. P. Hariharan, "Silver halide sensitized gelatin holograms: Mechanism of hologram formation," *Appl. Opt.*, Vol. 25, 2040–2042 (1986).
27. A. Fimia, I. Pascual, C. Vazquez, and A. Belendez, "Silver halide sensitized holograms and their applications," *Proc. SPIE*, Vol. 1136, 53–57 (1989).
28. H. Kiemle and W. Kreiner, "Lippmann-Bragg-Phasenhologramme mit hohem Wirkungsgrad," *Phys Lett. A*, Vol. 28, 425–426 (1968).
29. R. L. van Renesse, "Scattering properties of fine grained bleached emulsions," *Photogr. Sci. Eng.*, Vol. 24, 114–119 (1980).
30. W. Spierings, "Pyrochrome processing yields color-controlled results with silver halide materials," *Holosphere*, Vol. 10, 1–7 (1981).
31. R. K. Kostuk and J. W. Goodman, "Refractive index modulation mechanism in bleached silver halide holograms," *Appl. Opt.*, Vol. 30, 369–371 (1991).
32. G. Zhao and P. Mouroulis, "Diffusion model of hologram formation in dry photopolymer materials," *J. Mod. Opt.*, 41, 1929–1939 (1994).
33. R. K. Kostuk, "Factorial optimization of bleach constituents for silver halide holograms," *Appl. Opt.*, Vol. 30, 1611–1616 (1991).
34. N. Phillips, "Benign bleaching for healthy holography," *Holosphere*, Vol. 14, 21–22 (1986).
35. K. S. Pennington, J. S. Harper, and F. P. Laming, "New phototechnology suitable for recording phase holograms and similar information in hardened gelatin," *Appl. Opt.*, Vol. 18, 80–84 (1971).
36. W. R. Graver, J. W. Gladden, and J. W. Eastes, "Phase holograms formed by silver halide (sensitized) gelatin processing," *Appl. Opt.*, Vol. 19, 1529–1536 (1980).

37. A. Fimia, I. Pascual, and A. Belendez, "Silver halide sensitized gelatin as a holographic storage medium," *Proc. SPIE*, 952, 288–291 (1988).
38. F. P. Laming, S. L. Levine, and G. Sincerbox, "Lifetime extension of bleached holograms," *Appl. Opt.*, Vol. 10, 1181–1182 (1971).
39. P. Hariharan and C. S. Ramanathan, "Suppression of printout effect in photographic phase holograms," *Appl. Opt.*, Vol. 10, 2197–2199 (1971).
40. S. Kumar and K. Singh, "Stability improvement in bleached phase holograms," *Opt. Laser Technol.*, Vol. 23, 225–227 (1991).
41. H. J. Bjelkhagen and D. Vukicevic, "Color holography: A new technique for reproduction of paintings," *Proc. SPIE*, Vol. 4659, 83–90 (2002).
42. H. J. Bjelkhagen, T. H. Jeong, and D. Vukicevic, "Color reflection holograms recorded in an ultrahigh resolution single layer holographic emulsion," *J. Imaging Sci. Technol.*, Vol. 40, 134–146 (1996).
43. T. A. Shankoff, "Phase holograms in dichromated gelatin," *Appl. Opt.*, Vol. 10, 2101–2105 (1968).
44. R. K. Curran and T. A. Shankoff, "The mechanism of hologram formation in dichromated gelatin," *Appl. Opt.*, Vol. 9, 1651–1657 (1970).
45. B. J. Chang, "Dichromated gelatin holograms and their applications," *Opt. Eng.*, Vol. 19, 642–648 (1980).
46. S. Sjolinder, "Dichromated gelatin and the mechanism of hologram formalism," *Photogr. Sci. Eng.*, Vol. 25, pp. 112–118 (1981).
47. C. G. Stojanoff, O. Brasseur, S. Tropartz, and H. Schutte, "Conceptual design and practical implementation of dichromated gelatin films as an optimal holographic recording material for large format holograms," *Proc. SPIE*, Vol. 2042, pp. 301–311 (1994).
48. http://greatlakesgelatin.com. (accessed on 2 April 2019)
49. R. D. Rallison, "Control of DCG and non-silver holographic materials," *Proc. SPIE*, Vol. 1600, 26–37 (1992).
50. P. Kowaliski, *Applied Photographic Theory*, p. 365, John Wiley& Sons, Chichester, UK (1972).
51. T. J. Kim, *Optimization of Dichromated Gelating Film Coatings for Holographic Recordings*, M.S. Thesis, University of Arizona, Tucson, AZ (1991).
52. L. H. Lin, "Hologram formation in hardened dichromated gelatin films," *Appl. Opt.*, Vol. 8, 963–966 (1969).
53. R. G. Brandes, E. E. Francois, and T. A. Shankoff, "Preparation of dichromated gelatin films for holography," *Appl. Opt.*, Vol. 8, 2346–2348 (1969).
54. J. M. Russo, *Holographic Grating Over Lens Dispersive Spectrum Splitting for Photovoltaic Applications*, PhD. Dissertation, University of Arizona, Tucson, AZ (2014).
55. H. Shutte and C. G. Stojanoff, "Effects of process control and exposure energy upon the inner structure and the optical properties of volume holograms in dichromated gelatin films," *SPIE Proc.*, Vol. 3011, 255–266 (1997).
56. T. G. Georgekutty and H.-K. Liu, "Simplified dichromated gelatin hologram recording process," *Appl. Opt.*, Vol. 26, 372–376 (1987).
57. D. Zhang, M. Gordon, J. M. Russo, S. Vorndran, and R. K. Kostuk, "Spectrum splitting photovoltaic system using transmission holographic lenses," *J. Photonics Energy*, Vol. 3, 034597 (2013).
58. T. Kubota, T. Ose, M. Sasaki, and K. Honda, "Hologram formation with red light in methylene blue sensitized dichromated gelatin," *Appl. Opt.*, Vol. 15, 556–558 (1976).
59. T. Kubota and T. Ose, "Methods for increasing the sensitivity of methylene blue sensitized dichromated gelatin," *Appl Opt.*, Vol. 18, 2538–2539 (1979).
60. S. P. McGrew, "Color control in dichromated gelatin reflection holograms," *Proc. SPIE*, Vol. 215, 24–31 (1980).
61. K. Kurokawa, S. Koike, S. Namba, T. Mizumo, and T. Kubota, "Full color holograms recorded in Methylene blue sensitized dichromated gelatin," *Proc. SPIE*, Vol. 2577, pp. 106–111 (1995).
62. Prism Solar Technologies (www.prismsolar.com) has manufactured DCG holographic concentrators that have passed accelerated life testing for 25 years of operation. Since 2015 Prism Solar has discontinued the sale of their holographic concentrator modules. (accessed on 15 January 2013)
63. P. Gunter and J. P. Huignard, *Photorefractive Materials and Their Applications I*, Topics in Applied Physics, Vol. 61, Springer-Verlag, Berlin, Germany (1988).
64. H. J. Coufal, D. Psaltis, and G. T. Sincerbox, *Holographic Data Storage*, Springer-Verlag, Berlin, Germany (2000).

65. A. Ashkin, G. D. Boyd, J. M. Dziedzic, R. G. Smith, A. A. Ballman, J. J. Levinstein, and K. Nassau, "Optically-induced refractive index inhomogeneities in $LiNbO_3$ and $LiTaO_3$," *Appl. Phys. Lett.*, Vol. 9, 72 (1966).

66. D. L. Staebler and J. J. Amodei, "Thermally fixed holograms in $LiNbO_3$," *Ferroelectrics*, Vol. 3, 107–113 (1972).

67. P. Gunter and A. Krumins, "High sensitivity read-write volume holographic storage in reduced $KNbO_3$ crystals," *Appl. Phys.*, Vol. 23, 199–207 (1980).

68. R. L. Townsend, J. T. LaMacchia, "Optically induced refractive index changes in $BaTiO_3$," *J. Appl. Phys.*, Vol. 41, 5188 (1970).

69. J. Feinberg and R. W. Hellwarth, "Phase conjugating mirror with continuous wave gain," *Opt. Lett.*, Vol. 5, 519 (1980).

70. K. Megumi, H. Kozuka, M. Kobayashi, and Y. Furuhata, "High sensitivity holographic storage in Ce-doped SBN," *Appl. Phys. Lett.*, Vol. 30, 631–633 (1977).

71. M. Peltier and F. Micheron, "Volume hologram recording and charge transfer process in $Bi_{12}SiO_{20}$ and $Bi_{12}GeO_{20}$," *J. Appl. Phys.*, Vol. 48, 3683–3690 (1977).

72. M. B. Klein, "Beam coupling in undoped GaAs at 1.06 μm using the photorefractive effect," *Opt. Lett.*, Vol. 9, 350–352 (1984).

73. J. Strait and A. M. Glass, "Photorefractive four-wave mixing in GaAs using diode lasers operating at 1.3 μm," *Appl. Opt.*, Vol. 25, 338 (1986).

74. B. Mainguet, "Characterization of the photorefractive effect in InP:Fe by using two-wave mixing under electric fields," *Opt. Lett.*, Vol. 8, 657–659 (1988).

75. A. Yariv, S. Orlov, G. Rakuljic, and V. Leyva, "Holographic fixing, readout and storage dynamics in photorefractive materials," *Opt. Lett.*, Vol. 20, 1334 (1995).

76. A. Yariv, S. Orlov, and G. Rakuljic, "Holographic storage dynamics in lithium niobate: Theory and experiment," *J. Opt. Am. B*, Vol. 13, 2513 (1996).

77. D. H. Close, A. D. Jacobson, J. D. Margerum, R. G. Brault, and F. J. McClung, "Hologram recording on photopolymer materials," *Appl. Phys. Lett.*, Vol. 14, 159 (1969).

78. W. S. Colburn and K. A. Haines, "Volume hologram formation in photopolymer materials," *Appl. Opt.*, Vol. 10, 1636–1641 (1971).

79. B. L. Booth, "Photopolymer material for holography," *Appl. Opt.*, Vol. 11, 2994–2995 (1972).

80. R. H. Wopschall and T. R. Pampalone, "Dry photopolymer film for recording holograms," *Appl. Opt.*, Vol. 11, 2096–2097 (1972).

81. S. Calixto, "Dry polymer for hologram recording," *Appl. Opt.*, Vol. 26, pp. 3904–3910 (1987).

82. O. Beyer, I. Nee, F. Havermeyer, and K. Buse, "Holographic recording of Bragg gratings for wavelength division multiplexing in doped partially polymerized poly(methyl methacrylate)," *Appl. Opt.*, Vol. 422, 30–37 (2003).

83. D. A. Waldman, R. T. Ingwall, P. K. Dal, M. G. Horner, E. S. Kolb, H.-Y. S. Li, R. A. Minns, and H. G. Schild, "Cationic ring-opening photopolymerization methods for holography," *Proc. SPIE*, 2689, 127–141 (1996).

84. D. A. Waldman, H. Y. S. Li, and E. A. Cetin, "Holographic recording properties in thick films of ULSH-500 photopolymer," *Proc. SPIE*, 3291, 89–103 (1998).

85. W. K. Smothers, B. M. Monroe, A. M. Weber, and D. E. Keys, "Photopolymers for holography," *Proc. SPIE*, Vol. 1212, 20–29 (1990).

86. A. M. Weber, W. K. Smothers, T. J. Trout, and D. J. Mickish, "Hologram recording in DuPont's new photopolymer materials," *Proc. SPIE*, Vol. 1212, 30–39 (1990).

87. D. Jurbergs, F. K. Bruder, F. Deuber, T. Fäcke, R. Hagen, T. Hönel, T. Rölle, M. S. Weiser, and A. Volkov, "New recording materials for the holographic industry", *Proc. SPIE*, 7233, 72330K (2009).

88. https://www.films.covestro.com/Products/Bayfol.aspx. (accessed on 2 April 2019)

89. J. Marin-Saez, J. Atencia, D. Chemisana, and M.-V. Collados, "Characterization of volume holographic optical elements recorded in Bayfol HX photopolymer for solar photovoltaic applications," *Opt. Exp.*, Vol. 24, A720–A730 (2016).

90. P. A. Blanche et al. "Holographic three-dimensional telepresence using large area photorefractive polymer," *Nature*, Vol. 468, 80–83 (2010).

91. T. J. Bunning, L. V. Natarajan, V. P. Tondiglia, and R. L. Sutherland, *Ann. Rev. Mater. Sci.*, Vol. 30, 83–115 (2000).

92. W. E. Moerner, S. M. Silence, F. Hache, and G. C. Bjorklund, "Orientationally enhanced photorefractive effect in polymers," *J. Opt. Soc. B.*, Vol. 11, 320–330 (1994).

93. B. Kippelen, F. Meyers, N. Peyghambarian, and S. R. Marder, "Chromophore design for photorefractive applications," *J. Am. Chem. Soc.*, Vol. 119, 4559–4560 (1997).

94. P. M. Borsenberger, E. H. Magin, M. B. O'Regan, and J. A. Sinicropi, "The role of dipole moments on hole transport in triphenylamine-doped polymers," *J. Polym. Sci. B: Poly. Phys.*, Vol. 34, 317 (1996).

95. H. Bassler, "Charge transport in random organic photoconductors," *Adv. Mater.*, Vol. 5, 662 (1993).

96. B. Kippelen et al., "Infrared photorefractive polymers and their applications for imaging," *Science*, Vol. 279, 54 (1998).

97. www.norlandprod.com. (accessed on 2 April 2019)

98. N. K. Sheridan, "Production of blazed holograms," *Appl. Phys. Lett.*, Vol. 12, 316–318 (1968).

99. M. C. Hutley, "Blazed interference diffraction gratings for the ultraviolet," *Optica Acta*, Vol. 22 (1975).

100. M. J. Beesley and J. G. Castledine, "The use of photoresist as a holographic recording medium," Vol. 9(12), 2720–2724 (1970).

101. N. D. Lai, T. D. Huang, J. H. Lin, D. B. Do, and C. C. Hsu, "Fabrication of periodic nanovein structures by holography lithography technique," *Opt. Exp.*, Vol. 17, 3362–3369 (2009).

102. C.-W. Chien, Y.-C. Huang, J. H. Lin, D. B. Do, and C. C. Hsu, "Analysis of a two-dimensional photonic bandgap structure fabricated by and interferometric lithographic system," *Appl. Opt.*, Vol. 46, 3196–3204 (2007).

103. http://www.allresist.com/product-overview/products-experimental-sample/. (accessed on 2 April 2019)

104. H. Werlich, G. T. Sincerbox, and B. Yung, "Fabrication of high efficiency surface relief holograms," *IBM Res. Rpt.*, RJ3912 (1983).

105. J. C. Urbach and R. W. Meier, "Thermoplastic xerographic holography," *Appl. Opt.*, Vol. 5, 666–667 (1966).

106. T. L. Credelle and F. W. Spong, *RCA Rev.*, Vol. 33, 206 (1972).

107. W. S. Colburn and E. N. Tompkins, "Improved thermoplastic-photoconductive devices for holographic recording," *Appl. Opt.*, Vol. 13, 2934–2941 (1974).

108. A. A. Friesem, Y. Katzir, Z. Rav-Noy, and B. Sharon, "Photoconductor-thermoplastic devices for holographic nondestructive testing," *Opt. Eng.*, Vol. 19, 659–665 (1980).

109. T. Kreis, *Handbook of Holographic Interferometry: Optical and Digital Methods*, Wiley-VCH, Weinheim, Germany (2005).

110. M. Y. Bazenov, V. V. Grabovsky, and G. Zahaykevich, "Real-time Interferometric investigation of defects and stresses in solids," *SPIE Proc.*, 5024, 191–197 (2003).

111. G. Saxby, *Practical Holography*, Chapter 21, 3rd ed., IoP, Bristol, UK (2004).

112. R. Kashyap, *Fiber Bragg Gratings*, Academic Press, San Diego, CA (1999).

113. S. Kaim, S. Mokhov, I. Divliansky, V. Smirnov, J. Lumeau, B. Y. Zeldovich, and L. B. Glebov, "Saturation of multiplexed volume Bragg grating recording," *J. Opt. Soc. Am. A*, Vol. 32, 22–27 (2015).

114. H. Y. Abe, D. L. Kinser, R. A. Weeks, K. Muta, and H. Kawazoe, "Nature and origin of the 5 eV band in SiO_2-GeO_2 glasses," *Phys. Rev. B.*, Vol. 46, 11445–11451 (1992).

115. R. M. Atkins, V. Mizrahi, and T. Erdogan, "248 vacuum UV spectral changes in optical fibre preform cores: Support for a colour center model of photosensitivity," *Electron Lett.*, Vol. 29, 385–387 (1993).

116. G. P. Agrawal, *Nonlinear Fiber Optics*, p. 450, 2nd ed., Academic Press, San Diego, CA (1995).

117. J. M. Battiato, "*Fiber Bragg Gratings*," PhD. Dissertation, Chapter 3, University of Arizona, Tucson, AZ (1998).

118. V. Mizrahi, P. J. Lemaire, T. Erdogan, W. A. Reed, D. J. DiGiovanni, and R. M. Atkins, "Ultraviolet laser fabrication of ultrastrong optical fiber gratings and of Germania doped channel waveguides," *Appl. Phys. Lett.*, Vol. 63, 1727–1729 (1993).

119. S. D. Stookey, "Photosensitive glass," *Ind. Eng. Chem.*, Vol. 41, 856–861 (1949).

120. V. A. Borgman, L. B. Glebov, N. V. Nikonorov, G. T. Petrovski, V. V. Savvin, and A. D. Tsvetkov, "Photo-thermal refractive effect in silicate glasses," *Sov. Phys. Dokl.*, Vol. 34, 1011–1013 (1989).

121. L. B. Glebov, N. V. Nikonorov, E. I. Panysheva, G. T. Petrovskii, V. V. Savvin, I. V. Tunimanova, and V. A. Tsekhomskii, "New ways to use photosensitive glasses for recording volume phase holograms," *Opt. Spectrosc.*, Vol. 73, 237–241 (1992).

122. O. M. Efimov, L. B. Glebov, L. N. Glebova, K. C. Richardson, and V. I. Smirnov, "High-efficiency Bragg gratings in photothermorefractive glass," *Appl. Opt.*, Vol. 38, 619–627 (1999).
123. J. Lumeau and L. B. Glebov, "Modeling of the induced refractive index kinetics in photo-thermo-refractive glass," *Opt. Mat. Exp.*, Vol. 3, 95–104 (2013).
124. S. A. Ivanov, A. I. Ignat'ev, N. V. Nikonorov, and V. A. Aseev, "Holographic characteristics of a modified photohermorefractive glass," *J. Opt. Technol.*, Vol. 81, 356–360 (2014).

9

Holographic Displays

9.1 Introduction

Perhaps one of the most intriguing and appealing applications of holography are displays. The three-dimensional nature of the image provides a great deal of information to the observer and has also served as an art form both for showing precious artifacts and providing unique imagery. One of the challenges of holographic displays is to make the viewing retain the qualities of the three-dimensional object while keeping the process simple and accessible to many observers. While a hologram is usually formed with a temporally and spatially coherent optical source, it should be possible to reconstruct the image with a broadband or "white" non-laser source. However, as shown in Chapter 4, dispersion and chromatic aberration result when using broadband reconstruction sources and can cause serious image degradation. Fortunately, a number of techniques have been developed to minimize or eliminate these issues by controlling the hologram dispersion and by only correcting those factors that affect viewing. In this chapter, a review of some of the important holographic display methods is presented. A more comprehensive review of display holography can be found in the excellent book by Benton and Bove [1].

9.2 Reflection Display Holograms

In Chapter 1, the Lippmann photography process is described in which an image of an object is formed at the film plane with a highly reflective backing material. This configuration forms interference patterns across the film aperture that corresponds to both the spatial and spectral properties of the object. In the early 1960s, a Soviet physicist, Yuri N. Denisyuk, made the realization that he could record interference fringes similar to those in a Lippmann photograph by making a reflection hologram [2,3]. He made his initial recordings using a mercury arc lamp with limited coherence length and power, but the resulting holograms were sufficient to demonstrate the concept. The important contribution for holography was that the reflection holograms could be reconstructed with a white light source making hologram viewing much more assessable to the general public. Surprisingly, the idea was first met with skepticism by the Soviet science community. However, the Denisyuk reflection hologram only became widely accepted after other developments in holography were publicized in the mid-1960s. As a result, Yuri Denisyuk was awarded the Lenin Prize in 1969, which at that time was equivalent to the Nobel Prize in the West [4].

The Denisyuk reflection hologram is formed with a single recording beam that passes through the hologram to illuminate the object as shown in Figure 9.1. The incident beam acts as the reference, and light reflected by the object forms a counter-propagating beam producing interference fringes. The use of a single beam to form both the reference and object beams eliminates the optics required for a second optical path and makes the recording system more stable. In this case, the absorption of the emulsion becomes important as it determines how much light illuminates the object.

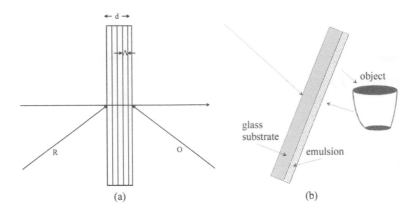

FIGURE 9.1 Basic reflection hologram with fringes parallel to the emulsion surface (a). A Denisyuk hologram with the object beam formed with light passing through the emulsion (b).

An important consideration when making reflection holograms are spurious gratings that form within the emulsion due to reflections from the glass and emulsion surfaces. One approach to reduce or eliminate these reflections is to use parallel or *p*-polarized light for the construction beam incident at the Brewster angle for the emulsion-air interface. If the emulsion and glass have similar refractive index values, there will be a small reflection at the glass-emulsion interface, but minimal or no reflection at the emulsion-air interface. There will be some loss in diffraction efficiency due to the *p*-polarized light as discussed in Chapter 5, but in general, the advantage of lower surface reflections and minimizing the formation of spurious gratings is much greater. A configuration suggested by Saxby [5] for recording high-quality single beam reflection holograms is illustrated in Figure 9.2. The neutral density filter controls the beam ratio and the object beam does not have to pass through the emulsion. Both the neutral density filter and the film are at the approximate Brewster angle (note that since the beam is expanding the Brewster angle cannot be exactly met for all diverging ray angles).

Full-color Denisyuk reflection holograms can also be formed by superimposing three holograms exposed with different wavelengths that span the visual response range of colors from the Internation Commission on Illumination (CIE) Color Chart. A configuration to record a hologram of this type is shown in Figure 9.3. Light from a red helium neon (HeNe) (633 nm) laser, a diode pumped frequency doubled neodymium yttrium aluminum garnet (ND:YAG) (532 nm) laser, and light from a diode pumped blue (457 nm) laser are combined with dichroic filters and then used to form the reflection hologram. The film selected for this type of hologram must be panchromatic. The Covestro Bayfol HX200 photopolymer and the Slavich PFG-03C silver halide film have a suitable spectral response for full-color reflection holography (see Chapter 8 for details on these materials).

The results obtained with full-color reflection holograms can be quite spectacular. Figure 9.4 below shows the recent work of Bjelkhagen [6] for a hologram reconstructed with a "white" light LED source.

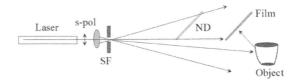

FIGURE 9.2 Configuration for recording single beam reflection holograms that does not require the beam to pass through the emulsion and uses an approximate Brewster angle of incidence to minimize surface reflections and the formation of spurious gratings.

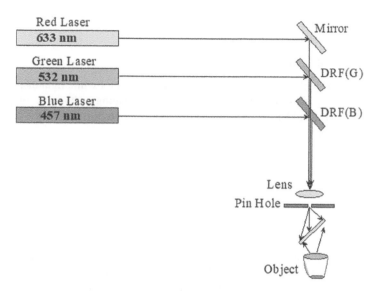

FIGURE 9.3 Schematic for constructing a full color reflection hologram with three lasers made to propagate along the same path by using dichroic reflection filters (DRF).

FIGURE 9.4 (See color insert.) Full color hologram of the 1911 Bay Tree Fabergé Easter Egg using an OptoClone recording system. (From Bjelkhagen, H.I. et al., *Proc. SPIE*, 9771, 2016. With permission from the SPIE.)

9.3 Transmission Display Holograms

9.3.1 Image Plane Holograms

One of the main problems with using transmission type holograms for white light display applications is dispersion and the resulting image blurring. From the paraxial imaging relations developed in Chapter 4, it was shown that the lateral and longitudinal image resolution are proportional to the spectral bandwidth of the reconstruction source $\Delta\lambda_2$:

$$\left|\Delta x_3\right| = \left(\frac{x_p}{z_p}\right) z_o \frac{\Delta\lambda_2}{\lambda_2}$$

$$\left|\Delta z_3\right| = z_o \frac{\Delta\lambda_2}{\lambda_2}.$$

(4.37)

However, it should also be noted that the image blur depends on the distance of the object from the film plane (z_o). This result implies that if the object field is projected within the film plane or a small distance from this plane then the image blur due to the spectral bandwidth of the reconstruction source can be greatly reduced. This idea is the motivation for the image plane hologram geometry illustrated in Figure 9.5 [7,8]. In this construction, an image of an object is straddled across the hologram plane and also illuminated with an off-axis reference beam. Objects with some depth (z_o) can be viewed with acceptable resolution using a broadband source ($\Delta\lambda$) as determined with Equation 4.37.

9.3.2 Rainbow (Benton) Holograms

As shown in the previous section, when a transmission hologram is reconstructed with a broadband source the diffracted light is dispersed and reduces image resolution. However, in the late 1960s, Stephen Benton realized that for viewing a three-dimensional image only horizontal parallax is important. He then incorporated this idea into "rainbow" transmission holograms that encode a slit aperture to restrict the wavelength band that is viewed [9]. During reconstruction, the dispersed light (rainbow) forms perpendicular to the length of the slit, hence the name "rainbow hologram" (Figure 9.6). The horizontal image forms with a narrow band of wavelengths with high lateral resolution and retains the depth perspective of the object. Several versions of the rainbow hologram have been developed to simplify construction and to expand its visual properties. These include the two-step rainbow hologram and the single-step image plane rainbow hologram and are described below. A complete review of rainbow holograms can be found in the book by Benton and Bove [1].

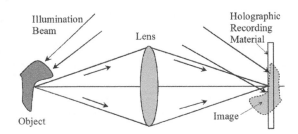

FIGURE 9.5 Geometry to form an image plane hologram. An image of an object is straddled across the recording material to limit the object distance (z_o).

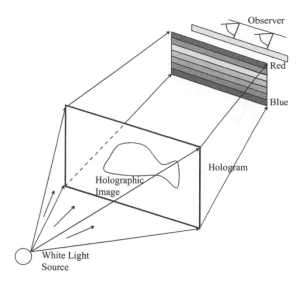

FIGURE 9.6 (See color insert.) Reconstruction of a rainbow hologram with a white light source showing the dispersive axis in the vertical direction and observation along a horizontal axis.

9.3.2.1 Two-Step Rainbow Hologram

In this design, a "master" transmission type hologram, H1, is formed with a slit across the film as shown in Figure 9.7a and the reference beam at an angle θ_r to the film normal. The master or H1 hologram is then rotated so that it is effectively reconstructed with a conjugate reference beam to form a real image. This image serves as the object beam for forming a second transmission hologram, H2 (Figure 9.7b). After processing, H2 is reconstructed with a conjugate reference beam and forms an image of the original object as well as an image of the slit used in H1. If an observer views the hologram through the image of the slit, he will see an image in a single color. If the head is moved vertically, the image will appear in a different color, and if the slit is viewed from a distance, the image will appear with a range of colors that change in the vertical direction from blue at the bottom to red at the top.

Selecting the value for the slit width is a tradeoff between having high resolution with a narrow (0.5–2.0 mm) slit and greater image brightness with a wider slit (10–20 mm). A narrow slit also causes a higher degree of image speckle. The slit width for a particular geometry can be found by making several H1 holograms with different slit width values until the desired imaging result is obtained. The spectral band of wavelengths observed through the slit can be estimated by using the grating equation and the geometry of the rainbow hologram. If the object angle is assumed to be normal to H2, then the grating period is: $\Lambda = \lambda/\sin\theta_r$. Using the dispersion relation for a thin grating (Equation 3.32) and the geometry of Figure 9.7, the spectral band across the slit is:

$$\Delta\lambda = \frac{\lambda}{\sin\theta_r}\frac{w}{z_{12}}\cos\theta_r. \tag{9.1}$$

Using typical values for the geometry of the rainbow hologram: $\lambda = 0.650$ μm, $w = 4$ mm, $\theta_r = 25°$, and $z_{12} = 400$ mm, the variation in wavelength across the slit in the vertical direction is approximately 14 nm. The availability of high power LED sources with small emission apertures, a range of operating wavelengths, and spectral bandwidths of 15–30 nm makes them a good option for rainbow hologram illumination.

Wyant extended the understanding and analysis of image blur in rainbow holograms [10]. He showed that the image degradation in a rainbow hologram is due to: (i) the spectral bandwidth of the reconstruction source, (ii) the source size; and (iii) diffraction from the slit. Taking each factor separately, the image blur resulting from this wavelength spread is:

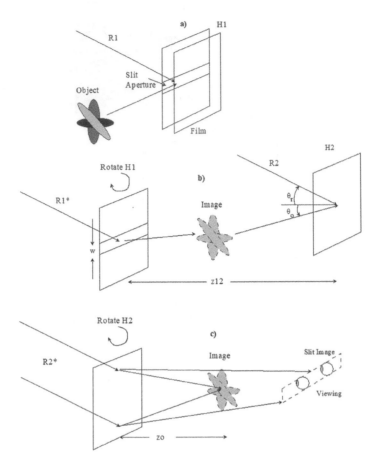

FIGURE 9.7 Two-step rainbow hologram. (a) Formation of the master hologram H1 with a slit across the film, (b) Reconstruction of a real image from H1 to form an object beam for the recording of H2, and (c) Reconstruction of the image and slit with the rainbow hologram H2.

$$\delta_{\Delta\lambda} = z_{2i}\frac{\partial\lambda}{\Lambda_{avg}}, \tag{9.2}$$

where z_{2i} is the distance from the hologram to the image, Λ_{avg} is the average grating period of the hologram, and d_{eye} is the diameter of the eye pupil. The effect of the source size on image blur can be expressed as:

$$\delta_{src} = \partial\omega_{src}z_{2i}, \tag{9.3}$$

with $\partial\omega_{src}$ equal to the angular divergence of the source. If the blur due to the source size is approximately equal to the blur resulting from the spectral bandwidth then:

$$\partial\omega_{src} = \frac{w + d_{eye}}{z_{12}}. \tag{9.4}$$

Finally, the approximate image blur due to diffraction effects can be expressed as:

$$\delta_{diff} = \frac{2\lambda\left(z_{2i} + z_{12}\right)}{w}. \tag{9.5}$$

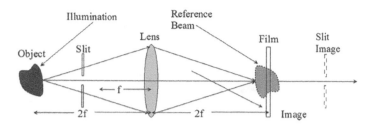

FIGURE 9.8 One-step rainbow hologram construction using an image plane hologram.

Another consideration is that the different wavelengths within the reconstructed slit image will be slanted with the upper red light closer to H2 than the blue light at the lower portion of the slit. This results from the fact that according to the grating equation, a hologram with a fixed grating period will diffract red light at a larger angle than blue light. This effect becomes important when designing full color or white light rainbow holograms and must be corrected. A detailed description of factors that affect the performance of rainbow holograms and methods to correct them are described in Chapter 15 of the book by Benton and Bove [1].

9.3.2.2 Single-Step Image Plane Rainbow Hologram

It is possible to simplify the recording of a rainbow hologram to a single step by using the image plane hologram described earlier. In this approach, a lens is used to form a 1:1 image of an object at the film plane as shown in Figure 9.8. A slit is placed in front of the lens in order to form a real image of the slit beyond the image of the object and the object and reference beams are set as shown. During reconstruction the hologram is illuminated with the reference beam, and the image is observed through the real image of the slit.

9.4 Composite Holographic Displays

Prior to holography, a variety of imaging techniques were invented to give different perspectives of a scene and a three-dimensional aspect to the display [11]. Perhaps the most common was the stereoscopic viewer that shows a sequence of scenes that were pre-recorded onto a disk [1]. Holography solved the problem of being able to reconstruct a three-dimensional image of a scene. However, it did not completely solve the problem of being able to view the changing perspectives of a scene. In order to address this problem, a number of techniques that combine earlier stereoscopic methods based in photography with holographic methods were developed. The following sections describe some of these techniques which have proven successful.

9.4.1 Holographic Stereograms

In general, a stereographic image provides different perspective views of a scene. A non-holographic stereogram can be formed by using a lenslet array in front of the film to image a different perspective of the scene onto the film. When the processed film is then viewed by an observer with the lenslet array in place, the different perspective views of the scene are simultaneously reconstructed providing the perception of a three dimensional (3D) image.

A similar process can be applied to the construction of a holographic stereogram [12,13]. One construction method is shown in Figure 9.9. First, different perspectives of a scene are photographed. Each photograph then serves as an object for recording a thin hologram strip. The process is repeated for each photograph. When the resulting hologram is viewed by an observer, the image formed by each eye provides two simultaneous 3D perspectives of the scene. In addition, as the observer moves laterally the views of the object will change.

An important variation of this process is to record photographs of slightly different positions of the object. The displaced object photographs are then used as the objects for recording the holographic strips in the stereogram (Figure 9.10). When an observer moves her head around, the resulting composite

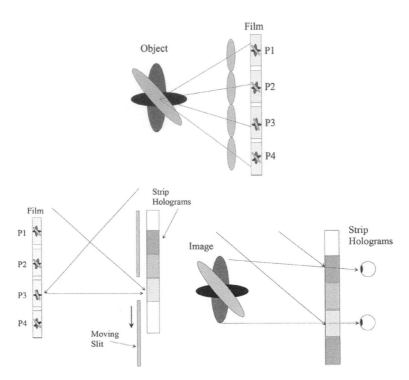

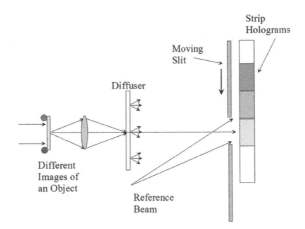

FIGURE 9.9 Method for forming a holographic stereogram. The top illustration shows the formation of the composite photograph of different views of the object. The lower left diagram shows the formation of the strip holograms, and the diagram at the lower right the reconstructed image.

FIGURE 9.10 System for forming a holographic stereogram of a changing scene. A different perspective of the object is holographically recorded on a different strip of the holographic recording material.

hologram image appears to move. Perhaps the most famous example of this type of hologram was that formed by Cross and Brazier of a woman seeming to blow a kiss and wink as the observer moves her head around the hologram. A video of the Cross-Brazier hologram can be seen in references 14 and 15.

9.4.2 Zebra Imaging Holographic Display

Perhaps one of the more impressive composite hologram techniques is the system developed by Zebra Imaging/Holotech [16]. In this approach, the complete display hologram aperture is divided into sub

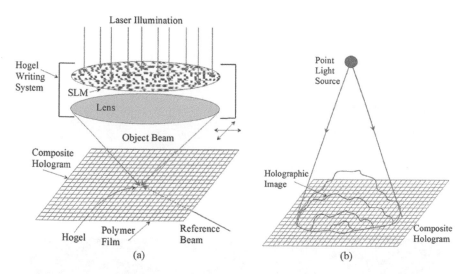

FIGURE 9.11 Illustration of the process used by Zebra Imaging for fabricating a composite display hologram (a). Different image scenes projected from the spatial light modulator are recorded at different coordinates on the film plane. The resulting hologram of the three-dimensional scene is reconstructed with a point source (b) similar to a conventional reflection hologram.

holograms that Zebra Imaging refers to as "hogels." The basic technique is illustrated in Figure 9.11a. The key aspect of their process is to correctly sample the graphical three-dimensional data that are to be displayed. Special software is used to sample and digitize the perspective image at different volumetric points across the scene. The sampled image data are then used to set a pattern on a spatial light modulator and form an object beam for recording a reflection type sub hologram at different locations on the aperture of the composite hologram. Each sub hologram or hogel gives a slightly different 3D reconstruction of the full volumetric data for the entire scene. The process is repeated thousands of times across the aperture forming an array of hogels that are recorded on holographic film. After completing the exposure process, the composite hologram is processed and then displayed with a point source (Figure 9.11b).

9.5 Updateable Holographic Displays

One of the most interesting and challenging types of holographic displays is a dynamic one that allows the images to be changed within a short period of time. This type of display requires a material that allows the formation, reading, and erasing of holograms in a reasonably short time period (i.e., seconds to minutes). A significant step toward achieving this performance was recently demonstrated by Blanche et al. [17]. Their approach is based on the development of an effective photorefractive photopolymer (PRPP) that can be fabricated on reasonably large apertures (100 × 100 mm). The PRPP is combined with a holographic stereography system to form a 3D image that is similar to that used by Zebra Imaging for static holographic displays.

The PRPP material used for updateable displays consists of a copolymer with a hole carrier transporting component and a carbaldehyde aniline group attached via an alkoxy linker [18]. The holographic display is formed by melting the polymer mixture and pouring it onto a glass plate coated with indium tin oxide to serve as an electrode. The polymer thickness is determined by using spacers with the desired thickness value. The structure is then covered with another piece of glass coated with indium tin oxide to act as the second electrode. The holograms are formed at 532 nm with an irradiance of approximately 1000 mW/cm². During the hologram exposure, an initial high voltage of ~9 kV is applied to increase the speed of the charge transport mechanism during the writing period to obtain diffraction efficiencies greater than 50% with the high voltage applied for times of approximately 0.5 sec. The voltage is then

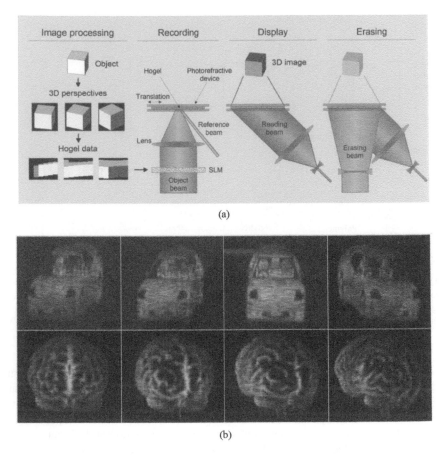

(a)

(b)

FIGURE 9.12   The schematic (a) shows the process for recording and reconstructing holographic displays in photorefractive photopolymers. The image (b) shows a full color holographic reconstruction of an object. (With kind permission from Taylor & Francis: *Optical Properties of Functional Polymers and Nano Engineering Applications*, 2014, Jainm V, and Kokil, A. Eds; Reprinted by permission from Macmillan Publishers Ltd. *Nature Lett.*, Tay, S. et al., 2008, Copyright 2008.)

lowered to a holding voltage of ~4 kV. The holograms are reconstructed with HeNe (632.8 nm) laser light and persist for several hours when reconstructed with red light. The holograms can be erased by turning off the voltage and illuminating with incoherent green illumination.

The holographic stereography process has been automated into a printer and display system as shown in Figure 9.12 [18]. First, different projections of a 3D object are produced from a subject of interest. The object can be a typical scene, medical imaging data (computerized axial tomography [CAT] scan, magnetic resonance imaging [MRI], optical), aerial or satellite recorded topography, 3D artwork, or other 3D subjects thus being useful for a variety of applications. The first step in the process consists of storing a set of 2D perspective images of the object. To record a sub hologram or hogel, data for a 2D perspective of the object are recalled and used to modulate the aperture of a spatial light modulator. The spatial light modulator aperture is then illuminated with a collimated and expanded green (532 nm) laser to form the object beam and combined with an off-axis reference beam. Next, the PRPP substrate is translated to a new position and another hogel is recorded with a different 2D perspective of the 3D object. During the recording phase, the voltage across the indium tin oxide contacts of the PRPP is set to a higher 9 kV potential to decrease the recording time. A typical time to record the array of hogels across the aperture of the PRPP is about 180 seconds (0.5 sec/hogel). For reconstruction, the voltage across the PRPP is reduced to the lower holding value of 4 kV, and the full area of the polymer is illuminated with an expanded red (632.8 nm) laser. The display can be viewed for several hours and then either refreshed or changed to view a new object.

9.6 Holographic Combiner Displays

A very useful application of holography are displays that combine different image information. In a typical system, a computer-generated display is overlaid onto a natural scene normally viewed by an observer. Holographic combiner displays can be categorized into two general groups: (i) head up and helmet mounted displays that project instrumentation data within the view of an observer [19] and (ii) augmented reality displays and eyewear that combine computer-generated animation with real scenery [20]. Holographic optical elements have a number of important attributes that are useful for these displays. These include the ability to: control the amount of reflected and transmitted light from the display; control the wavelength of the diffracted light; incorporate focusing power into the element; and form elements on relatively thin layers (a few microns) that can be applied to a variety of surfaces with different curvatures.

9.6.1 Head Up and Helmet Mounted Displays

Holographic elements are very effective in head up and helmet mounted displays (HUDs and HMDs). These systems display critical information to pilots of high-performance aircraft without requiring them to look away from a scene and greatly adds to their effectiveness and safety [21]. HUDs have also been used on surface vehicles to allow the operator to concentrate on driving and not be distracted by looking at instruments.

The basic idea for a HUD is illustrated in Figure 9.13. In this system, an image of an instrument display is relayed to an intermediate position. A reflection hologram then forms a virtual image of the display within the field of view of the pilot who can view the display information as long as the eyes are within the eye-box region. Since the instrument display often uses a single color or a narrow spectral band, the hologram produces an image without significant chromatic aberration. The reflection hologram also helps to select a narrow spectral band of wavelengths. Many head up displays are formed in dichromated gelatin as it has excellent optical quality, can be coated on either flat or curved surfaces, and is stable after sealing. When a curved surface such as a helmet visor is used, the curvature must be factored into the design of the holographic combiner.

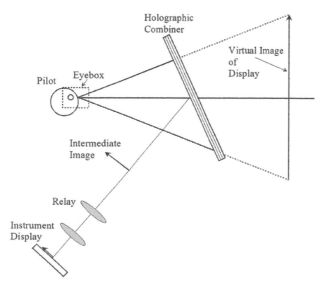

FIGURE 9.13 Holographic HUD using a reflection type holographic combiner. The virtual image of the instrument display is projected into the field of view of the observer/pilot.

9.6.2 Near Eye Augmented Reality Eyewear

In a typical augmented reality (AR) system, a head-mounted device projects a scene from a micro display into the field of view of an observer. This allows interactive computer images to be combined with real-world activities [22]. It is projected that AR will be useful in many applications including medical imaging, navigation, and entertainment [20,23]. It is similar to a HUD except that it provides interactive animation overlaid onto the real scene viewed by an observer.

It is possible to realize a system of this type with bulk optical components, but this leads to a large and heavy device that is cumbersome and awkward to use and restricts potential applications and markets. Therefore, a significant effort is underway to minimize the size of the AR system and to allow the projected images to be viewed through both eyes to minimize fatigue. The goal for many companies involved in AR development is to incorporate the system into form factor similar to that of a pair of normal eye glasses [24].

A useful design approach to achieve this goal is to use a small aperture micro display in combination with an exit pupil expander (EPE) [22]. The small display makes a variety of head or eyewear-mounted configurations possible while the EPE allows the display image to be projected over a larger field of view within a large eye motion box in front of both eyes.

One approach to realizing the EPE for both eyes can be accomplished with holographic elements combined with a folded or substrate mode optical system (Figure 9.14) [25]. In this arrangement, a reflective micro display is illuminated with a red-green-blue combination of light sources to provide full-color rendering. The reflected light is then collimated and sent through an input coupling transmission hologram. The input coupling hologram consists of either a multiplexed element or two cascaded holograms that diffract normally incident light to the left and right at angles that exceed the critical angle of the substrate. When the diffracted light in the waveguide reaches the output coupling reflection hologram, part of the light is diffracted with each successive interaction with the hologram. Focusing power can also be incorporated into the output coupling hologram and provides an extended field of view of the scene from the micro display. In this configuration, the micro display and light source can be mounted between the eyes on the eyewear.

However, there are several issues with implementing a design of this type with holographic elements. One is realizing a full-color image of the display. This can be accomplished by recording three holograms designed for operation with a red, green, and blue wavelength that span the visual range of CIE color chart to provide full color rendering of the display. This method was implemented by Sony Corporation [23] using the design shown in Figure 9.15. Note that in this arrangement, the intermediate wavelength (green) light is diffracted by a hologram in a separate waveguide to minimize crosstalk with the holograms diffracting red and blue light. However, a multiplexed hologram in one waveguide is used to diffract red and blue light since in this case the wavelength separation is sufficient to prevent significant cross coupling in diffraction efficiency for the two wavelength bands. Other groups working on different holographic near eye EPE systems are given in the references [20,24–26].

Another type of holographic near eye display is based on switchable Bragg gratings (SBGs) using polymer dispersed liquid crystals [27]. This approach has primarily been developed by DigiLens [28].

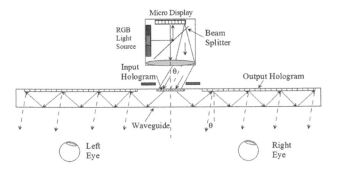

FIGURE 9.14 Waveguide holographic exit pupil expander for projecting an image from a micro display.

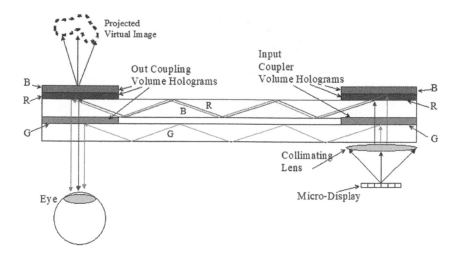

FIGURE 9.15 Configuration developed by the Sony Corporation for an AR near-eye display system using red-green-blue reflection volume holograms and a substrate mode waveguide. (From Mukawa, H. et al., *SID*, 89–92, 2008.)

As discussed in Chapter 8, the holographic polymer dispersed liquid crystal material is formed by combining a holographic polymer and a liquid crystal compound in a common solution. The material is deposited on a glass or plastic substrate and sealed with a cover glass. The cover glass and substrate are coated with transparent conductive electrodes. The assembly is then exposed to an object and reference beam to holographically encode a specific optical function such as focusing, beam expansion, or filtering as shown in Figure 9.16a. When the hologram is reconstructed without applying a voltage, the

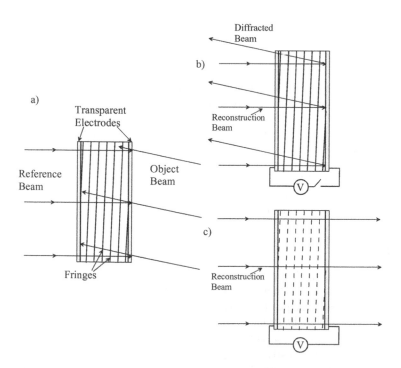

FIGURE 9.16 (a) Formation of a reflection hologram in a SBG material, (b) Reconstruction of the hologram without an applied voltage, and (c) Canceling the refractive index modulation with an applied voltage. (From Smith, R. et al., *Proc. SPIE*, 4207, 31–38, 2000.)

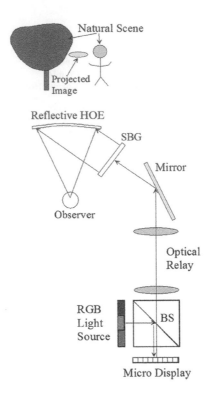

FIGURE 9.17 Schematic of a near-eye display using SBGs. The images formed by a micro display are combined with a real or natural scene viewed by an observer. (From Smith, R. et al., *Proc. SPIE*, 4207, 31–38, 2000.)

image beam is diffracted similar to a conventional volume hologram with a refractive index modulation (Figure 9.16b). However, when a voltage is applied to the conductive surfaces, the liquid crystal director within the droplets are aligned causing the refractive index to change and cancels the refractive index difference formed during the recording step (Figure 9.16c). This allows the encoded hologram to be turned on and off with the application of a voltage. The typical diffraction efficiency of an SBG can be varied from >99% to <0.1% using an electric field of 2–5V/μm within 10–50 μs. The low switching voltage makes it possible to use in wearable devices. Different optical functions such as focusing, beam expansion, and spectral filtering can be integrated into the SBG, and allow images from the micro optical display to move to different locations within the scene. The SBG also allows the possibility for eye tracking with the use of very small low power eye-safe lasers and sensors. One configuration for near-eye displays based on SBG devices is illustrated in Figure 9.17. In this system, the SBG, mirror, optics, micro display, and light sources can be combined into wearable eyewear [29].

PROBLEMS

1. An image plane hologram is reconstructed with an LED with a nominal wavelength of 600 nm and a spectral bandwidth of 30 nm. Light from the LED is essentially collimated and illuminates the hologram at an angle of 30° to the optical axis. If the object extends 0.5 cm from the hologram plane, compute the image spread in the lateral and axial directions.

2. A reflection hologram is formed with a single beam as with the Denisyuk method. The recording material is a silver halide film that absorbs about 20% of the incident beam. Compute the visibility of the resultant hologram. If the refractive index of the film is 1.47 and the illumination is at 632.8 nm, compute the maximum spatial frequency that can be recorded in the film.

3. It is desired to form two-step rainbow hologram with an H1 and H2 and the following parameters: source wavelength is 550 nm, source bandwidth is 40 nm, distance between H1 and H2 is 25 cm, image distance from H2 is 15 cm, average grating period is 0.7 μm, slit width is 4 mm, and the eye pupil diameter is 3 mm. Compute the image blur due to the source spectral bandwidth and diffraction assuming that the angular range due to the finite source size produces the same image blur caused by the spectral bandwidth of the LED.

REFERENCES

1. S. A. Benton and V. M. Bove Jr., *Holographic Imaging*, John Wiley & Sons, Hoboken, NJ (2008).
2. Y. N. Denisyuk, "Photographic reconstruction of the optical properties of an object in its own scattered radiation field," *Sov. Phys. Dokl.*, Vol. 7, 543–545 (1962).
3. Y. N. Denisyuk, "On the reproduction of the optical properties of an object by the wave field of its own scattered radiation," *Opt. Spectrosc.*, Vol. 18, 365–368 (1963).
4. S. A. Benton, "Holography reinvented," *Proc. SPIE*, 4737, 23–26 (2002).
5. G. Saxby, *Practical Holography*, p. 123, 3rd ed., Institute of Physics Publishing, Bristol, UK (2004).
6. H. I. Bjelkhagen, A. Lembessis, and A. Sarakinos, "Ultra-realistic imaging and OptoClones," *Proc. SPIE*, Vol. 9771, doi:10.1117/12.2207524 (2016).
7. L. Rosen, "Focused-image holography with extended sources," *Appl. Phys. Lett.*, Vol. 9, 337 (1966).
8. G. W. Stroke, "White-light reconstruction of holographic images using transmission holograms recorded with conventionally focused images and in-line background," *Phys. Lett.*, Vol. 23, 325 (1966).
9. S. A. Benton, "Hologram reconstructions with extended incoherent sources," *J. Opt. Soc. Am.*, Vol. 59, 1545–1546 (1969).
10. J. C. Wyant, "Image blur for rainbow holograms," *Opt. Lett.*, Vol. 1, 130–132 (1977).
11. T. Okoshi, *Three-Dimensional Imaging Techniques*, Academic Press, New York (1976).
12. S. A. Benton, "Survey of holographic stereograms," *Proc. SPIE*, Vol. 367, 15–19 (1982).
13. D. J. DeBitetto, "Holographic panoramic stereograms synthesized from white light recordings," *Appl. Opt.*, Vol. 8, 1740–1741 (1969).
14. http://www.jrholocollection.com/collection/cross.html. (accessed on 15 October 2018)
15. J. Hecht, "Holography and the laser," OPN, 34–41 (2010).
16. http://www.holotech3d.com/. (accessed on 15 October 2018)
17. P. A. Blanche, "Photorefractive polymers for 3D display application," In *Optical Properties of Functional Polymers and Nano Engineering Applications*, CRC Press, Boca Raton, FL (2014), V. Jain and A. Kokil, Eds., Chapter 3.
18. S. Tay et al., "An updatable holographic three-dimensional display," *Nature Lett.*, Vol. 451, 694–698 (2008).
19. J. Upatnieks, "Compact head-up display" U.S. Patent No. 4711512, December 8, 1987.
20. B. C. Kress and W. J. Cummings, "Towards the ultimate reality experience: HoloLens display architecture choices," SID Digest, Paper 11-1, 127–131 (2017).
21. R. B. Wood and M. J. Hayford, "Holographic and classical head up display technology for commercial and fighter aircraft," *Proc. SPIE*, 0883, 36–52 (1988).
22. T. Levola, "Diffractive optics for virtual reality displays," *J. SID*, Vol. 14, 467–475 (2006).
23. B. C. Kress and M. Shin, "Diffractive and holographic optics as optical combiners in head mounted displays," *Proceedings of the 2013 ACM Conference on Pervasive and Ubiquitous Computing Adjunct Publication*, pp. 1479–1482, Zurich, Switzerland, September 8–13, 2013.
24. B. C. Kress and T. Starmer, "A review of head-mounted displays (HMD) technologies and applications for consumer electronics," *Proc. SPIE*, Vol. 8720 A-1:A-13 (2013).
25. P. Ayras, P. Saarikko, and T. Levola, "Exit pupil expander with a large field of view based on diffractive optics," *J. SID*, Vol. 17, 659–664 (2009).
26. H. Mukawa, K. Akutsu, I. Matsumura, S. Nakano, T. Yoshida, M. Kuwahara, K. Aiki, and M. Ogawa, "A full color eyewear display using holographic planar waveguides," *SID*, 89–92 (2008).
27. T. J. Bunning, L. V. Natarajan, V. P. Tondigliam and R. L. Sutherland, "Holographic polymer dispersed liquid crystals (H-PDLCs)," *Ann. Rev. Mater. Sci.*, Vol. 30, 83–115 (2000).
28. R. Smith, M. Popovich, and S. Sagan, "Application specific integrated lenses and filters for microdisplays using electrically switchable Bragg grating technology," *Proc. SPIE*, Vol. 4207, 31–38 (2000).
29. J. Han, J. Liu, and Y. Wang, "Near-eye waveguide display based on holograms," *SID*, 4–6 (2016).

10

Holographic Interferometry

10.1 Introduction and Basic Principles

Holographic interferometry is a method of comparing optical fields that are transmitted or reflected from an object. One or both of the waves used in forming the interference pattern are produced by the hologram. It is a non-contact, non-destructive technique for measuring changes in the optical path length that occur when the physical properties of the object are altered in some way. Changes in the optical path length indicate variations to the thickness, surface topography, refractive index, or any combination of these parameters. Since changes in optical path length as small as 1/100th of the wavelength can be detected, the technique is very useful for sensing small changes or perturbations of a test object. The high sensitivity of the process places some restrictions on the total change in phase that can be measured. In addition, since this is an interferometric process, the temporal and spatial coherence properties of the light must be sufficient for the geometry of the setup. Nevertheless, the technique has proven very valuable for non-destructive testing of components and in manufacturing.

This chapter provides a basic introduction to holographic interferometry and how it is applied to surface measurements and deformations. More extensive discussion of this subject is covered extensively in the literature and in excellent texts by Kreis [1], Toal [2], and Hariharan [3].

The basic principle for measuring a phase difference using holographic interferometry can be evaluated by considering the coherent addition of two optical fields. The field reflected from the object in the initial state is:

$$\tilde{E}_1(x, y) = A_1(x, y)\exp(-j\varphi(x, y)),\tag{10.1}$$

where $A_1(x, y)$ is the amplitude, and $\varphi(x, y)$ the phase of the field on an observation plane at coordinates (x,y) on an observation plane. If the perturbation of the object is small, the phase will only vary by a small amount from the initial state. Therefore, the field reflected from the object in the perturbed state can be written as:

$$\tilde{E}_2(x, y) = A_2(x, y)\exp\{-j[\varphi(x, y) + \Delta\varphi(x, y)]\},\tag{10.2}$$

where $\Delta\varphi$ is the resulting phase difference between the two states. When the fields from the two states are coherently combined, the intensity across the observation plane is:

$$\begin{aligned}
I(x, y) &= |\tilde{E}_1 + \tilde{E}_2|^2 \\
&= |\tilde{E}_1|^2 + |\tilde{E}_2|^2 + 2|\tilde{E}_1|\cdot|\tilde{E}_2|\cos[\Delta\varphi(x, y)] \\
&= |A_1(x, y)|^2 + |A_2(x, y)|^2 \\
&\quad + 2|A_1(x, y)|\cdot|A_2(x, y)|\cos[\Delta\varphi(x, y)].
\end{aligned}\tag{10.3}$$

The function $I(x, y)$ is the interference pattern or interferogram that provides a map of the optical path difference that occurs for the object in the two states. Some issues exist that must be resolved before an accurate description of the object can be obtained. First, the cosine function is an even function which does not distinguish the sign of the phase value [i.e., $\cos(-\theta) = \cos(\theta)$]. Another problem is that the field amplitudes, A_1 and A_2, may vary over the aperture because of: non-uniform object illumination, non-uniform reflectivity of the object surface, speckle interference, or other degrading factors. To eliminate these problems and recover $\Delta\varphi$, several holographic interferometry exposure and interpretation methods have been developed such as phase shifting recording and phase unwrapping algorithms.

10.2 Methods for Forming Holographic Interference Patterns

A variety of techniques have been developed for generating holographic interference patterns for different measurement applications. The following sections provide an overview of two of the most commonly used methods for generating an interference pattern: double exposure holography and real-time interference monitoring. A discussion of interference patterns formed by deformation of an object surface, surface contouring, and variations in the refractive index of an object are also presented.

10.2.1 Double Exposure Holographic Interferometry

Double exposure holographic interferometry takes advantage of the multiplexing capability of holographic recording. In this process, two or more holograms are recorded of an object in different physical states. For instance, consider the recording of a holographic image of an object in its natural or relaxed state followed by the recording of the same object placed under compression by an outside force. The two holograms are sequentially recorded as shown in Figure 10.1 [1]. During the reconstruction process (Figure 10.2), the two image fields $\tilde{E}_1(x, y)$ and $\tilde{E}_2(x, y)$ interfere at the observation plane. If the amplitudes A_1 and A_2 are equal, the intensity at the observation plane becomes:

$$I(x, y) = 2I_{1,2}(x, y)\left[1 + \cos\left(\Delta\varphi(x, y)\right)\right], \tag{10.4}$$

where $I_{1,2}$ is the intensity of either object field, and $\Delta\varphi(x, y)$ is the phase difference between the two fields. When the phase difference is an even multiple of π, the fringe has maximum brightness and is a null when an odd multiple of π. The resulting fringe pattern indicates physical differences between the two states of the object that cause the optical phase of the recording beam to change. For a reflective object, this is typically a change in the depth or height of the object surface between the two recording states.

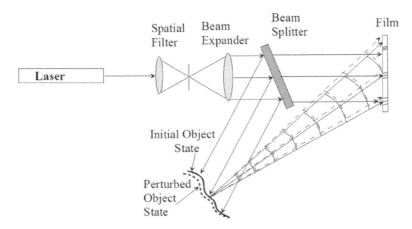

FIGURE 10.1 Double exposure holographic interferometry setup showing the fields recorded from the initial object state and the perturbed object state. Both holograms are recorded with the same reference beam.

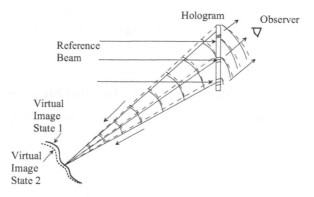

FIGURE 10.2 Reconstruction of the images from a double exposed hologram to form an interferogram.

10.2.2 Real-Time Holographic Interferometry

Another useful type of holographic interferometry compares changes to a surface or its refractive index in real-time and is useful for non-destructive testing applications. In this case, a hologram is made of the object in its initial or reference state. After the object is subjected to some physical change or wear, it is re-inserted in the exact initial position as it was when the hologram was recorded. Then both the hologram and the object are illuminated as shown in Figure 10.3. Light from the reconstructed hologram

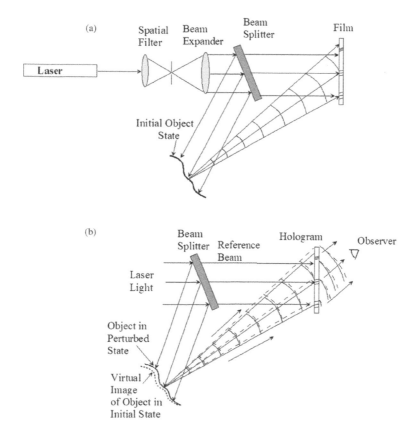

FIGURE 10.3 (a) Recording a hologram for use in a real-time interferometry setup and (b) Reconstruction of the hologram of the initial object superimposed with the field from the object in a perturbed state.

and the object interfere resulting in a fringe pattern that shows the differences between the two states of the object. However, returning the object to the exact original position in the optical system is difficult and places some restrictions on this method.

10.3 Measuring Surface Displacements with Holographic Interferometry

Holographic interferometry is useful for measuring small changes in the surface of an object that may result from applying a force, changing the temperature, or wear. The resulting change in optical phase due to surface change can be determined using the configuration illustrated in Figure 10.4. The vector $\vec{l}(x,y,z)$ shows the displacement of a surface point that occurs as the object is perturbed between the two states.

Assuming that the object is in air ($n = 1.00$), the difference in optical path length for rays propagating from a light source to the object in the two different states can be expressed as:

$$\delta = \vec{S}_1 + \vec{R}_1 - \left(\vec{S}_2 + \vec{R}_2\right)$$
$$= S_1\hat{s}_1 + R_1\hat{r}_1 - (S_2\hat{s}_2 + R_2\hat{r}_2),$$

(10.5)

where $\vec{S}_1$ and $\vec{S}_2$ are the distance vectors from the light source to the object in the two perturbed states, and $\vec{R}_1$ and $\vec{R}_2$ are the distance vectors from the object points to the observation point. Since the surface displacement is on the order of a few wavelengths of light while the distance from the source and the observer positions to the surface are typically 10's of centimeters, the unit vectors can be approximated as: $\hat{s}_1 \approx \hat{s}_2 \approx \hat{s}$; $\hat{r}_1 \approx \hat{r}_2 \approx \hat{r}$. Using this approximation, the optical path difference can be simplified to:

$$\delta = \left(\hat{r} - \hat{s}\right)\vec{l},$$

(10.6)

and the resulting phase difference is:

$$\Delta\delta\left(x,y\right) = \frac{2\pi}{\lambda}(\hat{r} - \hat{s})\vec{l}\left(x,y,z\right) = \vec{l}\left(x,y,z\right)\vec{V}_s.$$

(10.7)

$\vec{V}_s$ is often referred to as the sensitivity vector and indicates the direction in which the maximum phase difference occurs [4].

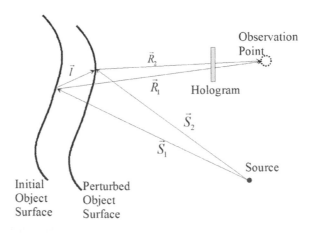

FIGURE 10.4 Geometry for measuring a transient surface displacement using holographic interferometry.

10.4 Surface Contouring with Holographic Interferometry

Another very useful application of holographic interferometry is measuring the surface profile or contour of an object. There are two common approaches to surface contouring using holographic interferometry: (a) the two-wavelength method and (b) the two-point source technique. The two approaches are similar in that they both vary the phase of the optical field and record holograms of the resulting interference patterns [5]. The surface profile is then reconstructed from the fringe patterns. The holograms can either be recorded in film type materials or by using a camera to form a digital hologram.

10.4.1 Multiple Wavelength Surface Contouring

The two-wavelength holographic contouring approach consists of recording two holograms on the same recording medium with two different wavelengths λ_a and λ_b [6]. The two wavelengths are typically obtained using a tunable laser [7]. A typical arrangement for recording the holograms is shown in Figure 10.5.

A beam splitter is used to set the direction of illumination parallel to the object depth direction (z-axis). The light not reflected by the beam splitter is used to form an off-axis reference beam at the hologram recording plane. Holograms are recorded at both wavelengths in either a film type material for a permanent analog recording or using a camera with the digital holography process as described in Chapter 7. The object distance for both holograms is z_o and the two wavelengths used differ by only small amounts. In this case, during reconstruction with one of the wavelengths (i.e., λ_a), two images will form. The image corresponding to λ_a will be highly resolved and have an axial position of z_o, while the image corresponding to the recording with λ_b will be displaced and somewhat aberrated due to the slightly different reconstruction wavelength. The axial position of the image from the hologram formed with λ_b can be determined with the paraxial image relation:

$$z_3 = \frac{z_p z_o z_r}{z_o z_r + \mu z_p z_r - \mu z_p z_o}$$

$$= \frac{z_r^2 z_o}{z_o z_r + \mu z_r^2 - \mu z_p z_o},$$

(4.28)

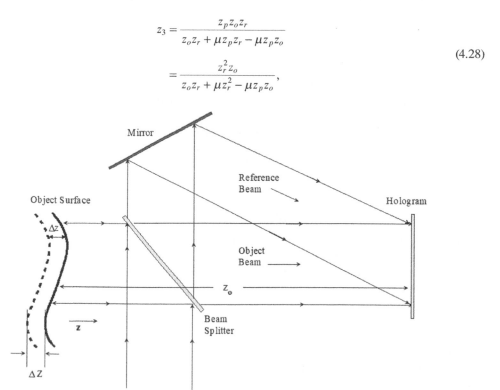

FIGURE 10.5 Holographic setup for measuring surface contours with holographic interferometry.

with $\mu = \lambda_a/\lambda_b$. Furthermore, since planar reference and reconstruction beams are used, $z_r = z_p = \infty$:

$$
\begin{aligned}
z_{3,b} &= \frac{z_p z_o z_r}{z_o z_r + \mu z_p z_r - \mu z_p z_o} \\
&= \frac{z_r^2 z_o}{z_o z_r + \mu z_r^2 - \mu z_r z_o} \\
&= \frac{z_o}{\mu} = \frac{\lambda_b z_o}{\lambda_a},
\end{aligned}
\tag{10.8}
$$

The difference in axial positions of the images formed by the two holograms reconstructed with λ_a are:

$$
\Delta z = z_a - z_b = z_o \frac{|\lambda_a - \lambda_b|}{\lambda_a}. \tag{10.9}
$$

The phase difference corresponding to the difference in optical path length to and from the object and to the hologram plane $(2\Delta z)$ for the two images is:

$$
\Delta\varphi(x,y) = \frac{2\pi}{\lambda_b} 2\Delta z = 4\pi \frac{|\lambda_a - \lambda_b|}{\lambda_a \lambda_b}. \tag{10.10}
$$

Since two images are simultaneously being formed, the phase difference defines a set of interference fringes and depends on the distance z_o to the hologram location. During the reconstruction, an image of the object forms with an overlay of contour fringes where a fringe indicates a surface with the same height. The difference in image height between two adjacent fringes is:

$$
\Delta Z = \frac{\lambda_a \lambda_b}{2|\lambda_a - \lambda_b|} = \frac{\Lambda}{2}, \tag{10.11}
$$

where the period between adjacent fringes, Λ, is sometimes referred to as an equivalent or synthetic wavelength. However, the accuracy of this method is limited since one of the holograms is reconstructed at a slightly different wavelength from the recording wavelength causing chromatic aberrations as discussed in Chapter 4.

One way around this shortcoming is to use digital holography. As in the previous method, the holograms are formed with two different wavelengths λ_a and λ_b, however, since the reconstruction is done computationally, the matched wavelength can be used for reconstructing holograms a and b. This eliminates the aberration and the inaccuracy introduced using a permanent recording material.

10.4.2 Contouring with Two Point Sources

Surface contouring with holographic interferometry can also be accomplished using two spatially separated point sources in the construction of a double exposure hologram [4]. This produces a phase difference similar with that produced with the two-wavelength process. The basic geometry is illustrated in Figure 10.6. Two point sources, P_1 and P_2, are displaced by a separation vector $\vec{l}$:

$$
\vec{l} = \vec{p}_2 - \vec{p}_1. \tag{10.12}
$$

The source points, P_1 and P_2, are only separated by a very small amount compared to the distance to the object. Therefore, on a geometrical optics scale the two points are essentially at the same location (P_s) with:

$$
\vec{p}_1 \approx \vec{p}_2 \approx \vec{p}. \tag{10.13}
$$

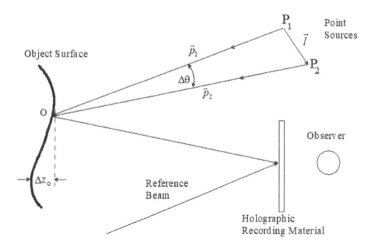

FIGURE 10.6 Holographic contouring of an object surface using two shifted point sources to forma a double exposure hologram.

With this approximation, the difference in optical path length from the source points to the object and the corresponding phase difference can be expressed as:

$$\delta = \vec{p} \cdot \vec{l} ,$$ (10.14)

and

$$\Delta \varphi = \frac{2\pi}{\lambda} \vec{p} \cdot \vec{l} .$$ (10.15)

The phase difference describes a set of interference fringes that overlap with the image of the object surface. The fringes form a group of hyperboloids with the two foci being the two point sources at P_1 and P_2. If the distance from the sources points to the object surface is large compared to the object height variations (Δz_o), then the fields propagating from these sources can be considered plane waves. In this case the separation between adjacent fringes indicates a change in height at the object surface given by:

$$\Delta d_{fringe} = \frac{\lambda}{2\sin(\Delta\theta / 2)},$$ (10.16)

where $\Delta\theta$ is the angle between the two point sources and the object point.

10.5 Holographic Interferometry Measurement of Refractive Index Variations

As shown in the previous sections, the geometrical length can be varied to measure either a surface deformation or contour. In a similar way, an optical phase difference will occur if the length is held constant and the refractive index is varied during the recording of a hologram [4]. If it is assumed that the length is along the z direction, the change in phase resulting from a variation in refractive index can be expressed as:

$$\Delta\varphi(x, y) = \frac{2\pi}{\lambda_o} \int_{d_1}^{d_2} [n(x, y, z) - n_i]dz,$$ (10.17)

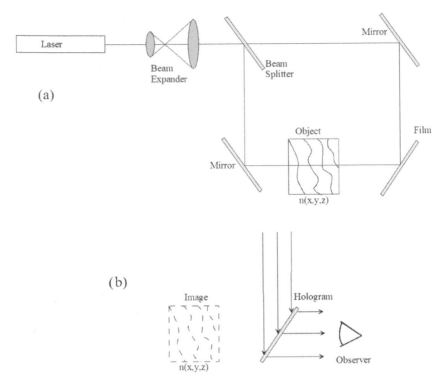

FIGURE 10.7 Construction (a) and reconstruction (b) of a hologram of a phase object that can be used to analyze the refractive index variation of the object.

where $n(x,y,z)$ is the spatially varying refractive index, and n_i is the initial index of the medium. A holographic setup for recording variations of a phase object is shown in Figure 10.7a and reconstruction in Figure 10.7b. The beam that passes through the object records the phase variation resulting from the refractive index distribution of the object. An interference profile can be formed by using double exposure or real-time holography as previously described. If the rays essentially propagate along the z-axis of the optical system, the resulting interference pattern from a double exposed hologram recording is given by:

$$I(x,y) = 2I_o(x,y)\left[1 + \cos\{\Delta\varphi(x,y)\}\right],\qquad(10.18)$$

where it is assumed that the reference and object beams have an equal intensity of I_o. Although it is possible to record phase variations in this manner, problems can arise due to scatter from dust particles and from aperture constraints on the field of view. If a diffuser is placed before the object during recording, the effects of these problems are essentially eliminated and is often used in practice.

10.6 Phase Shifting Holographic Interferometry

The process of changing or shifting the phase and recording a set of holograms can be used to eliminate ambiguities in the holographic interferometry technique [2,4,8]. Recording multiple holograms of the same fields with a different phase shift between each recording gives additional equations for the interference patterns to uniquely solve for the phase. The approach is similar to that described in Chapter 7 for on-axis digital holography.

As an example of this method, consider recording three interference patterns with a phase shift angle δ and 2δ between subsequent recordings:

$$I_1(x,y) = I_a(x,y) + I_b(x,y)\cos\left[\Delta\varphi(x,y)\right]$$

$$I_2(x,y) = I_a(x,y) + I_b(x,y)\cos\left[\Delta\varphi(x,y)+\delta\right] \quad (10.19)$$

$$I_3(x,y) = I_a(x,y) + I_b(x,y)\cos\left[\Delta\varphi(x,y)+2\delta\right].$$

The set of equations can be solved to determine $\Delta\varphi$ without ambiguity provided the phase shift angle δ is known. A known phase shift angle can be introduced by moving either the reference or object beam on a piezoelectric transducer or by having one of the beams propagate through a fiber stretcher [4]. If four phase shifted recordings are made instead of three, the phase difference can be recovered without knowing the precise value of δ provided that the value is a constant for all recordings. In this case:

$$I_1(x,y) = I_a(x,y) + I_b(x,y)\cos\left[\Delta\varphi(x,y)\right]$$

$$I_2(x,y) = I_a(x,y) + I_b(x,y)\cos\left[\Delta\varphi(x,y)+\delta\right]$$

$$I_3(x,y) = I_a(x,y) + I_b(x,y)\cos\left[\Delta\varphi(x,y)+2\delta\right] \quad (10.20)$$

$$I_4(x,y) = I_a(x,y) + I_b(x,y)\cos\left[\Delta\varphi(x,y)+3\delta\right].$$

The value for the phase difference can then be found by eliminating δ, I_a, and I_b from the set of equations resulting in:

$$\Delta\varphi = \tan^{-1}\left\{\frac{\sqrt{I_1+I_2+I_3}\cdot\sqrt{3I_2-3I_3-I_1+I_4}}{I_2+I_3-I_1-I_4}\right\}, \quad (10.21)$$

where the phase difference and intensity values are functions of x and y. Additional information on phase shifting digital holography and interferometry can be found in references [9–12].

10.7 Analysis of Holographic Interference Patterns

Once the phase difference is measured, it is converted into a spatial map of the optical path difference from the object under investigation [4]. However, an ambiguity exists in the cosine function that appears in the intensity equation for the interference function (Equation 10.3), namely, that:

$$\cos(\Delta\varphi) = \cos(\Delta\varphi + m2\pi), \quad (10.22)$$

where m is an integer. As shown in the previous section on phase shifting, the phase difference is computed using the $\tan^{-1}$ function. Therefore, phase discontinuities occur at values of $\pm\pi$ and must be resolved in order to determine the actual phase difference function across the area under analysis. To address this problem, a number of algorithms have been developed to perform the unwrapping operation [13,14]. One approach is to sample the phase of adjacent pixels (m) and take the difference:

$$\Delta\varphi(m+1) - \Delta\varphi(m). \quad (10.23)$$

(In this expression, a one-dimensional phase difference function is used to simplify the discussion, but can readily be extended to two dimensions.) If the phase difference at the $(m + 1)$th pixel is $< -\pi$, then 2π is added to phase difference value for pixels greater than the $(m + 1)$th pixel. If the phase difference is $> +\pi$, then the phase difference value is reduced by 2π for pixels greater than $(m + 1)$. If the value at the sampling

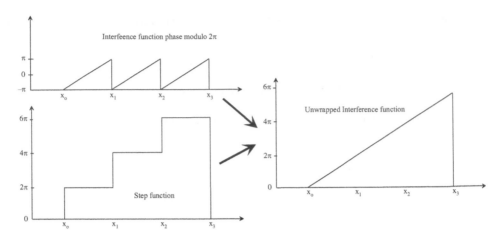

FIGURE 10.8   Illustration of the phase unwrapping process.

point is neither $<-\pi$ or $>+\pi$, then the value is not changed at that location. The sampling process can be started at any point in the interference pattern by making phase difference comparisons to the right and left of the starting point. Areas that are occluded due to holes in the object or masking must have additional complexity built into the algorithm to scan along the edge of those regions. A graphical illustration of the process is shown in Figure 10.8 and several other examples of algorithms are given in references [13] and [14].

PROBLEMS

1. What potential issues might arise when making a double exposure hologram in a photopolymer material like the Bayer/Covestro material described in Chapter 8? How should the exposure energies be adjusted to equalize the efficiency of the recordings?

2. Consider the interferometer setup to illuminate an opaque object as shown in Figure 10.9 below. If $\lambda = 532$ nm, $d_1 = 1$ mm, and $d_2 = 0.001$ mm, determine the phase difference that occurs in the object wave at the point x_1. Write an expression for the object field amplitude at the hologram plane assuming z_{12} satisfies the near-field diffraction propagation requirement. If the object reflects 20% of the incident light, what transmission-reflection beam ratio beam splitter will give the highest contrast fringes at the hologram?

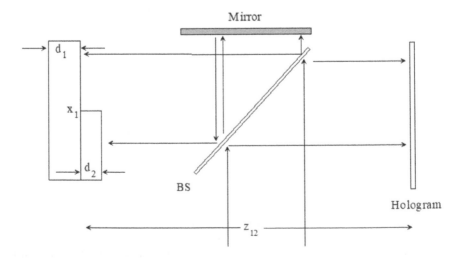

FIGURE 10.9   Configuration for Problem 2.

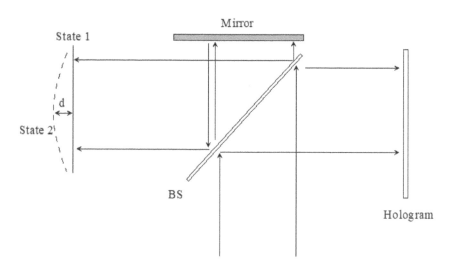

FIGURE 10.10 Double exposure hologram setup for Problem 3.

3. It is desired to form a double exposure hologram with 532 nm light of a reflective membrane as shown in Figure 10.10. In the first state, the membrane is flat, and in the second, the membrane is deflected by a distance d at the center of the aperture $D = 2$ cm forming a spherical surface. What deflection d will result in five ring fringes within the aperture. Assume that each fringe corresponds to a 2π phase difference.

4. Consider a reflective surface that is contoured using the two-wavelength holographic interferometry technique. A tunable laser diode is used with a nominal wavelength of 650 nm. What wavelength tuning range is required to scan a height of 1.5 μm? What is the corresponding *synthetic wavelength*?

REFERENCES

1. T. Kreis, *Handbook of Holographic Interferometry*, Wiley-VCH, Weinheim, Germany (2005).
2. V. Toal, *Introduction to Holography*, Chapter 10, CRC Press, Boca Raton, FL (2011).
3. P. Hariharan, *Optical Holography*, Chapter 14, Cambridge University Press, Cambridge, UK (1984).
4. U. Schnars and W. Jueptner, *Digital Holography*, Springer-Verlag, Berlin, Germany (2005).
5. B. Hildebrand and K. Haines, "Multiple-wavelength and multiple–source holography applied to contour generation," *J. Opt. Soc. Am.*, Vol. 57, 155–162 (1967).
6. A. A. Friesem and U. Levy, "Fringe formation in two-wavelength contour holography,"*Appl. Opt.*, Vol. 15, 3009–3020 (1976).
7. A. Wada, M. Kato, and Y. Ishii, "Multiple-wavelength digital holographic interferometry using tunable laser diodes," *Appl. Opt.*, Vol. 47, 2053–2060 (2008).
8. K. Creath and J. C. Wyant, "Direct phase measurement of aspheric surface contours," *SPIE Optical Manufacturing, Testing, and Aspheric Optics*, Vol. 645, 101–106 (1986).
9. I. Yamaguchi and T. Zhang, "Phase shifting digital holography," *Opt. Lett.*, Vol. 22, pp. 1268–1270 (1997).
10. S. Lai, B. King, and M. A. Neifeld, "Wavefront reconstruction by means of phase-shifting digital in-line holography," *Opt. Comm.*, Vol. 169, pp. 37–43 (2000).
11. I. Yamaguchi, J. Kato, S. Ohta, and J. Mizuno, "Image formation in phase-shifting digital holography and applications to microscopy," *Appl. Opt.*, Vol. 40, 6177–6186 (2001).
12. I. Yamaguchi, T. Matsumura, and J. Kato, "Phase shifting color digital holography," *Opt. Lett.*, Vol. 27, pp. 1108–1110 (2002).
13. J. Munoz, G. Paez, and M. Strojnik, "Two-dimensional phase unwrapping of subsampled phase-shifted interferograms," *J. Mod. Opt.*, Vol. 51, 49–63 (2004).
14. K. Itoh, "Analysis of the phase unwrapping algorithm," *Appl. Opt.*, Vol. 21, 2470 (1982).

11

Holographic Optical Elements and
Instrument Applications

11.1 Introduction

As described in earlier chapters, holography is a method of recording the wavefront of an optical field. This property is important for the control, transformation, and manipulation of an incident optical beam and increases options for optical design. Holograms can be formed in either transmission or reflection and with on- or off-axis geometries. In addition, the diffraction efficiency characteristics of holographic gratings can be tailored to control the spatial and spectral aspects of the fields after interacting with the hologram. The polarization properties of the diffracted light can also be controlled by varying the properties of the holographic grating allowing the formation of polarization elements. Finally, more than one optical function or transformation can be multiplexed in the same recording material allowing several bulk optical elements to be replaced with a single thin film element. These attributes make holograms ideal candidates for use as optical elements that offer considerable versatility and cost savings in optical systems ranging from medical imaging to solar energy conversion. This chapter will investigate some of the important design and performance characteristics of holographic optical elements and show how they can be used in a variety of applications.

11.2 Holographic Lenses

Holographic lenses have been the subject of considerable study [1–10]. This is due to in part to the potential for making physically thin elements with high optical power. As a result holographic lenses have been used in a number of applications including signal processing [11], head-up displays [12–14], laser collimators [15], and other applications that use a narrow band reconstruction source. However, holographic lenses can be used with broadband reconstruction illumination for certain imaging and for non-imaging applications such as solar energy concentrators [16]. When used with broadband illumination applications, the dispersive properties of the holograms must also be integrated into the design procedure. In this section, a methodology is presented that integrates the important factors that should be included in the design of a holographic lens. For this discussion, it is assumed that the lens is formed as a volume hologram to achieve high diffraction efficiency in a single diffraction order.

The main steps in the design and analysis of a holographic lens include:

1. Ray analysis of the diffracted light
2. Diffraction efficiency analysis across the lens aperture
3. Evaluation of the polarization properties of the diffracted light
4. Consideration of spectral bandwidth properties of the illumination source used for reconstruction and the dispersion properties of the hologram.

Several complicating factors also enter into the design such as:

1. The variation in the grating vector across the hologram aperture
2. Differences between the construction and reconstruction wavelength

3. Shrinkage or swelling of the thickness of recording material
4. Change in the average refractive index of the recording material.

In order to illustrate the design process, consider a holographic lens used as a solar concentrator. In this case, the hologram must diffract the incident broadband, randomly polarized solar illumination to one or more solar cells with high diffraction efficiency. One approach for realizing the lens is illustrated in Figure 11.1. In this case, the holographic lens diffracts and focuses light to the boundary between a wide bandgap (responsive to short wavelengths) and a narrow bandgap (responsive to long wavelengths) photovoltaic cell. This design uses the dispersive property of the lens to separate the spectrum between the two photovoltaic (PV) cells. This is accomplished by designing the hologram with a "transition" wavelength with a value set between the spectral response ranges for the two PV cells. Light at this wavelength is focused to the boundary between the two cells when the holographic lens is illuminated with normally incident light. As shown in Figure 11.1, light incident at other wavelengths will focus along an arc. According to the grating equation (Equation 3.9), longer wavelength light will come to focus above the PV plane and then expand on the narrow bandgap PV cell, while shorter wavelength light focuses behind the PV plane. This effectively separates the spectral bands at the PV plane and directs them to PV cells that have a higher response to those spectral components.

The geometry for the lens must also be laid out so that high diffraction efficiency can be achieved across the entire lens aperture. In order to accomplish this, the hologram "Q" parameter (Equation 5.109) should exceed a value of about 10 where the "Q" parameter is given by:

$$Q = \frac{2\pi\lambda d}{n\Lambda^2} \geq 10, \tag{5.109}$$

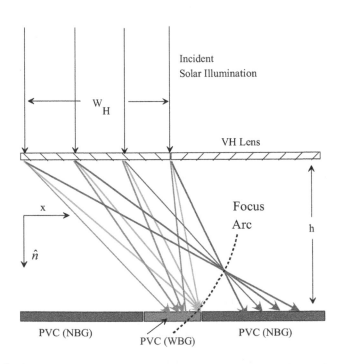

FIGURE 11.1 (See color insert.) A volume holographic lens used in a spectrum splitting solar energy conversion system. A wide bandgap (WBG) photovoltaic cell (PVC) is surrounded by a narrow bandgap (NBG) PVC. The "green" rays indicate the light paths for illumination with the "transition" wavelength. (From Vorndran, S. et al., *SPIE Proc.*, 9937, 99370K1:8, 2016. With permission from the SPIE.)

with λ the reconstruction wavelength, d the grating thickness, n the average refractive index of the grating, and Λ the volumetric grating period. As shown in Chapter 5, the volumetric grating period is determined from the K-vector relation:

$$\vec{K}(x) = \vec{k}_1(x) - \vec{k}_2(x)$$

$$\left|\vec{K}(x)\right| = \frac{2\pi}{\Lambda(x)}, \tag{11.1}$$

where $\vec{k}_1(x)$ and $\vec{k}_2(x)$ are the propagation vectors for forming the hologram at a particular position on the hologram aperture. Note that for simplicity in this example, the grating vector is assumed to vary only in the x-direction, however, it can also be extended to the orthogonal direction if necessary. For the lens illustrated in Figure 11.1, the hologram width (W_H), position and widths of the PV cells, and the hologram thickness (d) are arranged so that the ray closest to the location separating the wide and narrow bandgap cells satisfies Equation 5.109. Other positions on the hologram aperture will automatically satisfy this condition because of their larger inter-beam angle and smaller grating period.

The resulting diffraction efficiency of the lens can be evaluated using the approximate coupled wave relation for a phase type transmission hologram (Equation 5.70):

$$\eta = \frac{\sin^2\left(v^2 + \xi^2\right)^{1/2}}{\left(1 + \xi^2/v^2\right)}, \tag{5.70}$$

where the parameters v and ξ defined earlier in Chapter 5 are:

$$v = \frac{\pi n_1 d}{\lambda c_r c_s}; \xi = \frac{\vartheta d}{2c_s}. \tag{5.69}$$

The spectral bandwidth, $\Delta\lambda$, can be approximated using the detuning parameter that causes the diffraction efficiency given by Equation 5.70 to change from a peak value at $\xi = 0$ to the first null at $\xi \approx 2.7$. Using this condition, the full spectral bandwidth can then be approximated as:

$$\Delta\lambda \approx 5.4n\frac{\Lambda(x)^2}{d}\cos\theta_o, \tag{11.2}$$

where θ_o is the Bragg angle, n is the average refractive index of the hologram, d is the hologram thickness, and $\Lambda(x)$ is the grating period at a particular location x within the aperture. Since the period generally varies across the hologram aperture, the spectral bandwidth also varies. However, materials like dichromated gelatin can achieve a large refractive index modulation and allows the thickness to be thin resulting in a large spectral bandwidth, while still maintaining high diffraction efficiency.

For spectrum splitting applications, the system tracks the sun so that the incident direct illumination remains normal to the hologram surface. However, to maximize the solar conversion energy yield, diffuse light must also be collected. A volume holographic element allows for large off normal, non-Bragg incident angle light to pass through the hologram and illuminate the PV cells. This characteristic allows conversion of diffuse illumination and increases the total energy yield of the system.

A final consideration in the lens design is the polarization dependence of the diffracted beam across the hologram aperture. When the hologram is illuminated with randomly polarized light as in this case with sunlight, the incident polarization can be considered to have equal amounts of light in and perpendicular to the plane of incidence. As shown in Chapter 5, the coupling coefficient is a function of the Bragg and grating slant angles:

$$\kappa_\parallel = -\kappa_\perp \cos 2\left(\theta_o - \varphi\right), \tag{11.3}$$

where $\kappa_\perp$ and $\kappa_\parallel$ refer, respectively, to the coupling coefficients for light polarized perpendicular and parallel to the plane of incidence, θ_o to the Bragg angle, and φ to the angle of the grating vector relative to the z or propagation axis.

As discussed previously, the angle and magnitude of the grating vector for a holographic lens can vary considerably across the aperture. The resulting diffraction efficiency for perpendicular and parallel field components can be made more similar by over exposing the hologram for the perpendicular case. This process will result in some loss in efficiency for the perpendicular component, but can provide an overall improvement in total diffraction efficiency.

An example of a volume holographic cylindrical lens formed in a 20 μm thick film of dichromated gelatin is illustrated in Figure 11.2. The hologram was exposed using a 532 nm diode pumped laser with the geometry shown in the figure. The reference beam angle was offset to compensate for material swelling during processing and allow reconstruction at normal incidence with the 663 nm transition wavelength. The two PV cell materials were indium gallium phosphide (InGaP) for the wide bandgap material (688 nm bandgap wavelength) and silicon (Si) for the narrow bandgap material (1100 nm bandgap wavelength). The figure also shows the diffraction efficiency across the aperture after using the method described above. For this example, $\kappa_\parallel \approx 0.82\kappa_\perp$ at the right edge and $\kappa_\parallel \approx 0.53\kappa_\perp$ at the left edge and contributed to variation in the diffraction efficiency. This can be further compensated by adjusting the hologram exposure energy. Figure 11.3 shows the spectrum diffracted by the hologram to the wide and narrow bandgap PV cells. The figure also shows the raytrace through adjacent volume holographic lenses to the PV cells at the transition wavelength and at wavelengths above and below the transition wavelength. The techniques shown in this example can be adapted to a variety of holographic lens designs for both imaging and non-imaging applications.

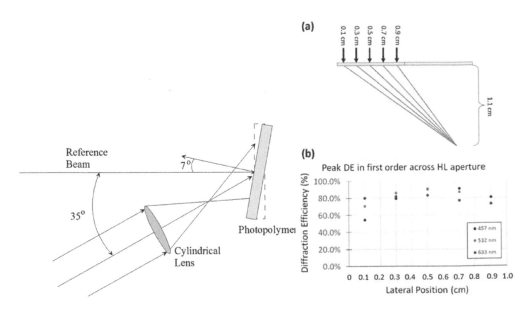

FIGURE 11.2 The construction setup for an off-axis holographic lens (a) and the corresponding experimental diffraction efficiency (b). The hologram was formed in a 20 μm thick DCG film at 532 nm and the reconstructed diffraction efficiency was measured at three different wavelengths (457, 532, and 633 nm). (From Vorndran, S. et al., *Appl. Opt.*, 55, 7522–7529, 2016. With permission of the Optical Society of America.)

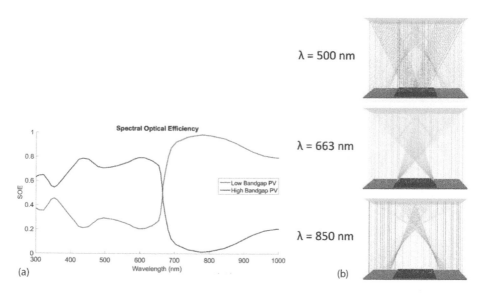

FIGURE 11.3 (See color insert.)   The spectral filtering properties (a) and raytrace (b) through the spectrum splitting volume holographic lens described in the text for a combination of silicon and indium gallium phosphide photovoltaic cells. (From Vorndran, S. et al., *Appl. Opt.*, 55, 7522–7529, 2016. With permission of the Optical Society of America.)

11.3 Holographic Spectral Filters

Spectral filters are important optical components used to control the spectral bandwidth in a variety of optical systems. Low or high pass filters (Figure 11.4a and b) are often needed for spectroscopy applications to block reflected light from high power laser excitation sources and not over power the desired weak signals. Filters of this type are used in both Raman and fluorescence spectroscopy systems where the desired signals can be 3–4 orders of magnitude lower than the excitation power. Low and high pass spectral filters as well as bandpass filters (Figure 11.4c) are also used to control the path of different spectral bands filtered from sunlight in spectrum splitting photovoltaic energy systems [17,18]. Another type of spectral filter is a notch or band rejection filter (Figure 11.4d) that reflects a very narrow spectral band and transmits light at other wavelengths with very little loss. Notch filters are often used in laser systems and astronomy [19]. Some desirable features of spectral filters are high reflection efficiency, sharp transitions between high and low transmittance wavelengths, insensitivity to polarization, and good optical quality.

Reflective spectral filters are typically formed by depositing alternating layers of high and low refractive index materials on glass substrates [20,21]. The desired spectral properties are achieved by using the reflective and interference effects produced by index differences between layers within the stack. This material system is similar to the modulation profile obtained in a volume reflection hologram. Both interference filters and reflection holograms can be analyzed using coupled wave models [22,23]. Reflection holographic filters can be formed with the grating vector ($\bar{K}$) either normal (as in an interference filter) or slanted with respect to the film surface (Figure 11.5). The slanted grating vector of a hologram can be used as a design parameter to diffract a specific wavelength or wavelength range at a particular angle. This is more difficult to achieve with interference filters formed by evaporating materials as the entire process must be adjusted for the off-axis angle of incidence.

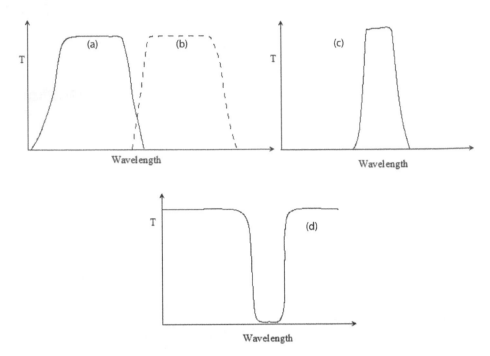

FIGURE 11.4 Different types of optical filters: (a) low pass, (b) high pass, (c) band pass, and (d) notch filter characteristics.

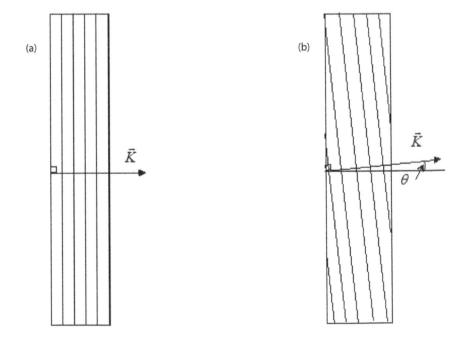

FIGURE 11.5 Reflection holographic filters with the grating vector, $\vec{K}$, normal (a) and slanted (b) with respect to the grating surface.

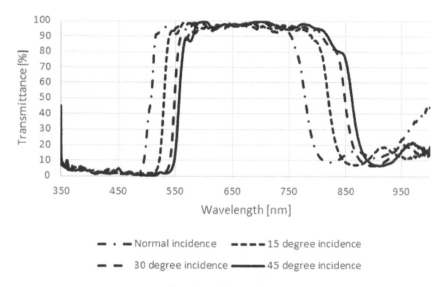

FIGURE 11.6 Measured spectral transmittance if a dichroic bandpass filter as a function of angle of incidence. The filter is designed for a 45° angle of incidence. At lower angles of incidence, the band edge shifts to shorter wavelengths at both the upper and lower band edge. (From Zhang, D. et al., *J. Phot. Energy*, 4, 2014. With permission from the SPIE.)

An interesting characteristic of reflection type interference filters is the "blue shift" of the band edge wavelength of the reflected spectrum when the angle of incidence increases. This shift can be determined using the Bragg condition and is given by:

$$\lambda(\theta) = 2\Lambda\sqrt{n^2 - \sin^2\theta}, \tag{11.4}$$

where n is the average refractive index of the medium, θ is the angle of incidence in air, and Λ is the grating period. Figure 11.6 shows the blue shift in the transmittance for a dichroic reflection filter (Thorlabs DMLP505L) when the angle of incidence is changed from the design angle of 45° to smaller angles of incidence. The blue shift also occurs for holographic reflection filters when reconstructed at angles that differ from the designed angle of incidence. This effect can have a significant impact on optical system performance and should be considered in systems that are sensitive to the band-edge wavelength such as fluorescence excitation and spectrum splitting solar energy systems.

Volume holographic reflection filters can be fabricated using one of the geometries shown in Figure 11.7. The first configuration (Figure 11.7a) shows an interferometer with two counter propagating construction beams. The grating vector in this case lies along the optical axis, however, slanted gratings can also be constructed by having one or both beams at an angle to the axis. The Lippmann geometry shown in Figure 11.7b incorporates a high reflectance mirror index matched to the back surface of the hologram substrate. The prism used in the hologram construction shown in Figure 11.7c allows light to be coupled into the film/substrate at angles that exceed the total internal reflection (TIR) angle of the film. The hologram substrate is index matched to the prism with a fluid, and the TIR beam reflected from the film surface forms the second beam for the interference recording. The angle of incidence in this configuration can be adjusted to change the peak diffraction efficiency wavelength. The ability to control the peak diffracted wavelength and the high stability of the prism configuration makes this a very desirable arrangement for forming reflection hologram filters.

Adjusting the exposure density and thickness of the hologram can be used to control the spectral bandwidth of the reflection hologram. If the material is exposed within its linear response range, the refractive index will be modulated with a sinusoidal function. However, when the material is exposed beyond the saturation level of the recording material, the refractive index modulation will have other functional forms and result in different diffraction efficiency profiles. For example, consider the case

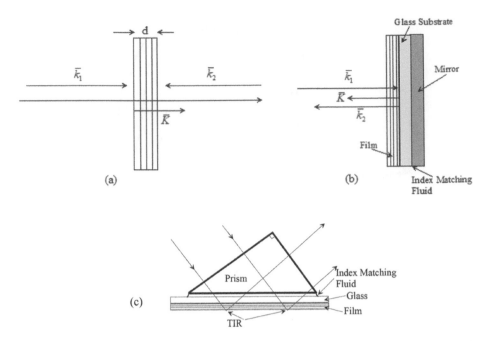

(a)

(b)

(c)

FIGURE 11.7 Configurations for forming spectral filters with reflection volume holograms. (a) Standard interferometer, (b) Lipmann geometry, and (c) Prism coupler and TIR construction.

shown in Figure 11.8 where the refractive index modulation varies from the linear to non-linear range. The dashed curve in this figure shows the diffraction efficiency for a 10 μm thick reflection hologram with a refractive index modulation amplitude of 0.045. If the amplitude is increased to 0.07, the diffraction efficiency has significantly broader reflected spectral bandwidth response (dash-dot curve in Figure 11.8). This change in refractive index modulation is achieved by increasing the exposure density during hologram recording.

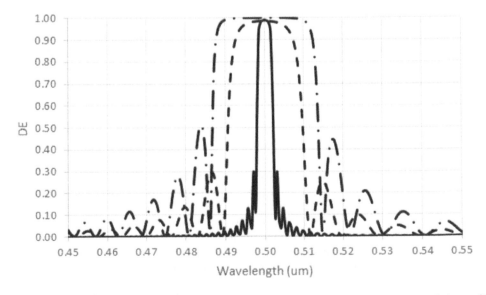

FIGURE 11.8 Diffraction efficiency of a reflection hologram with a sinusoidal modulated refractive index profile and constant grating period throughout the thickness. Solid line shows a hologram with $d = 50$ μm and $n_1 = 0.010$; dashed curve has $d = 10$ μm and $n_1 = 0.045$; and dash-dot curve a hologram with $d = 10$ μm and $n_1 = 0.07$.

The spectral bandwidth can also be varied by controlling the hologram thickness. The solid curve in Figure 11.8 shows the diffraction efficiency for a 50 μm thick hologram and a refractive index modulation amplitude of 0.01. This results in near 100% diffraction efficiency and a much narrower spectral bandwidth. The plots shown in Figure 11.8 were obtained using approximate coupled wave analysis as discussed in Chapter 5 and is a good first step for analyzing holographic filter designs.

Further control of the holographic filter spectral characteristics can be achieved by controlling the holographic material composition and development process. This is especially true for materials like dichromated gelatin (DCG) which can be formulated and modified to suit a variety of requirements for different applications. For instance, it is possible to make both the amplitude of the refractive index modulation and the grating period vary as a function of the depth to produce the following functions:

$$n_1(z) = n_1(d) - \Delta n_1 (1 - z/d)^a$$

$$\Lambda(z) = \Lambda(0) - \Delta\Lambda(z/d)^b ,$$

(11.5)

where d is the total hologram thickness, z is the distance from the hologram/substrate interface, and a and b are coefficients determined from experimental data for the recording material. The variation of index modulation and grating period with depth obtained in this manner can be used to increase the spectral bandwidth by greater amounts than can be obtained by simply increasing the exposure density and decreasing the thickness. Figure 11.9 shows the spectral transmission bandwidth for a 17 μm thick DCG film with $n_1(d) = 0.125$, $\Delta n_1 = 0.0437$; $\Lambda(0) = 0.41$ μm; $\Delta\Lambda = 0.0193$ μm; $a = 3$; and $b = 2$ [24]. The hologram was formed using the prism construction method shown in Figure 11.7c and has a full width half maximum (FWHM) spectral bandwidth of over 200 nm.

The formation of more complex hologram grating profiles by controlling the material and processing properties can also be used to reduce the diffraction efficiency of the sidelobes that occurs with uniform

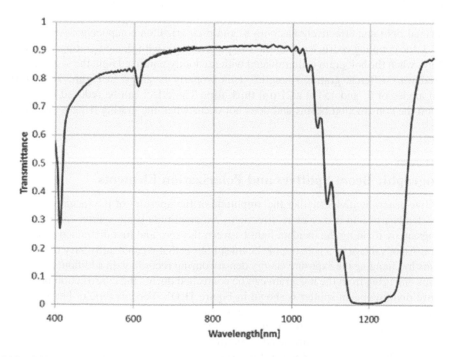

FIGURE 11.9 Diffraction efficiency for a reflection volume hologram formed in a 17 μm thick DCG film with the grating period and refractive index modulation amplitude varied with depth into the hologram to increase the spectral bandwidth and to decrease the sidelobes. (From Zhang, D. et al., *Opt. Exp.*, 20, 14260–14271, 2012. With permission from the Optical Society of America.)

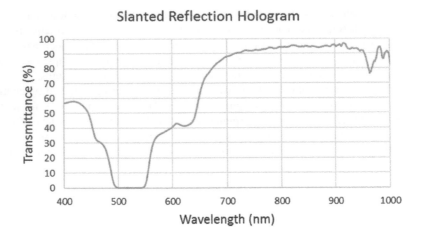

FIGURE 11.10 Diffraction efficiency for a slanted reflection hologram formed with 0° and 45° construction beams in a 21 μm thick film for operation at 550 nm. (Courtesy of Yuechen Wu, University of Arizona PhD graduate student, 2018.)

sinusoidal modulated gratings. This effectively extends the useful bandwidth range and makes the efficiency profile more uniform. However, one consequence of non-linear modulation is the occurrence of higher diffracted orders. In Figure 11.9 the "dips" in the transmittance functions near 610 and 410 nm indicate the presence of second and third diffraction orders corresponding to the primary order near 1200 nm. The magnitude of these orders can be reduced somewhat by varying the exposure energy. In many cases, the second and third order wavelengths are significantly displaced from the spectrum of interest so as not to be a major problem.

Another issue that can occur with slanted reflection holograms is the formation of a "side-hump" in the transmittance characteristic of the filter. This results when the hologram is reconstructed with randomly polarized light and effectively has both s- and p-polarization components. As shown earlier in Equation 11.3, the coupling coefficients for the two polarizations with slanted grating vectors are different. Therefore, when the hologram is illuminated with randomly polarized light the s- and p-components are diffracted with different grating strengths. This property is shown in Figure 11.10, formed with construction angles of 0° and 45° in a 21 μm thick film. The effect can be reduced by decreasing the angle between the construction beams and does not occur when the grating fringes are parallel to the film surface.

11.4 Holographic Beam Splitters and Polarization Elements

A beam splitter generally divides either the amplitude or the aperture of the incident beam and separates the divided beams into two or more directions. An amplitude beam splitter can be formed with a volume hologram by dividing the incident light between the zero and first diffracted order as shown in Figure 11.11a. An advantage of a holographic beam splitter is that it can readily vary the power between the two beams by changing the exposure energy density during recording. In addition, the angle between the two beams emerging from the hologram can be controlled during the construction process.

An aperture dividing beam splitter is shown in Figure 11.11b. For this type of beam splitter, two or more different holograms are formed within the aperture of the hologram. Ideally, the efficiency of each hologram should be 100%, but if this is difficult to achieve, the zero or non-diffracted order can be directed away from the desired diffracted beam directions. Holographic aperture beam splitters are useful in digital holographic microscope systems (Figure 11.12) [25,26]. The limited spatial resolution of cameras used in digital holographic microscopes (~2 μm) constrains the inter-beam angle (θ) between the object and reference beams. Setting up an optical system with conventional optical components to achieve this is difficult due to their physical size and mounting requirements. However, the small

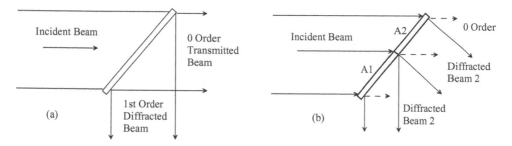

FIGURE 11.11 (a) Configuration for a holographic amplitude dividing beam splitter and (b) a configuration for an aperture dividing holographic beam splitter.

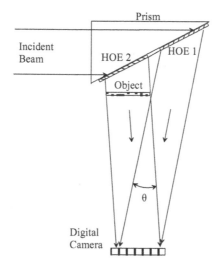

FIGURE 11.12 An aperture holographic beam splitter used to form the object and reference beams for a digital holographic microscope. (From Rostykus, M. et al., *Opt. Exp.*, 25, 4438–4445, 2017. With permission from the Optical Society of America.)

inter-beam angle can readily be achieved with an aperture dividing beam splitter on a single piece of holographic film. The prism in this configuration is index matched to the hologram and reduces the Fresnel reflection losses that otherwise would occur at the relatively large angles of incidence to the holograms.

It is also possible to use volume holograms to control the polarization of an optical beam. As shown in Chapter 5, the diffraction efficiency of a volume hologram illuminated with s- and p-polarized light can be written as:

$$\eta_s = \sin^2(C)$$

$$\eta_p = \sin^2[C \cdot \cos(2\theta_o)].$$

(5.108)

Therefore, if the grating strength parameter $C = \pi \rightarrow \eta_s = 0$ and s-polarized light is transmitted by the hologram without diffraction. The diffraction efficiency of the same hologram for p-polarized light is a maximum when $C \cdot \cos(2\theta_o) = \pi/2$ and occurs when $\cos 2\theta_o = 1/2 \rightarrow \theta_o = 30°$ inside the grating medium. This effect can be used to make a polarization beam splitter where s-polarized light is transmitted through

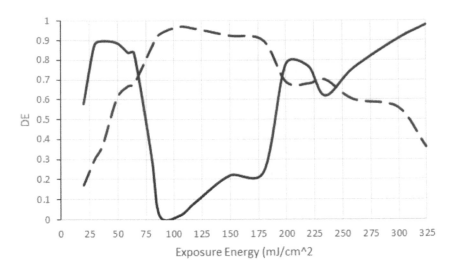

FIGURE 11.13 Experimental holographic polarization beam splitter formed in DCG. The solid curve indicates the efficiency of s-polarized light, and the dashed line shows the diffraction efficiency of p-polarized light.

the hologram and p-polarized light is diffracted at an angle away from the zero order. An example of a polarizing beam splitter formed in dichromated gelatin is shown in Figure 11.13. For this plot, a series of holograms with different exposure values were formed in 8.5 μm thick samples of dichromated gelatin emulsions. The reconstruction angles are at $\pm 30°$ to the surface normal inside the film with a reconstruction wavelength of 632.8 nm. In this figure, the solid curve shows the diffraction efficiency for s-polarized light, and the dashed curve is for p-polarized light. The s-polarized light reaches a maximum at an exposure near 35 mJ/cm² and then starts to decrease in a manner consistent with Equation 5.108 for η_s. The s-polarized light passes through a minimum with an exposure energy density near 85 mJ/cm². At this exposure energy density, the p-polarized reaches a maximum diffraction efficiency value exceeding 90%. Therefore, a hologram fabricated with these characteristics can be used as a polarization beam splitter with s-polarized light in the zero order and p-polarized light in the first diffraction order.

Other polarization elements such as retardation plates can also be fabricated with volume holograms [27–29]. When the grating period is reduced, the angle of the diffracted beam increases for a given wavelength. Eventually, the diffracted beam angle will exceed 90° and no longer propagates. At this point, the holographic optical element acts as an effective medium and an incident field with polarization along the grating planes experiences a different refractive index than a field polarized perpendicular to the grating planes. This results in form birefringence [30,31]. The dielectric permittivity in the two orthogonal directions can be described with effective medium theory as [29,31]:

$$\varepsilon_{\parallel} = \varepsilon_o$$

$$\varepsilon_{\perp} = \sqrt{\varepsilon_o^2 - \varepsilon_1^2} \, , \tag{11.6}$$

where ε_o is the average permittivity of the holographic film, ε_1 is the permittivity modulation amplitude, and $\varepsilon_{\parallel}$ and $\varepsilon_{\perp}$ are, respectively, the permittivity values in a direction perpendicular to and parallel to the grating K-vector. The resulting form birefringence Δn is:

$$\Delta n = n_{\parallel} - n_{\perp} = \sqrt{\varepsilon_{\parallel}} - \sqrt{\varepsilon_{\perp}}$$

$$= \sqrt{\varepsilon_o} - \sqrt[4]{\varepsilon_o^2 - \varepsilon_1^2} \tag{11.7}$$

$$\cong \frac{1}{4} \frac{\varepsilon_1^2}{n_o^3} = \frac{n_1^2}{n_o} ,$$

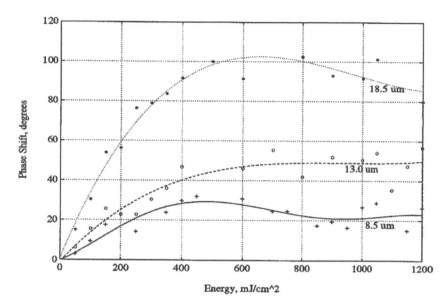

FIGURE 11.14 Plots of the phase shift at a wavelength of 632.8 nm in dichromated gelatin film at different exposure levels and thickness values. (From Kim, T.J., *Opt. Lett.*, 20, 2030–2032, 1995. With permission from the Optical Society of America.)

where $n_o, n_1, n_\parallel n_\perp$ are the refractive index parameters corresponding to the permittivity values. The resulting birefringence can be used to produce a phase difference between the two polarization components. When the hologram is illuminated with linearly polarized light with the polarization direction at 45° to the grating fringes, the resulting phase difference becomes:

$$\Delta\varphi = \frac{2\pi\,\Delta n d}{\lambda}, \tag{11.8}$$

where d is the hologram thickness, and λ is the wavelength. Figure 11.14 shows the phase variation as a function of exposure for different hologram thickness values [29]. For these plots, the holograms have a grating period $\Lambda = 0.286\,\mu m$ and are illuminated with 632.8 nm wavelength light. The results show that a quarter wave plate can readily be formed in dichromated gelatin film with moderate thickness (18.5 μm) and exposure values (~400 mJ/cm²).

11.5 Folded Holographic Optical Elements

Folding or guiding an incident optical beam within an optical substrate has many applications in optical communications, interconnects, displays, and illuminators. The basic arrangement for a component of this type is illustrated in Figure 11.15. In this case, an incident beam is diffracted by a transmission hologram at an angle that exceeds the critical angle of the substrate and undergoes total internal reflection until it reaches an output coupling hologram. It is not a true waveguide as the beam propagation is described by geometrical optics rather than physical or wave optics. For instance, an image can be transferred in the substrate mode system, but not in a waveguide. (Note that a fiber image guide can transfer an image, but for that device an array of fibers is used with each fiber sampling the light distribution.)

Many types of substrate mode holograms are possible [32–34]. For instance, the input hologram can be multiplexed and diffract incident light into different directions within the substrate [32]. A hologram of this type is designed by first assuming a specific polarization state of the reconstruction beam. Then the grating strength parameters for the different holograms, v_m, are divided into two sets,

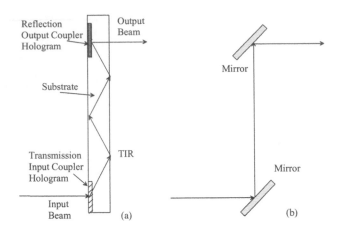

FIGURE 11.15   A folded or substrate mode hologram (a) and an equivalent folded system with mirrors (b).

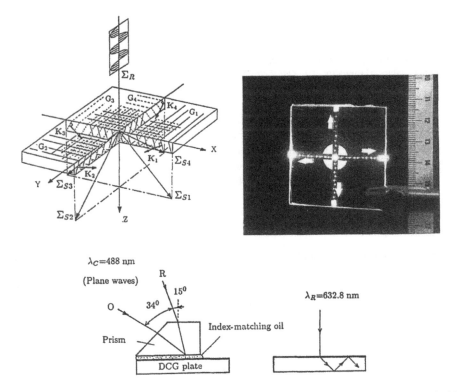

FIGURE 11.16   A multiplexed substrate mode hologram formed in dichromated gelatin that diffracts an incident HeNe laser beam into four orthogonal directions. (From Kato, M. et al., *JOSA A.*, 7, 1441–1447, 1990. With permission from The Optical Society of America.)

one corresponding to s-polarized light and the other to p-polarized light as illustrated in Figure 11.16 and specified in Equation 11.9:

$$v_m = \frac{\kappa_m d}{\left(c_{Sm} c_R\right)^{1/2}} = \frac{\pi n_m d}{\lambda \left(c_{Sm} c_R\right)^{1/2}}, \qquad m = 1,3$$

$$v_m = \frac{\kappa_m d}{\left(c_{Sm} c_R\right)^{1/2}} \left(\hat{r} \cdot \hat{s}_m\right) = \frac{\pi n_m d}{\lambda \left(c_{Sm} c_R\right)^{1/2}} \left(\hat{r} \cdot \hat{s}_m\right), \quad m = 2,4$$

(11.9)

where κ_m are the coupling coefficients for the four different gratings ($m = 1$–4), n_m are the refractive index modulation values for the four gratings, d is the hologram thickness, $\hat{r}$ and $\hat{s}_m$ are, respectively, unit vectors in the directions of the reconstruction beam and the m diffracted beams, and c_{Sm} and c_R are, respectively, the cosine factors for the diffracted beams and the reconstruction beam. The final step in the design is to adjust the individual exposures to equalize the diffraction efficiency values of the four holograms.

Frequently, holograms are formed at a wavelength where the recording material is sensitive and reconstructed at a different operating wavelength. The refractive index modulation required to achieve maximum diffraction efficiency is a function of the reconstruction wavelength as shown in Equation 5.69. This requirement adds complexity to the design since the light must also satisfy the TIR condition for the substrate. However, by making a few trial exposures with the material system, the diffraction efficiencies for each hologram can be equalized after a few iterations [32].

The corresponding diffraction efficiencies of the individual gratings can be calculated using a modification to the approximate coupled wave equations given by:

$$\eta_{Si} = \frac{v_i^2 \sin^2 \left(\sum_{m=1}^{4} v_m^2 \right)^{1/2}}{\sum_{m=1}^{4} v_m^2}; \quad (i = 1,...,4). \tag{11.10}$$

An example of a multiplexed substrate mode hologram formed in dichromated gelatin film that diffracts normally incident light into four nearly equal efficiency guided beams is shown in Figure 11.16. Holograms of this type have been used in a variety of optical interconnect applications [35,36].

In addition to distributing the power in an incident beam to different locations on a substrate, it is also possible to realize different polarization control operations with substrate mode holographic components [37]. An example of an element that switches the direction of an output beam based on its polarization state of an incident beam is shown in Figure 11.17. In this device, a ferroelectric liquid crystal half wave plate is used to change the state of the incident polarization. A polarization insensitive input coupling hologram diffracts both states of polarization with equal diffraction efficiency into the substrate. This is accomplished by making $\kappa_\parallel \approx \kappa_\perp$ as described in Equation 11.3. A polarization selective hologram is placed at the output location to separate the two states of polarization for the detection process. This hologram is also formed by controlling the coupling coefficient, but with an exposure that makes the diffraction efficiency of the parallel polarization component approximately zero for light incident at 45°.

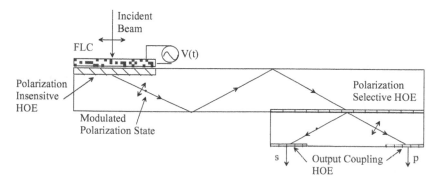

FIGURE 11.17 Cascaded substrate mode holographic system for switching the position of an optical beam based on its polarization. (From Kostuk, R.K., *Appl. Opt.*, 29, 3848–3854, 1990. With permission from the Optical Society of America.)

11.6 Volume Holographic Imaging

As described in Chapter 7, digital holography is often used for microscopy applications. This approach, while quite useful, requires the subject to remain stable during the recording. It also requires reconstruction of the image through a computational algorithm. These requirements make digital holography difficult to use for real time visualization applications such as during medical endoscopy procedures.

As an alternative, pre-recorded volume holographic elements with special filtering properties can be used to enhance conventional microscopy imaging. Volume holographic imaging (VHI) is one example of this approach and not limited by the stability requirements during the viewing process. The method acquires the image in real time without the need for computation to reconstruct the image. In VHI, the holographic optical element acts as a highly functionalized filter that incorporates the Bragg selectivity, multiplexing, and the spectral properties of volume holograms into the imaging process.

Barbastathis and Brady [38] initially described the theoretical framework for volume holographic imaging. Later they also showed that a volume hologram can be used in place of a pinhole to effectively form a confocal imaging system [39] that selects information from a narrow axial section of the object. An illustration of the confocal imaging effect and background rejection with a volume hologram is illustrated in Figure 11.18. In this case, a hologram is recorded with an off-axis reference wave and an on-axis object point source displaced by a distance of Δz from the focus of a lens. The wavefront of the object wave has a slight curvature, and the resulting volume hologram has a K-vector that varies across the aperture similar to that with a large f-number (f/#) lens. During reconstruction if the hologram is illuminated with a point source at the same location as used during construction, the wavefront satisfies the Bragg condition across the entire aperture of the hologram and diffracts the field with high efficiency. However, if the hologram is reconstructed with a collimated plane wave, the incident field will only be Bragg matched in a small region of the aperture near the optical axis, and the total diffraction efficiency across the aperture will be significantly lower.

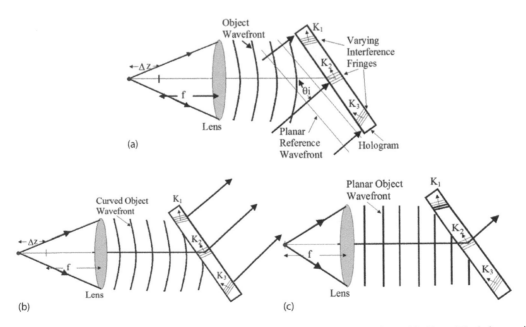

FIGURE 11.18 Schematic showing the wavefront selectivity properties of volume holographic filters. The hologram is recorded with a slightly curved wavefront (a). When reconstructed with the same wavefront, (b) the diffraction efficiency is matched across the full aperture of the hologram. When the wavefront is changed, (c) only a small region near the center of the aperture is Bragg matched.

The depth resolution of the VHI system is a measure of how well it can resolve features along the optical axis, and reject light from other depths and form high contrast images. The FWHM depth (z) resolution of the VHI system can be approximated as [40]:

$$\Delta z_{FWHM} = \frac{C \lambda f_{lens}^2}{r_{lens} \, d \tan \theta_s},$$ (11.11)

where f_{lens} and r_{lens} are, respectively, the focal length and aperture radius of the microscope objective, d is the hologram thickness, θ_s is the angle of the reconstruction beam relative to the axis, λ is the wavelength, and C is a constant obtained by fitting to an exact integral form for the axial power distribution [40]. Therefore, for high background rejection, it is important to have thick holographic elements and to use large numerical aperture lenses.

Optical methods such as confocal microscopy (CM) [41] and optical coherence tomography (OCT) [42] are capable of imaging features beneath the tissue surface. This makes them useful for diagnosing the onset of diseases such as cancer. CT and OCT systems typically accomplish depth imaging with complex point-by-point scanning techniques and numerical reconstruction of the image. The confocal selectivity property of VHI can be used to evaluate subsurface tissue features as well. This is accomplished by multiplexing holograms with each hologram collecting light from a different tissue depth. During reconstruction, the full object volume is illuminated and the multiplexed hologram filters out light from different depths and diffracts each image to a different position on a camera as shown in Figure 11.19.

During reconstruction, if the object is illuminated with a narrow spectral band laser source, only a narrow line image will form at the camera (Figure 11.19, left). The width of the image field can be extended by using a broad spectral bandwidth source with the wavelengths laterally dispersed across the object. Recall that multiple wavelengths can be Bragg matched to the same grating vector if the incident and diffracted angles are also changed as shown in Figure 11.20. Therefore, a different wavelength images a different lateral spatial region of the object (Figure 11.19, right). This property of volume holograms effectively increases the field of view of the microscope. The number of depth sections that can be simultaneously imaged depends on the field of view of each image and the width of the camera aperture. Figure 11.21 shows the result of projecting five depth sections obtained with five multiplexed holograms with depths separated by 75 μm [43]. The LED used in this case has a spectral bandwidth of 30 nm and produces a lateral field of view of approximately 250 μm for each image. For this multiplexed element, the diffraction efficiency of the individual holograms was varied from 20% for the surface detecting grating to 50% for the depth grating. This was accomplished by varying the exposure for each hologram and helps compensate for the difference in signal strength between the surface and depth images.

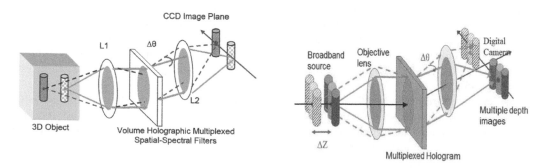

FIGURE 11.19 Basic configurations for VHI. The system shown selects light from two different depths using two multiplexed holograms. The configuration on the left shows imaging with narrow band illumination, and the system on the right shows the system operation with dispersed illumination. (From Luo, Y. et al., *Opt. Lett.*, 33, 566–568, 2008. With permission from the Optical Society of America.)

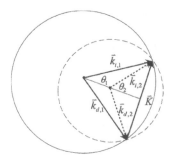

FIGURE 11.20 Bragg circle diagram showing that two different propagation vectors can be matched to the same grating vector provided that the angles for the propagation vectors are changed. This principle allows the field of view (FOV) of the VHI system to be extended.

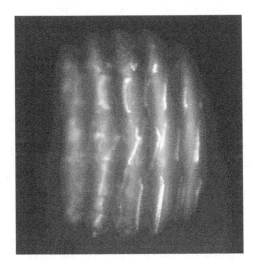

FIGURE 11.21 Degenerate mode volume holographic imaging using a volume hologram with five multiplexed gratings. The hologram was formed in a 1.8 mm thick pq-MMA polymer. The object is illuminated with an LED that has a 30 nm spectral bandwidth. The depth samples are separated by 75 μm. (From Luo, Y. et al., *Opt. Lett.*, 33, 566–568, 2008. With permission from the Optical Society of America.)

However, a problem occurs with the VHI system when using a broadband illumination source if the light is not laterally dispersed across the object. In this case, multiple wavelengths exist at each point within the object volume. The images still form, but the depth selectivity and background rejection significantly degrade. This condition is referred to as degenerate volume holographic imaging. The degeneracy can be removed either by using a second grating to disperse broadband illumination [44,45] or by scanning a narrow spectral band of wavelengths from a source along one dimension [45].

If the sample is treated with a fluorescent dye and excited with a laser, the fluorescence emission from the object will be dispersed in the image. The imaging in this case is degenerate since multiple wavelengths are emitted from each point in the object, however, image contrast can be increased for different features in the object using background subtraction methods. In addition, new types of fluorescent biomarkers can indicate the presence and location of tissue and cell abnormalities. A comparison of reflectance and fluorescence mode volume holographic images obtained from normal and cancerous ovarian tissue with standard histological sections is shown in Figure 11.22 [46]. The volume holographic images show similar feature content as the traditional histological sectioning method, but can be obtained without having to wait days for the staining process to be completed.

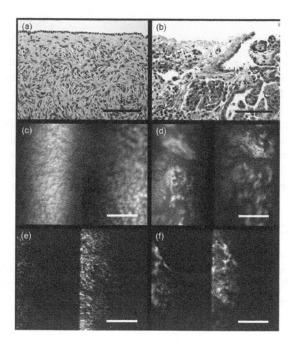

FIGURE 11.22 Comparative images of normal (left) and cancerous (right) ovarian tissue samples. Figures *a* and *b* histological sections, *c, d* reflectance mode volume holographic imaging of the tissue samples shown in (*a, b*), and *e, f* are fluorescence mode volume holographic images of the tissue samples shown in *a, b*. (From Orsinger, G.V. et al., *J. Biomed. Opt.*, 19, 036020, 2014. With permission from the SPIE.)

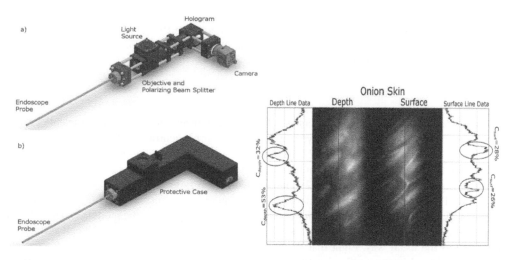

FIGURE 11.23 (a) Handheld VHI endoscope with a 30 cm long, 3.8 mm diameter gradient index (GRIN) probe. (b) Handheld VHI endoscope with case. Images obtained with the endoscope are shown on the right. (From Howlett, I.D. et al., *SPIE Proc.*, 8927, 2014. With permission from the SPIE.)

The volume holographic imaging approach is sufficiently robust to allow miniaturization and packaging into a handheld laparoscope for use in surgical procedures. Figure 11.23 shows a VHI endoscope that uses a gradient index rod probe to relay an image from an object to the holographic wavefront sensing system and image two object depths [47,48]. The system is used with a standard trocar sheath, conforms to clinical sterilization requirements, and has been used in surgical procedures for detecting ovarian cancer.

11.7 Holographic Optical Elements in Solar Energy Conversion Systems

PV solar energy conversion is an increasingly important clean energy source that can reduce the effects of climate change. However, at the present time these systems only provide a small fraction of the world's total energy production. In order to become a more significant energy source, the PV conversion efficiency must be increased with lower cost systems. Optics can play a role in satisfying these requirements by performing light management operations that improve the efficiency of PV systems. Holographic optical elements are particularly well suited for this task since they can be used to realize highly functional components in thin film materials and manufactured using large volume production methods to reduce cost [49]. In addition, it has been shown that, once sealed holographic optical elements can survive accelerated life testing which is indicative of 25–30 years of operational lifetimes [50].

11.7.1 Holographic Concentrators

One way to reduce the cost of a PV system is to collect light with a low cost optical element and concentrate it on a high efficiency PV cell. This approach reduces the amount of expensive PV cell material needed and can potentially reduce the overall system cost while still capturing the same amount of sunlight. The geometrical concentration ratio is:

$$CR_G = \frac{A_{opt}}{A_{PV}}, \tag{11.12}$$

where A_{opt} is the area of the optical collector, and A_{PV} is the area of the PV cell. If the collector focuses light to a line, it is considered a 2-dimensional concentrator with:

$$CR_{2D} = n/\sin\alpha, \tag{11.13}$$

where n is the refractive index between the optical element and the PV cell, and α is the angle of incidence (half angle). When the collector focuses light to a point, the configuration is considered a 3D concentrator, and the concentration ratio becomes:

$$CR_{3D} = n^2/\sin^2\alpha. \tag{11.14}$$

In both cases, the collection angle is inversely related to the concentration ratio, and the collector must accurately track the sun in order to achieve high levels of solar concentration at the PV cell.

Bloss [51] and Ludman [52] had first shown in the early 1980s that holographic optical elements can serve as solar concentrators. One configuration for a broadband holographic planar concentrator (HPC) is shown in Figure 11.24 [53]. This is a 2D, low concentration ratio concentrator with a concentration ratio of ~2X. The lower concentration ratio allows for a larger acceptance angle (±48.6°) providing significant energy capture without the need for tracking. In this arrangement, the module is populated with alternating strips of holographic elements and PV cells. Light that illuminates the PV cell surface (D_{PV}) is converted in the usual way. Light that illuminates the holographic element (D_h) is diffracted either directly or after total internal reflection to the PV cell surface. The grating fringe planes are arranged in a direction that is roughly parallel to the trajectory of the sun during the course of the day as shown in Figure 11.24 (right). The seasonal variation of the declination angle of the sun changes the angle of incidence (±23.45°) in a direction that is perpendicular to the grating fringe planes and must also be considered in the design. The peak diffracted wavelength will vary with the changing angle of incidence and must be designed to fall within the spectral response range of the particular PV cell used in the system. For silicon, the spectral response is between 350 and 1100 nm. Achieving high diffraction efficiency over large angular and spectral bandwidth requires thin recording materials with high refractive index modulation. This type of performance can be obtained with DCG films which also have good long term durability after sealing to prevent the incursion of moisture. Analysis has shown that by cascading two

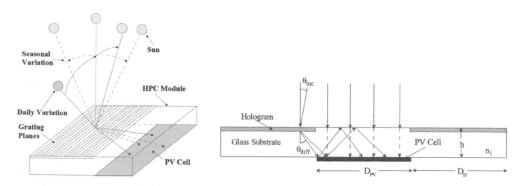

FIGURE 11.24 Schematic for a HPC (left). Regions D_h are areas filled with holographic elements that diffract light to a PV cell surface (D_{PV}) either by direct diffraction or after TIR within the substrate. The figure on the right shows sunlight being diffracted as the sun changes position during the day and with seasonal variations. (From Kostuk, R.K. and Rosenberg, G., *Proc. SPIE*, 7043, 2008. With permission from the SPIE.)

FIGURE 11.25 (See color insert.) A HPC module produced by Prism Solar Technologies for rooftop and building integrated applications. (From Prism Solar Technologies. With permission.)

hologram layers of DCG, a spectral bandwidth of 300 nm can be obtained and that nearly 50% of the available solar illumination can be collected with a fixed HPC module [54]. Figure 11.25 shows a commercial HPC module produced by Prism Solar Technologies [50].

11.7.2 Light Trapping Holographic Optical Elements

Another approach to lower the cost of PV cells is to reduce the amount of expensive PV cell material by decreasing the cell thickness. However, this can also decrease light absorption and conversion efficiency. To offset the loss in light collection, light trapping methods can be used to increase the effective optical path length within the cell and the probability of absorption. An approach that has been used for many years to accomplish this is to texture the top and bottom surface of the PV cell. This method increases the effective optical path length by a factor of $4n^2$ [55]. Therefore, for silicon with a refractive index of 3.5, the path length is increased by approximately 50 times. It has also been shown that the optical path length can be further increased to $4n^2/(\sin\theta_a)^2$ using a Rugate filter where θ_a is the collection angle of the filter [56,57].

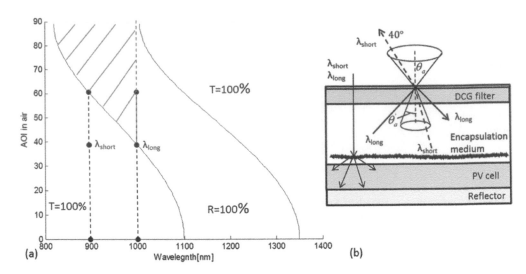

FIGURE 11.26 Holographic enhanced light trapping filter. Plot (a) shows the hologram reflectance as a function of the angle of incidence and wavelength. Diagram (b) is a schematic of the device showing the relative locations of the PV cell, diffusing surface, and reflection hologram. (From Zhang, D. et al., *Opt. Exp.*, 20, 14260–14271, 2012. With permission from the Optical Society of America.)

However, filters of this type are difficult and expensive to manufacture in large quantities as would be needed for PV energy conversion systems. However, the Bragg selectivity and spatial and spectral filtering properties of reflection type holographic elements have similar properties as Rugate filters. When used in a configuration as shown in Figure 11.26, they can significantly increase light trapping in a thin PV cell [58]. For this application, the reflection hologram is designed to diffract light at longer wavelengths near the bandgap of the PV cell and to transmit shorter wavelength light that is not as efficiently converted by the PV cell. The reflection hologram is also designed to diffract light at large angles of incidence and to transmit light at smaller angles. Incident light near normal incidence angles is not Bragg matched and passes through the hologram. The light is then scattered in both forward and backward directions by a textured surface (diffuser) positioned above the PV cell. The forward scattered light enters the cell and the backward scattered light at longer wavelengths is reflected by the hologram back to the PV cell where it has another chance for absorption. The increased absorption over what can be obtained with just using a diffuser is shown in Figure 11.27 and indicates an increase of approximately 15%.

11.7.3 Holographic Spectrum Splitting Systems

The theoretical limit for the conversion efficiency of a single bandgap PV cell is approximately 33% as determined by detailed energy balance between the temperature of the sun and the PV cell [59]. In order to increase conversion efficiency, multiple PV cells with different energy bandgaps are needed that span the solar emission spectrum. This can be accomplished using a broadband concentrator and a stacked or tandem multi-junction PV cell. While a number of systems of this type have been demonstrated [60], the individual cells are series connected which limits the total output current to the lowest output cell. In addition, the cells are typically manufactured by epitaxial growth methods that deposit one bandgap material on top of the other. This requires precise lattice matching and limits the choice of materials that can be used. Another issue is that the high cost of the cells per unit area limits the size of the cell, and they are typically used with large high concentration ratio collectors. As discussed earlier, this limits the acceptance angle of the concentrator and prevents the collection of diffuse illumination that makes up a significant fraction of the available solar energy.

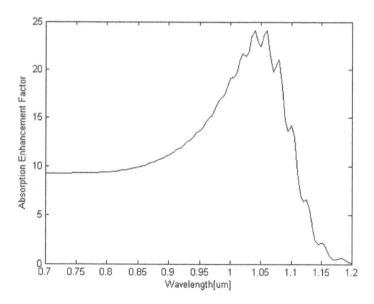

FIGURE 11.27 Absorption enhancement factor as a function of wavelength produced by the holographic light trapping filter shown in Figure 11.26. (From Zhang, D. et al., *Opt. Exp.*, 20, 14260–14271, 2012. With permission from the Optical Society of America.)

An alternative approach to realizing a multiple bandgap PV system is to spatially divide the incident solar spectrum into spectral components, and direct each component to a single bandgap cell with a response that matches the incident spectral component [17,61]. Figure 11.28 shows the spectral response for silicon (narrow bandgap) and gallium arsenide (GaAs wide bandgap) PV cells as well as the solar illumination spectrum. The optics for an ideal spectrum splitting system transfers 100% of the short wavelength light to the wide bandgap cell and 100% of longer wavelength light to the narrow bandgap cell.

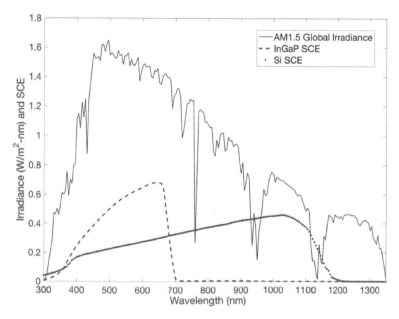

FIGURE 11.28 The spectral irradiance for the air mass (AM) 1.5 splar spectrum along with the spectral conversion efficiencies for silicon and indium gallium phosphide type PV cells. (From Vorndran, S. et al., *Appl. Opt.*, 55, 7522–7529, 2016. With permission of the Optical Society of America.)

A measure of the spectral-spatial filtering properties of an optical system can be quantified using the spectral optical efficiency defined by:

$$SOE(\lambda) = \sum_i \frac{\Phi_{PVC,i}(\lambda)}{\Phi_o(\lambda)}, \quad (11.15)$$

where $\Phi_{PVC,i}(\lambda)$ is the optical spectral power density incident on PV cell i, and $\Phi_o(\lambda)$ is the solar spectral power density incident on the spectrum splitting system. The spectral management process can be accomplished in either reflection or transmission geometries as shown in Figure 11.29. For the reflection configuration, the incident broadband illumination is successively filtered and reflected to different PV cells (Figure 11.29a). Reflection holograms can be used in this arrangement and have been shown to produce system efficiencies of nearly 90% of the value reached with an ideal spectral filter [62].

Transmission holographic optical elements can also be effectively used in spectrum splitting systems to spatially separate different spectral bands for conversion by different bandgap PV cells. The dispersion properties of transmission holograms are well suited to this application. However, it has been shown that they work best when the diffracted beam is focused [63] either with a separate lens [64] or by including focus power in the hologram [65]. A volume holographic lens can be used to form a very compact spectrum splitting module [66]. As described in the section on holographic lenses earlier in this chapter, a spectrum splitting volume holographic lens can be formed with an off-axis cylindrical lens in the object beam path to focus light at a transition wavelength to a line that separates PV cells with two different bandgaps (Figure 11.1 in Section 11.2). The resulting spectral optical efficiency is shown in Figure 11.3. While not ideal, the volume holographic lens in combination with GaAs wide bandgap and silicon narrow bandgap cells form a spectrum splitting system that has nearly 20% higher efficiency than just using the higher efficiency GaAs cell alone [66]. In addition, diffuse light entering at non-normal (non-Bragg) incidence can be converted into electricity by the PV cells resulting in higher overall energy yield. For the cylindrical volume holographic lens in a GaAs/silicon PV cell spectrum splitting system, the energy yield was 17% more than just using GaAs alone [66]. An image of the dispersive spectrum

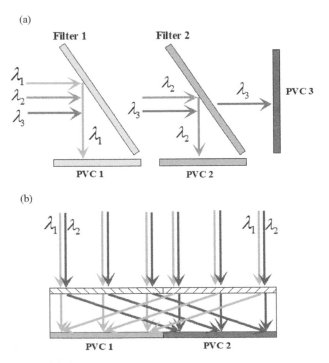

FIGURE 11.29 Schematic representations for a reflection (a) and transmission (b) type spectrum splitting systems.

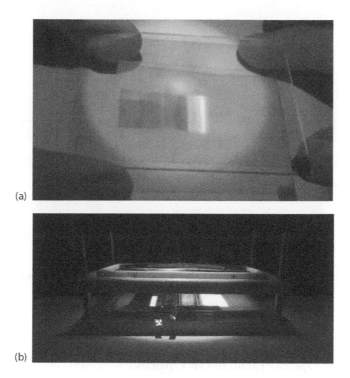

(a)

(b)

FIGURE 11.30 (See color insert.) A volume holographic cylindrical lens fabricated in a photopolymer material showing the dispersed spectrum (a) for use in a spectrum splitting system with GaAs and silicon PV cells (b). (From Chrysler, B. et al., *SPIE Proc.*, 10368, 10368G, 2017. With permission from the SPIE.)

splitting cylindrical lens and a prototype spectrum splitting system that combines the lens with a GaAs and silicon cell is shown in Figure 11.30.

11.7.4 Other Solar Applications of Holographic Optical Elements

The spectral and spatial properties of holographic optical elements make them useful for several passive solar applications as well. For instance, the spectral content of light that enters a building can be transferred and distributed by using holographic elements in windows and skylights. This capability can modify both the illumination and heating properties of buildings to passively control living and working environments. As shown in previous examples, volume holograms can be made to diffract a specific spectral band when the sun is incident over a certain field of view. This mechanism can be used to diffract light entering a window to the ceiling of a room, where it is then scattered to illuminate areas farther from the window [67,68] (Figure 11.31).

It was shown by Ludman that if half of a 0.5 m² window is covered with a 50% efficient transmission type holographic element, that it can replace the equivalent of 750 W of lighting with conventional light bulbs [67]. In addition, the transmission type holographic daylighting element can be combined with a reflection hologram that diffracts a portion of the solar spectrum during peak illumination hours during the summer to reduce radiative heating of the building [15].

Another use of holographic elements and sunlight is for the purification of water and for chemical processing of various compounds. Water purification can be accomplished by focusing ultra-violet (UV) light to a line focus fixed on a UV transparent glass tube containing water with a photo catalyst suspension [69,70]. The UV light (350–390 nm) in sunlight is sufficient to activate the photo catalyst and break down toxic organic compounds [68]. In addition to water purification, Stojanoff [15] has shown that spectral selective reflection holograms laminated onto curved sections of glass can be used to form solar concentrators for photochemical processing of different compounds.

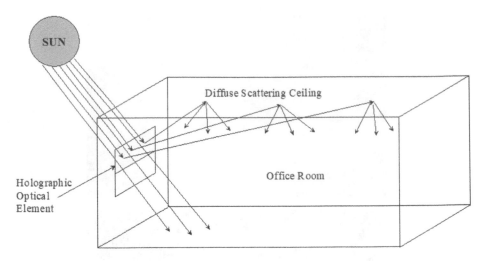

FIGURE 11.31   A diagram showing a configuration for holographic daylighting of the interior of a room.

11.8 Holographic Optical Elements in Optical Interconnects and Communications Systems

11.8.1 Optical Interconnects

During the early 1980s, the rapid advances in integrated circuit (IC) manufacturing enabled a vast industry of personal computers and electronic devices. However, it was recognized that electrical signal communication on and off and within the IC could become a serious problem to the rate and capacity of information transfer [71]. Electronic devices are excellent for switching, however, they experience capacitive and other electrical loading effects that reduce signal transmission capability. The converse is true for optic systems. They have excellent transmission capability, but more restrictive switching requirements (it should be noted that the rapid advances in silicon photonics [72] is making this less of an issue). For on-chip applications, the complexity of interconnect pathways and the need for both point-to-point and broadcast operations limit the use of waveguide optics. However, the multiplexing and wavefront recording capability of holographic optical elements make them attractive for this application [73]. In addition, the use of folded planar optics [35,36,74] helps reduce alignment and space requirements [75]. Figure 11.32 is an example of the application of holography for interconnects. It shows the free-space implementation of a holographic clock distribution system [76,77] and the corresponding planar hologram counterpart.

11.8.2 Optical Communications

The advent of low loss (<0.15 dB/km at 1.55 μ) single mode optical fibers led to the development and commercial feasibility of long haul fiber optic communication systems. In addition, the use of dense wavelength division multiplexing (DWDM) methods allows many independent signals to travel through the same fiber and greatly increases the communication capacity of the system. One component that has enabled the development of DWDM systems is the fiber Bragg grating. It was initially discovered by accident by K. O. Hill [78] while illuminating a single mode fiber with 488 nm light from an argon laser. They noticed that the amount of back reflected light in the fiber increased as a function of exposure. They concluded that a reflection grating was being formed as a result of reflected light interfering with

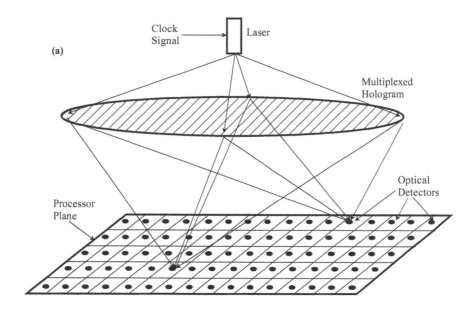

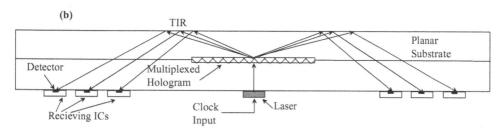

FIGURE 11.32 Holographic optical interconnects for clock signal distribution on a plane. (a) Free-space holographic distribution and (b) planar holographic distribution. (From Clymer, B.D. and Goodman, J.W., *Opt. Eng.*, 25, 1103–1108, 1986. With permission from SPIE.)

the light launched into the fiber and a weak photosensitivity response in the glass fiber. It was later found by Lam and Garside that grating formation at 488 nm was actually the result of a two photon process in combination with a stronger photosensitive effect at 244 nm [79]. The photosensitivity of the fiber is due to color centers that are formed when the core is produced by doping with germanium to increase refractive index. Exposure with a frequency doubled argon laser at 244 nm produces reflection holograms with significantly shorter exposure times. This allows the fabrication of gratings using conventional interferometer setups in combination with satisfying the Bragg matching conditions at communication wavelengths (1530–1570 nm). For this case, the holographic exposure system is set up for a transmission hologram, however, the fiber is placed as shown in Figure 11.33 to record a reflection hologram in the fiber. The Bragg diagram in the inset of the figure shows how the propagation vectors at 244 nm are used to form the $\vec{K}$ vector and Bragg matching for the reflection grating at the end use wavelength of 1550 nm. The end use wavelength can also be determined from the grating equation. It can be shown (see homework problem 11.6) that the end use wavelength within the fiber is:

$$\lambda_2 = \frac{n_1 n_2 \lambda_1}{\sin \theta},$$

(11.16)

where λ_1 and λ_2 are the construction and reconstruction wavelengths, respectively, n_1 is the refractive index of the fiber at λ_1, and n_2 is the effective mode index within the fiber at λ_2.

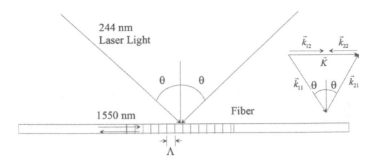

FIGURE 11.33 Interferometric setup for forming a fiber Bragg grating at a UV wavelength for use at a communications wavelength within the fiber. The inset shows the corresponding Bragg diagram.

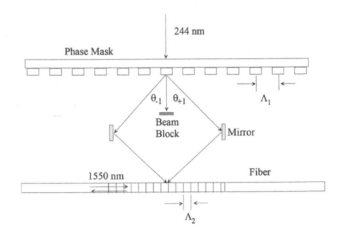

FIGURE 11.34 Construction setup for forming a fiber Bragg grating using a phase mask grating and the + and − diffraction orders.

When a large number of fiber Bragg gratings with similar properties are required, it is often more cost effective to use a computer-generated hologram (phase mask) for the exposure process. In this case, the computer-generated hologram (CGH) phase mask is designed to diffract the +1 and −1 orders with equal diffraction efficiency at the recording wavelength when illuminated at normal incidence as shown in Figure 11.34. In this case, the zero order is blocked and the two diffracted orders are reflected from a symmetric pair of mirrors to set up the interference pattern which then exposes the fiber. Since the + 1 and −1 diffraction orders are used for the exposure, the grating period of the two gratings are related according to:

$$\Lambda_{fiber} = \frac{\lambda_1}{2\sin\theta} = \frac{\Lambda_{PM}}{2}, \tag{11.17}$$

where Λ_{fiber} is the required period of the fiber Bragg grating ($\lambda_2/2n_2$), and Λ_{PM} is the period of the CGH phase mask.

11.8.3 Optical Code Division Multiple Access Waveguide Holograms

Holographic optical elements can also be used in a waveguide format to directly interface with fiber optic communications systems. One example is the formation of cascaded gratings that diffract different wavelengths with an optical delay between each diffraction grating. The combination of different wavelengths and phase delays can be used to form a matched code that is used at the transmission and receiving locations in the system [80]. A diagram of a simple two grating device formed in a 1.8 mm thick layer of

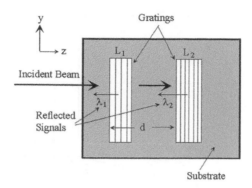

FIGURE 11.35 Code division multiple access device formed with an edge illuminated hologram in pq-MMA. The grating lengths in part determine the bandwidth of the diffracted spectra. (From Mossberg, T.W. and Raymer, M.G., *Opt. Photon. News*, 12, 50–54, 2001.)

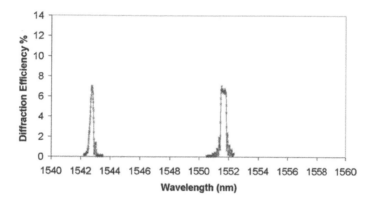

FIGURE 11.36 Spectra diffracted from the cascaded edge illuminated hologram shown in Figure 11.34. (From Mossberg, T.W. and Raymer, M.G., *Opt. Photon. News*, 12, 50–54, 2001.)

phenanthrenequinone-doped poly methyl methacrylate (pq-MMA) is shown in Figure 11.35 [81]. In this case, the grating periods determine the wavelengths that are selected, and the separation d and the refractive index of the polymer determine the phase delay between the diffracted components. Figure 11.36 shows the two spectra within the fiber communication band that were diffracted back into the launch fiber. It was also shown by Russo [82] that the wavelengths selected can be temperature tuned at a rate of 0.03 nm/°C for additional coding diversity.

PROBLEMS

1. Consider a cylindrical holographic lens formed as shown in Figure 11.37. Assume that the construction and reconstruction light has a wavelength of 488 nm, $D = 20$ mm, $h = 5$ mm, $s = 15$ mm, $n = 1.50$, $d = 15$ μm. Assume s-polarized incident light. Determine the volume hologram Q parameter (Equation 5.109) at the bottom middle and top of the aperture. What is the minimum distance (h) that will satisfy the Q factor for the volume hologram regime at the bottom of the aperture? At this condition, what refractive index modulation is needed at the bottom to match a diffraction efficiency of 90% at the top of the aperture?

2. For the cylindrical lens described in Problem 1 and the conditions when the diffraction efficiency for s-polarized light is 90%, compute the difraction efficiency for p-polarized incident light at the bottom, middle, and top of the aperture.

3. A reflection hologram is formed in a 10 μm thick recording material with a refractive index of 1.50 at a wavelength of 532 nm as shown in Figure 11.38 below. Compute and plot the

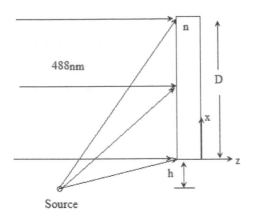

FIGURE 11.37 Construction geometry of a cylindrical lens for Problem 11.1.

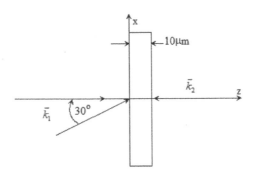

FIGURE 11.38 Geometry for the formation of a reflection hologram for Problem 11.3 with $\vec{k}_1$ at 0° and at 30° to the z-axis.

diffraction efficiency as a function of wavelength for s-polarized light with the construction vector at 0° and at 30° to the z-axis and $\Delta n = 0.035$. For the case with $\vec{k}_1$ at 30° to the z-axis, compute the diffraction efficiency as a function of wavelength for p-polarized light and plot with s-polarized diffraction efficiency.

4. Referring to Figure 11.18, assume that $\Delta z = 30$ μm for the shift of the on axis object beam, a wavelength of 650 nm, and that the objective lens has an $NA = 0.55$. The relation for the amount of wavefront defocus for the axial shift of the point source is given by [83]:

$$\Delta W_{df} = -\frac{1}{2} \frac{\Delta z \cdot (NA)^2}{\lambda}.$$

  a. Hologram is formed with the axial shifted object wave and a reference wave at 45° to the optical axis. As shown in Figure 11.18, the grating vector $\left(\vec{K}\right)$ changes as a function of the x-distance from the axis. If it is assumed that the wavefront of the object beam is planar at the axis, compute the index modulation required to achieve a diffraction efficiency of 45% in a 2.0 mm thick photopolymer with an average refractive index of 1.52 at this location.

  b. Assume that the same refractive index modulation as found in part (a) exists at the edge of the hologram aperture. If it is also assumed that the object wave is tilted, an additional $\lambda \cdot \Delta W_{df}$ with respect to the on-axis wavefront at the edge of the aperture, determine the change in the diffraction efficiency at this location.

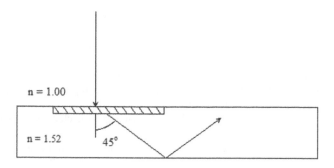

FIGURE 11.39 A substrate mode hologram for use in Problem 11.5 that diffracts a normally incident plane wave at 45° into the substrate with a refractive index of 1.52.

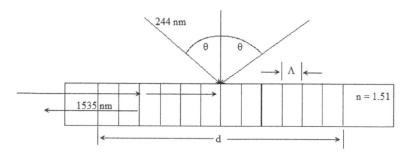

FIGURE 11.40 Construction of a fiber Bragg grating at 244 nm that diffracts incident light at a wavelength of 1535 nm (Problem 11.7).

5. It is desired to form a substrate mode hologram as shown in Figure 11.39 that diffracts a normally incident beam in air to an angle of 45° in the glass substrate at a wavelength of 650 nm. The exposing wavelength is 480 nm and holographic material has the same refractive index as the glass substrate. Determine the construction angles for the hologram if a prism is not used. What is the longest wavelength that can be used to make the recording that does not require a prism coupler?

6. Using the grating equation (Equation 3.9), derive Equation 11.16 for the diffracted wavelength within a fiber:

$$\lambda_2 = \frac{n_1 n_2 \lambda_1}{\sin\theta},$$

where λ_1 and λ_2 are the construction and reconstruction wavelengths, respectively, n_1 is the refractive index of the fiber at λ_1, and n_2 is the effective mode index within the fiber at λ_2.

7. Using the figure below (Figure 11.40), determine the construction angles (θ) required to form the grating at 244 nm and reconstruct it at 1535 nm within the fiber using the Bragg condition. Assume that the fiber is surrounded by air. Assuming that the fiber Bragg grating can be treated as a reflection grating, determine the grating length (d) that is needed in order to have a spectral bandwidht of 1 nm.

REFERENCES

1. J. N. Latta, "Computer based analysis of holography using ray tracing," *Appl. Opt.*, Vol. 10, pp. 2698–2710 (1971).
2. J. N. Latta, "Computer based analysis of hologram imagery and aberrations II: Aberrations induced by a wavelength shift," *Appl. Opt.*, Vol. 10, 609–618 (1971).

3. J. N. Latta, "Analysis of multiple hologram optical elements with low dispersion and low aberrations," *Appl. Opt.*, Vol. 11, 1686–1696 (1972).

4. W. T. Welford, "A vector raytracing equation for hologram lenses of arbitrary shape," *Opt. Comm.*, Vol. 14, 322–323 (1975).

5. W. T. Welford, "Isoplanatism and holography," *Opt. Comm.*, Vol. 8, 239–243 (1973).

6. H. W. Holloway and R. A. Ferrante, "Computer analysis of holographic systems by means of vector ray tracing," *Appl. Opt.*, Vol. 20, 2081–2084 (1981).

7. D. G. McCauley, C. E. Simpson, and W. J. Murbach, "Holographic element for visual display applications," *Appl. Opt.*, Vol. 12, 232–242 (1973).

8. K. Winick, "Designing efficienct aberration-free holographic lenses in the presence of a construction-reconstruction wavelength shift," *J. Opt. Soc. Am. A*, Vol. 5, 702–712 (1982).

9. Y. Amitai, A. A. Friesem, and V. Weiss, "Designing holographic lenses with different recording and readout wavelength," *J. Opt. Soc. Am. A*, Vol. 7, 80–86 (1990).

10. M. R. Latta and R. V. Pole, "Design techniques for forming 488 nm holographic lens with reconstruction at 633 nm," *Appl. Opt.*, Vol. 18, 2418–2421 (1979).

11. R. C. Fairchild and J. R. Fienup, "Computer-originated aspheric holographic optical elements," *Opt. Eng.*, Vol. 21, 133–140 (1982).

12. R. B. Wood and M. J. Hayford, "Holographic and classical head up display technology for commercial and fighter aircraft," *Proc. SPIE*, Vol. 0883, 36–52 (1988).

13. H. Peng, D. Cheng, J. Han, C. Xu, W. Song, L. Ha, J. Yang, Q. Hu, and Y. Wang, "Design and fabrication of a holographic head-up display with asymmetric field of view," *Appl. Opt.*, Vol. 53, H177–H185 (2014).

14. C. M. Bigler, P.-A. Blanche, and K. Sarmi, "Holographic waveguide heads-up display for longitudinal image magnification and pupil expansion," *Appl. Opt.*, Vol. 57, 2007–2013 (2018).

15. C. G. Stojanoff, "Engineering applications of HOEs manufactured with enhanced performance DCG films," *Proc. SPIE*, Vol. 6136 (2006).

16. S. D. Vorndran, B. Chrysler, B. Wheelwright, R. Angel, Z. Holman, and R. K. Kostuk, "Off-axis holographic lens spectrum-splitting photovoltaic system for direct and diffuse solar energy conversion," *Appl. Opt.*, Vol. 55, 7522–7529 (2016).

17. A. Mojiri, R. Taylor, E. Thomsen, and G. Rosengarten, "Spectral beam splitting for efficient conversion of solar energy—A review," *Renew. Sustain. Energy Rev.*, Vol. 28, 654–663 (2013).

18. A. Barnett et al., "Very high efficiency solar cell modules," *Prog. Photovoltaics Res.Appl.*, Vol. 17, 75–83 (2009).

19. C. Moser, L. Ho, E. Maye, and F. Havermeyer, "Fabrication and applications of volume holographic optical filters in glass," *J. Phys. D: Appl. Phys.*, Vol. 41, 224003 (2008).

20. https://www.thorlabs.com/navigation.cfm?guide_id=2210. (accessed on 15 October 2018).

21. H. A. Macleod, *Thin-Film Optical Filters*, 5th ed. (Series in Optics and Optoelectronics), CRC Press, Baca Raton, FL (2018).

22. M. G. Moharam and T. K. Gaylord, "Chain-matrix analysis of arbitrary-thickness dielectric reflection gratings," *J. Opt. Soc. Am.*, Vol. 72, 187–190 (1982).

23. D. Zhang et al., "Optical performance of dichroic spectrum-splitting filters," *J Photon Energy*, Vol. 4 (2014).

24. D. Zhang, S. Vorndran, J. M. Russo, M. Gordon, and R. K. Kostuk, "Ultra light-trapping filters with broadband reflection holograms," *Opt. Exp.*, Vol. 20, 14260–14271 (2012).

25. M. Rostykus, F. Soulez, M. Unser, and C. Moser, "Compact lensless phase imager," *Opt. Exp.*, Vol. 25, 4438–4445 (2017).

26. M. Rostykus, M. Rossi, and C. Moser, "Compact lensless subpixel resolution large field of view microscope," *Opt. Lett.*, Vol. 43, 1654–1657 (2018).

27. L. H. Cescato, E. Gluch, and N. Streibl, "Holographic quarter wave plates," *Appl. Opt.*, Vol. 29, 3286–3290 (1990).

28. G. Campbell and R. K. Kostuk, "Effective-medium theory of sinusoidally modulated volume holograms," *J. Opt. Soc. A*, Vol. 12, 1113–1117 (1995).

29. T. J. Kim, G. Campbell, and R. K. Kostuk, "Volume holographic phase retardation elements," *Opt. Lett.*, Vol. 20, 2030–2032 (1995).

30. R. C. Enger and S. K. Case, "Optical elements with ultrahigh spatial-frequency surface corrugations," *Appl. Opt.*, Vol. 22, 3220–3228 (1983).

31. M. Born and E. Wolf, *Principles of Optics*, p. 705, 6th ed., Pergamon Press, New York (1983).

32. M. Kato, Y.-T. Huang, and R. K. Kostuk, "Multiplexed substrate-mode holograms," *JOSA A.*, Vol. 7, 1441–1447 (1990).
33. S. K. Case, "Coupled-wave theory for multiply exposed thick holographic gratings," *J. Opt. Soc. Am.*, Vol. 65, 724–729 (1975).
34. R. Alferness and S. K. Case, *J. Opt. Soc. Am.*, Vol. 65, 730–739 (1975).
35. J.-H. Yeh and R. K. Kostuk, "Free-space holographic optical interconnects for board-to-board and chip-to-chip interconnections," *Opt. Lett.*, Vol. 21, 1274–1276 (1996).
36. J.-H. Yeh, R. K. Kostuk, and K.-Y. Tu, "Hybrid free-space optical bus system for board-to-board interconnections," *Appl. Opt.*, Vol. 35, 6354–6364 (1996).
37. R. K. Kostuk, M. Kato, and Y. T. Huang, "Polarization properties of substrate-mode holographic interconnects," *Appl. Opt.*, Vol. 29, 3848–3854 (1990).
38. G. Barbastathis and D. J. Brady, "Multidimensional tomographic imaging using volume holography," *Proc. IEEE*, Vol. 87, 2098–2120 (1999).
39. G. Barbastathis, M. Balberg, and D. J. Brady, "Confocal microscope with a holographic filter," *Opt. Lett.*, Vol. 24, 811–813 (1999).
40. A. Sinha, W. Sun, T. Shih, and G. Barbastathis, "Volume holographic imaging in transmission geometry," *Appl. Opt.*, Vol. 43, 1533–1551 (2004).
41. T. Wilson, Ed., *Confocal Microscopy*, Academic Press, London, UK (1990).
42. D. Huang et al., "Optical coherence tomography," *Science*, Vol. 254, 1178–1181, (1991).
43. Y. Luo, P. J. Gelsinger, J. K. Barton, G. Barbastathis, and R. K. Kostuk, "Optimization of multiplexed holographic gratings in pq-PMMA for spectral-spatial imaging filters," *Opt. Lett.*, Vol. 33, 566–568 (2008).
44. J. M. Castro, P. J. Gelsinger-Austin, J. K. Barton, and R. K. Kostuk, "Confocal-rainbow volume holographic imaging system," *Appl. Opt.*, Vol. 50, 1382–1388 (2011).
45. W. Sun and G. Barbastathis, "Rainbow volume holographic imaging," *Opt. Lett.*, Vol. 30, 976–978 (2005).
46. G. V. Orsinger et al., "Simultaneous multiplane imaging of human ovarian cancer by volume holographic imaging," *J. Biomed. Opt.*, Vol. 19, 036020 (2014).
47. I. Howlett, "Wavelength coded volume holographic imaging endoscope for multi-depth imaging," *J. Biomed. Opt.,* Vol. 22, 100501 (2017).
48. I. Howlett, W. Han, M. Gordon, P. Rice, J. K. Barton, and R. K. Kostuk, "Volume holographic imaging endoscopic design and construction techniques," *J. Biomed. Opt.*, Vol. 22, 056010 (2017).
49. E. D. Aspnes, J. E. Castillo, R. D. Courreges, P. S. Hauser, G. Rosenberg, and J. M. Russo, "Volume holographic replicator for transmission type gratings," U.S. Patent No. 8,614,842 B2, Issued December 24, 2013.
50. Image from Prism Solar Technologies, www.prismsolar.com with permission. (accessed on 15 October 2018).
51. W. H. Bloss, M. Griesinger, and E. R. Reinhardt, "Dispersive concentrating systems based on transmission phase holograms for solar applications," *Appl. Opt.*, Vol. 21, 3739–3742 (1982).
52. J. E. Ludman, "Holographic solar concentrators," *Appl. Opt.*, Vol. 21, 3057–3058 (1982).
53. G. A. Rosenberg, "Device for concentrating optical radiation," U.S. Patent No. 5,877,874, Issued March 2, 1999.
54. J. M. Castro, D. Zhang. B. Myer, and R. K. Kostuk, "Energy collection efficiency of holographic planar solar concentrators," *Appl. Opt.*, Vol. 49, 858–870 (2010).
55. E. Yablonovitch and G. D. Cody, "Intensity enhancement in textured optical sheets for solar cells," *IEEE Trans. Electron Devices*, 29, 300–305 (1982).
56. S. Fahr, C. Ulbrich, T. Kirchartz, U. Rau, C. Rockstuhl, and F. Lederer, "Rugate filter for light-trapping in solar cells," *Opt. Express*, 16, 9332–9343 (2008).
57. C. Ulbrich et al., "Directional selectivity and ultra-light-trapping in solar cells," *Phys Status Solidi A*, 205, 2831–2843 (2008).
58. D. Zhang, S. Vorndran, J. M. Russo, M. Gordon, and R. K. Kostuk, "Ultra light-trapping filters with broadband reflection holograms," *Opt. Exp.*, Vol. 20, 14260–14271 (2012).
59. W. Shockley and H. J. Queisser, "Detailed balance limit of efficiency of *p-n* junction solar cells," *J. Appl. Phys.*, Vol. 32, 510 (1961).
60. K. Ghosal, D. Lilly, J. Gabriel, and S. Burroughs, "Semprius field results," *AIP Conf. Proc.*, Vol. 1616, 272–275 (2014).
61. A. G. Imenes and D. R. Mills, "Spectral beam splitting technology for increased conversion efficiency in solar concentrating systems: A review," *Sol. Energy Mater. Sol. Cells*, Vol. 84, 19–69 (2004).

62. S. D. Vorndran, S. Ayala Pelaez, Y. Wu, J. M. Russo, R. K. Kostuk, J. T. Friedlein, S. E. Shaheen, and C. K. Luscombe, "Holographic spectral beamsplitting for increased organic photovoltaic conversion efficiency," *Proc. SPIE*, Vol. 9184 (2014).

63. J. M. Russo, S. Vorndran, Y. Wu, and R. K. Kostuk, "Cross-correlation analysis of dispersive spectrum splitting techniques for photovoltaic systems," *J. Photonics Energy*, Vol. 5 (2015).

64. J. M. Russo, D. Zhang, M. Gordon, S. Vorndran, Y. Wu, and R. K. Kostuk, "Grating-over-lens concentrating photovoltaic spectrum splitting systems with volume holographic optical elements," *Proc. SPIE*, Vol. 8821 (2013).

65. D. Zhang, M. Gordon, J. M. Russo, S. Vorndran, and R. K. Kostuk, "Spectrum-splitting photovoltaic system using transmission holographic lenses," *J. Photonics Energy*, Vol. 3, 034597 (2013).

66. S. Vorndran, B. Chrysler, B. Wheelwright, R. Angel, and R. K. Kostuk, "Off-axis holographic lens spectrum-splitting photovoltaic system for direct and diffuse solar energy conversion," *Appl. Opt.*, Vol. 55, 7522–7529 (2016).

67. J. E. Ludman, J. R. Riccobono, G. D. Savant, J. L. Jannson, E. W. Campbell, and R. Hall, "Holographic daylighting," *SPIE Proc.*, Vol. 2532, 436–446 (1995).

68. H. F. O. Muller, "Application of holographic optical elements in buildings for various purposes like daylighting, solar shading and photovoltaic power generation," *Renew. Energ.*, Vol. 5(5–8), 935–941 (1994).

69. J. A. Quintana, P. G. Boj, J. Crespo, M. Pardo, and M. A. Satorre, "Line focusing holographic mirrors for solar ultraviolet energy concentration," *Appl. Opt.*, Vol. 36, 3689–3693 (1997).

70. D. F. Ollis, E. Pelizzeti, and N. Serpne, "Photocatalyzed destruction of water contaminants," *Environ. Sci. Technol.*, Vol. 25, 1523–1529 (1991).

71. J. W. Goodman, F. J. Leonberger, S. Y. Kung, and R. A. Athale, "Optical interconnections for VLSI systems," *Proc. IEEE*, Vol. 72, 850–866 (1984).

72. P. R. Prucnal and B. J. Shastri, *Neuromorphic Photonics*, CRC Press, Boca Raton, FL (2017).

73. H. Lee, X.-G. Gu, and D. Psaltis, "Volume holographic interconnections with maximal capacity and minimum crosstalk," *J. Appl. Phys.*, Vol. 65, 2191–2194 (1989).

74. J. Jahns and A. Huang, "Planar integration of free-space optical components," *Appl. Opt.*, Vol. 28, 1602–1605 (1989).

75. R. K. Kostuk, J. W. Goodman, and L. Hesselink, "Design considerations for holographic optical interconnects," *Appl. Opt.*, Vol. 26, 3947–3953 (1987).

76. B. D. Clymer and J. W. Goodman, "Optical clock distribution to silicon chips," *Opt. Eng.*, Vol. 25, 1103–1108 (1986).

77. B. D. Clymer and J. W. Goodman, "Timing uncertainty for receivers in optical clock distribution for VLSI," *Opt. Eng.*, Vol. 27, 944–954 (1988).

78. K. O. Hill, Y. Fujii, D. C. Johnson, and B. S. Kawasaki, "Photosensitivity in optical waveguides: Application to reflection filter fabrication," *Appl. Phys. Lett.*, 32, 647 (1978).

79. D. K. W. Lam and B. K. Garside, "Characterization of single mode optical fiber filters," *Appl. Opt.*, Vol. 20, 440–445 (1981).

80. T. W. Mossberg and M. G. Raymer, "Optical code-division multiplexing: The intelligent solution," *Opt. Photon. News*, Vol. 12, 50–54 (2001).

81. R. K. Kostuk, W. Maeda, and C.-H. Chen, "Cascaded holographic polymer reflection grating filters for optical code division multiple access applications," *Appl. Opt.*, Vol. 44(35), 7581–7586, 2005.

82. J. M. Russo and R. K. Kostuk, "Temperature dependence properties of holographic gratings in phenanthrenequinone doped poly methyl methacrylate photopolymers," *App. Opt.*, Vol. 46, 7494–7499 (2007).

83. J. C. Wyant and K. Creath, "Basic wavefront aberration theory for optical metrology," *Applied Optics and Optical Engineering*, Vol. XI, Academic Press, Boston, MA (1992).

84. S. Vorndran, et al., "Comparison of holographic lens and filter systems for lateral spectrum splitting," *SPIE Proc.*, Vol. 9937, 2016.

85. I. D. Howlett, M. Gordon, J. W. Brownlee, J. K. Barton, and R. K. Kostuk, "Volume holographic reflection endoscope for In-Vivo ovarian cancer clinical studies," *SPIE Proc.*, 8927, 2014.

86. R. K. Kostuk and G. Rosenberg, "Analysis and design of holographic solar concentrators," *Proc. SPIE*, 7043, 2008.

87. B. Chrysler et al., "Volume holographic lens spectrum splitting photovoltaic system for high energy yield with direct and diffuse solar illumination," *SPIE Proc.*, 10368, 2017.

12

Holographic Data Storage

12.1 Introduction

A number of different types of storage media and systems exist including magnetic, electronic, and optical recording methods. Some of the important characteristics of data storage systems include the storage density, write time, data access time, read time, and data persistence. Most data storage systems write and read each data bit in serial fashion. Conventional optical storage systems have migrated to disk-based formats and great strides have been made in increasing storage densities. The current storage density of Blu-ray disk technologies is about 12.5 Gbit/in^2 and the theoretical limit is projected to be 75 Gbit/in^2 [1]. However, even these impressive storage densities are not expected to keep pace with the data storage demands of a growing technological society.

During the past 20 years, a variety of techniques have also been developed to increase the data transfer and access rates of data for removable data storage systems. These systems are widely used for portable consumer video applications. A large percentage of these systems are based on rotating media formats such as compact disks, digital video disks, and high bit density Blu-ray disks. However, despite these advances the transfer rates of serial data are a bottleneck to taking full advantage of higher storage densities within practical access times and transfer rates. For instance, transfer rates for current Blu-ray systems are still only about 15–20 MB/s and limits practical storage capability to about 100 GB per disk [1].

Holography by nature is a parallel recording and reconstruction process. This characteristic was recognized by Dennis Gabor [2] shortly after he invented holography and was later described more formally in a classic paper by van Heerden in 1963 [3]. In that paper, van Heerden points out that the theoretical optical storage density in a volume (V) of recording material is proportional to V/λ^3, where λ is the wavelength of light. Therefore, in a volume of 1 cm^3 and a wavelength of 1 μm, a terabit of information could in theory be stored. In fact, page oriented holographic storage systems have demonstrated the highest storage density of any removable media (>700 Gb/in^2) and are theoretically capable of up to 40 Tb/in^2 [4]. Equally important is the fact that an entire "page" of data can be stored and accessed at one time, which greatly increases transfer rates. These factors make holographic storage very attractive for data storage applications.

Work on holographic data storage systems (HDSSs) started with the advent of the laser in the 1960s, however, the lack of suitable optical interface devices such as spatial light modulators (SLMs) and camera arrays as well as holographic recording materials with the right characteristics prevented development. By the mid-1990s, this situation began to change. High performance SLMs and camera arrays as well as dry-processed photopolymers are now available for removable write once read many storage applications. HDSS research and development programs were started by government and industries in the United States, Japan, and Korea and are continuing to the present day. Two excellent books are available that provide an in depth review of HDSS technology and history and are suggested for a complete review of the subject [see references 1 and 5]. The purpose of the following chapter is to provide a summary of some of the main features of HDSS and in particular the demands on the holographic material, recording, and reconstruction process.

12.2 Holographic Data Storage System Configurations

A basic HDSS has several components as shown in Figure 12.1. First, during the encoding process, large blocks or arrays of data are converted into an optical object wave. This is done by electronically modulating a two-dimensional (2D) array of pixels within the aperture of a spatial light modulator. The pixel configuration corresponds to an array of electronic data that will be transferred and stored by the system. The pixels of an SLM can be altered to modify either the reflectance or transmittance of an

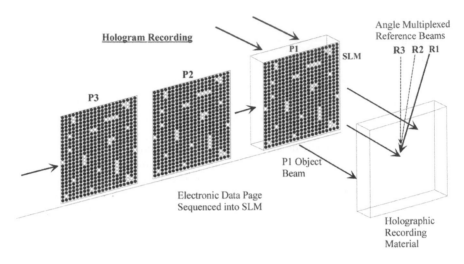

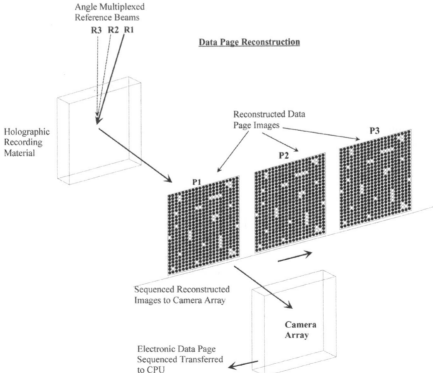

FIGURE 12.1 Schematic showing the encoding and reconstruction of data in a holographic data storage system. A page of data $(P_1, P_2, P_3, \ldots)$ is encoded using a SLM, illuminated, and then used as an object beam for a holographic exposure with a coded reference beams $(R_1, R_2, R_3, \ldots)$.

optical beam to indicate a "1" or "0" bit at a specific location in the array of electronic data. When the surface of the SLM is illuminated, the reflected or transmitted beam represents a full page of encoded electronic data. Each object wave must then be matched or encoded with a distinct reference beam for recording a hologram. The encoding is done so that a specific page can later be recalled with the desired data without ambiguity.

The recording process is repeated thousands of times to maximize the storage density of the medium. For reconstruction, an encoded optical reference beam illuminates the hologram and reconstructs the desired data page. The reconstructed object beam representing the data page is then projected onto a pixelated camera aperture to convert the optical signal back to an electronic format for subsequent use by the computer processing system.

Several different geometries are possible for recording holograms in a HDS system. Two configurations that have been the most successful are the image plane and Fourier transform hologram recording geometries. For the image plane geometry shown in Figure 12.2, an image of the SLM aperture is formed on the hologram recording material and serves as the object beam for the holographic recording. This configuration is useful for recording scenes as in mapping applications, but the recording density is not as high as can be achieved with a Fourier transform (FT) geometry. In the FT hologram arrangement (Figure 12.3), the SLM aperture is placed at the front focal plane of a lens and the holographic recording material at the back focal plane.

The optical FT of the SLM aperture minimizes the size of the object beam and the required hologram recording area on the medium surface. In addition, since the FT geometry records the spatial frequency content of the data page, non-essential spatial frequencies can be discarded provided their elimination

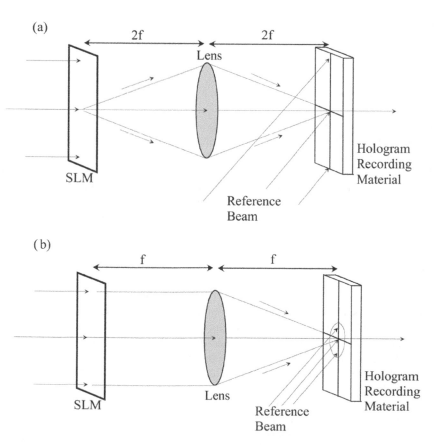

FIGURE 12.2 Holographic data storage recording geometries: (a) image plane geometry with the SLM imaged onto the surface of the recording material and (b) FT configuration with the SLM placed in the front focal plane and the recording material placed in the back focal plane of a lens.

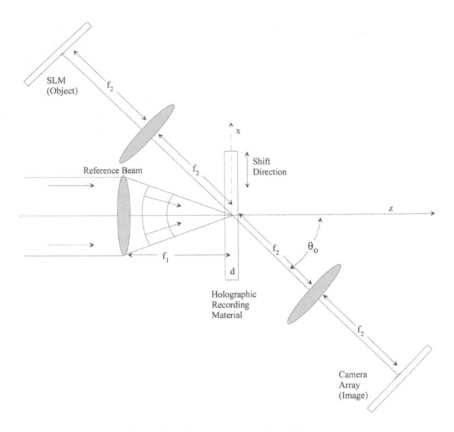

FIGURE 12.3   Illustration of a recording and readout system for shift multiplexing.

preserves adequate data content fidelity. Error correction schemes are also used to reduce the bit error rate of the information to an acceptable level. This approach allows further reduction in the size of the hologram area and increases the storage capacity of the medium.

The FT configuration produces an exact transform at the back focal plane of the lens when a planar object is placed at the front focal plane of the lens [6]. The object for the HDS system is an SLM surface which is essentially planar. However, the recording material for an HDS system will generally have significant thickness, and therefore the placement of Fourier plane within the material must be optimized for the recording and multiplexing method that is used.

12.3 Hologram Multiplexing Techniques

Typically, the image resolution of an optical system used to holographically record and reconstruct a page of information will be lower than a serial "bit-wise" system. This is due to the fact that the bit-wise system images an on-axis point and will generally have lower aberration than an off-axis imaging configuration. Therefore, in order for the HDS system to have an advantage in storage density, many holograms must be multiplexed in the same region of the recording material. This can be achieved using any variety of multiplexing schemes.

Multiplexing several holograms in the same volume of the recording material is useful provided that the data can be recovered with sufficient fidelity to unambiguously retrieve a data page. A number of multiplexing methods have been developed and demonstrated for holographic data storage systems that take advantage of the volume selectivity, directionality, and wavefront matching properties of the holographic process. The methods that have achieved some degree of success include angle, wavelength, phase shift, peristrophic, and polytopic holographic multiplexing [1,5]. Combinations of these methods have also been used to maximize storage density of HDS systems. The basic features of these different hologram multiplexing methods are presented in the following sections.

12.3.1 Angle Multiplexing

One of the most straightforward methods of hologram multiplexing is to change the angle of the object or reference beam (or both) between sequential exposures of the recording material. For this method, the storage density that can be achieved primarily depends on the angular Bragg selectivity of the volume hologram. Assuming that only the reference beam angle is changed between successively recorded holograms, the minimum separation between reference beam angles, $\Delta\theta$, can be estimated using the angular difference between the peak and first null of the diffraction efficiency profile for an individual hologram. $\Delta\theta$ can be determined using the detuning parameter, ϑ, from the coupled wave analysis described in Chapter 5. The diffraction efficiency profile of a transmission type phase hologram using the approximate coupled wave model is given by:

$$\eta = \frac{\sin^2\left(v^2 + \xi^2\right)^{1/2}}{\left(1 + \xi^2/v^2\right)},\tag{5.70}$$

with $v = \frac{\pi n_1 d}{\lambda c_r c_s}$; $\xi = \frac{\vartheta d}{2c_s}$.

Maximum diffraction efficiency occurs when $v = \pi/2$ and $\xi = 0$. The first null occurs when $\xi \approx 2.7$ (see Figure 5.11) and results in an angular selectivity of:

$$\Delta\theta = \frac{2.7}{\pi}\frac{\cos\theta_o}{\sin(\varphi - \theta_o)}\left(\frac{\Lambda}{d}\right).\tag{12.1}$$

This relationship shows that the angular selectivity is directly proportional to the grating period Λ and inversely proportional to the grating thickness d. For a grating with a period of 0.50 μm and a thickness of 10 μm, the angular selectivity is about 2.4° and 0.24° with a 100 μm thickness. This indicates that significant benefit in angle multiplexed hologram storage density can be gained by increasing the effective thickness of the recording material.

Angular multiplexing is relatively easy to implement and is often used in combination with other multiplexing techniques to achieve high data storage density in holographic recording materials [7]. Several of these methods are discussed in the following sections.

12.3.2 Wavelength Multiplexing

In this technique, holograms are recorded in the material at different wavelengths rather than different angles and is also based on the Bragg selectivity properties of the hologram. The wavelength separation, $\Delta\lambda$, between successive holographic recordings is the wavelength difference from the Bragg wavelength for the initial hologram that causes a drop in diffraction efficiency from a maximum value to the first minimum. This again occurs when $\xi \approx 2.7$ in the diffraction efficiency relation as described above. When Bragg detuning is only a function of wavelength, the wavelength selectivity is:

$$\Delta\lambda \approx 5.4 n \frac{\Lambda^2}{d}\cos\theta_o,\tag{12.2}$$

where n is the average refractive index of the material, Λ is the grating period, d is the grating thickness, and θ_o is the Bragg angle. In this case with $\Lambda = 0.5$ μm and a grating thickness of 100 μm, the wavelength selectivity $\Delta\lambda \sim 19$ nm. Although the availability of tunable wavelength lasers has been improving in recent years, a tuning range of several hundred nm would be required for these recording conditions. For this reason, other forms of hologram multiplexing have proven to be more practical than those relying on wavelength change.

12.3.3 Shift Multiplexing

In most multiplexing methods, a plane wave reference beam is used to record individual holograms. However, for the shift multiplexing method, a more complex phase function is used for the reference beam [8]. For example, if a spherical reference wave is used, the propagation vector and corresponding spatial frequency varies over the aperture of the hologram. This variation can be considered a change in angle of incidence of the reference beam across the hologram aperture.

When the hologram is shifted by δ_x in the x-direction, the angular spectrum of the reference beam also shifts and reduces angular Bragg matching across the hologram aperture. This in turn reduces the total diffraction efficiency of the hologram. The approximate amount of lateral shift δ_x required to accomplish this is given by:

$$\delta_x = \frac{\lambda f_1}{d \tan \theta_o} + \lambda f_{ref}^{\#}, \qquad (12.3)$$

where d is the hologram thickness, f_1 is the focal length of the lens used to form the spherical reference beam, θ_o is the angle of the object beam (Figure 12.3), and $f_{ref}^{\#}$ is the f-number of the reference beam. The first term in Equation 12.3 indicates the Bragg selectivity and the second term the diffraction limited uncertainty of the focus position of the spherical reference beam.

A convenient method to implement shift multiplexing is by using a disk coated with a holographic recording material and rotated by the shift value to record and read out data for individual holograms. This configuration is shown in Figure 12.4 and also allows shift multiplexing to be combined with other multiplexing methods such as angle and peristrophic to further increase the data storage density of the disk.

12.3.4 Peristrophic Multiplexing

In this form of multiplexing, the medium is rotated about the normal to the material surface between each exposure as shown in Figure 12.5 [9]. Peristrophic multiplexing can be performed in a thin recording material that has low Bragg selectivity, however, the method is more effective in thicker and more selective materials. The rotation causes several things to occur that allows more holograms to be recorded. First, the rotation translates the reconstructed image away from the aperture of the detector array allowing for another image to be read out. Next, for thick media, the angle shift caused by the rotation detunes the hologram from the Bragg condition, reducing the diffraction efficiency of the initial hologram. Finally, when a polymer is rotated, additional monomer is available for diffusion. This effectively

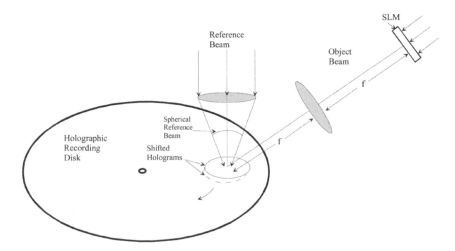

FIGURE 12.4 Shift multiplexing implemented in a disk format. The disk has a layer of holographic recording material for writing and reading holograms.

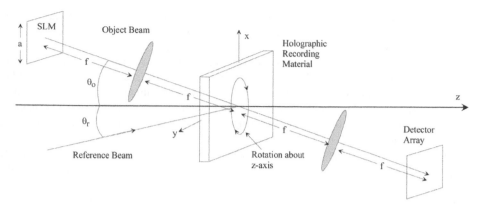

FIGURE 12.5 Configuration for peristrophic hologram multiplexing.

extends the dynamic range of the polymer and allows additional holograms to be recorded in the same region of recording material.

As shown in Figure 12.5, the rotation is about the normal to the film plane (*z*-axis). If the length of the SLM is *a*, then the angular rotation that prevents overlap of the images when recorded in the Fourier transform geometry is:

$$\delta\theta \geq \frac{a/f}{\sin\theta_o + \sin\theta_r},$$ (12.4)

where *f* is the focal length of the FT transform lenses, θ_o is the angle that the object beam makes with the rotation axis, and θ_r is the angle that the reference beam makes with the rotation axis.

12.3.5 Polytopic Multiplexing

With conventional angle multiplexing methods, a set of holograms are recorded in an overlapping volume of recording material as shown in Figure 12.6a. The holograms are typically recorded using the Fourier transform hologram geometry (Section 3.5.3) to minimize the lateral area required for a set of

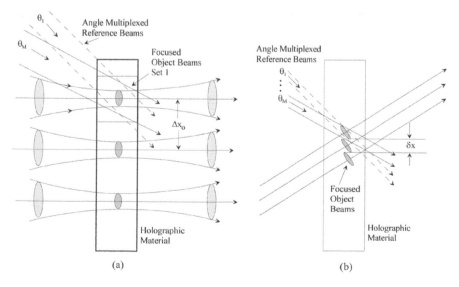

FIGURE 12.6 (a) Conventional angle multiplexed set of holograms in a volume material with sets of holograms separated by Δx_o and (b) polytopic angle multiplexing of sets of holograms with separation δx.

multiplexed holograms. After recording one set of multiplexed holograms, the material is displaced allowing another set of holograms to be recorded in unexposed material. The displacement between different sets of multiplexed holograms is typically made large enough so that the exposed areas do not overlap and reduce crosstalk between holographic images. Using this basic approach, angle multiplexing allows the recording of several thousand holograms in volume materials like $LiNbO_3$ photorefractive crystals [7]. This method has also been used to obtain ~200 Gbyte storage capacities on 13 cm diameter disks with holographic photopolymer recording materials [10]. As impressive as these results are, they are still not sufficient to compete with magnetic storage techniques. In order to increase the storage capacity of holographic memories, Anderson and Curtis proposed a new approach to angle multiplexing called "polytopic" multiplexing [11]. This method has been shown to increase holographic storage capacity by an order of magnitude over conventional angle multiplexing methods [10], thus providing a significant advantage.

In polytopic multiplexing, the focus regions of the object beams for each set of multiplexed holograms are separated during the recording step. However, unlike in conventional angle multiplexing (Figure 12.6a), where the expanded construction beams are also separated on the recording material, polytopic multiplexing allows partial overlap of these regions (Figure 12.6b). As a result, during the reconstruction process with polytopic multiplexing, some unwanted images from adjacent sets of multiplexed holograms also form. However, the unwanted holographic images can be filtered out using an aperture.

For polytopic multiplexing, the required spatial separation between neighboring sets of recorded holograms (δx) is given by [1]:

$$\delta x = a_F \cdot \cos\theta_o + a_F \cdot \sin\theta_o \cdot \tan(\theta_o + \varphi), \tag{12.5}$$

where a_F is the diameter of the focused object beam, θ_o is the center angle of the object beam relative to the surface normal of the recording material, and φ is the half angle corresponding to the numerical aperture of the Fourier transform lens. The aperture area used to filter adjacent sets of reconstructed holograms should be greater than the Nyquist area (A_{Nyq}) occupied by the spatial frequencies of the data given by:

$$A_{Nyq} = \left[\frac{\lambda f_{FT}}{\Delta_p} \right]^2, \tag{12.6}$$

where Δ_p is the width of the SLM pixels, and f_{FT} is the focal length of the Fourier transform lens used in the recording and reconstruction system. A system for recording and reconstructing holograms using polytopic multiplexing is shown in Figure 12.7. In this illustration, the holograms are reconstructed with conjugate reference beams. The adjacent reconstructed data page is at a slight angle to the optical axis between the lenses, whereas the desired page is parallel to the optical axis and passes through the filter [1].

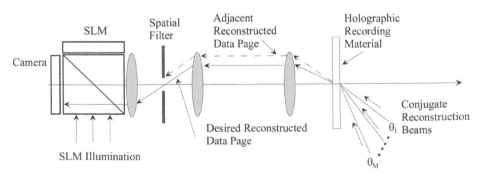

FIGURE 12.7 Schematic diagram for implementing a polytopic hologram multiplexing system in combination with angle multiplexing.

12.4 Recording Material Considerations

The holographic recording material for a HDS system must satisfy several requirements. First, it must allow the recording of thousands of data pages in the same region of material. This is required to achieve high storage density and implies the need for a large material dynamic response range. A high multiplexing requirement also necessitates the need for a thick material that allows high angular and wavelength selectivity. The reconstructed holograms must also be of sufficient optical quality to allow data recovery with low bit error rates. The material should have minimal shrinkage or swelling during processing to reduce complications of data recovery, and holograms must be stable for long periods of time that are consistent with the storage application needs (which could be for 10's of years). Finally, an ideal material should allow writing and reading of holograms in a short time frame and not require complex processing procedures. At the current time, the materials that have come closest to satisfying these requirements are photorefractive crystals and photopolymers. However, the format flexibility and lower cost of photopolymer compounds have led to their use in several recent HDS systems [1,5]. The following sections describe the metrics established for assessing materials for HDS applications and specific material issues related to the multiplexing and recovery of large numbers of colocated holograms. The discussion mostly focuses on photopolymer materials since these are becoming the preferred medium, however, the metrics and other considerations can be applied to other material systems.

12.4.1 Holographic Material Dynamic Range for Multiplexing

A useful parameter for assessing the performance of a recording material for holographic data storage applications is the $M^{\#}$. It provides a numerical value for comparing different materials and represents the dynamic response of the material to holographic exposure. The $M^{\#}$ is based on the approximate coupled wave expression for the diffraction efficiency of a transmission phase hologram reconstructed at the Bragg condition:

$$\eta_i = \sin^2 (v_i),\tag{12.7}$$

with the grating strength parameter given by:

$$v_i = \frac{\pi \, n_i d}{\lambda \cos\theta},\tag{12.8}$$

where n_i is the refractive index modulation, d is the grating thickness, λ is the wavelength, and θ is the reconstruction angle for the hologram. When multiplexing a large number of holograms, the index modulation for each hologram will be small. In this case, an expression for a single hologram (i) can be written as:

$$\sqrt{\eta_i} = \sin(v_i) \approx v_i,\tag{12.9}$$

and a cumulative grating strength parameter for M multiplexed holograms is:

$$v_M = \sum_{i=1}^{M} \sqrt{\eta_i}\,.\tag{12.10}$$

The maximum cumulative grating strength is called the $M^{\#}$ of the material or:

$$M^{\#} = \sum_{i=1}^{M} \sqrt{\eta_i}\,.\tag{12.11}$$

Typical $M^\#$ values for photopolymers used for holographic data storage range from ~2 for 38 μm thick Dupont HRF-150 [12] to 17 for 200 μm thick Aprilis ULSH-500 [13]. The individual hologram diffraction efficiency for recording ~1000 holograms is on the order of 10^{-4}–10^{-3}.

12.4.2 Hologram Exposure Scheduling

If equal exposure energy density is used to record each hologram in a set of M multiplexed holograms, the resulting diffraction efficiency of the holograms changes and depends on the particular position in the exposure sequence. Figure 12.8 from Pu et al. [12] shows the diffraction efficiency for 90 holograms recorded in a Dupont HRF-150 photopolymer with a constant 1 mJ/cm² of exposure energy for each hologram. Notice that hologram formation does not take place until the exposure energy density exceeds 4–5 mJ/cm², and that for holograms recorded later in the sequence, the diffraction efficiency is significantly lower than the peak value. These characteristics are undesirable and are a direct result of the non-linear characteristic of the cumulative grating strength of the recording material. In order to record a 1000 or more holograms in the same region of the recording material, it is necessary to use the entire dynamic response range and to have near equal diffraction efficiency for each hologram. Fortunately, it is possible to achieve this by using an exposure compensation method called "exposure scheduling" [12].

The first step in developing an exposure schedule is to characterize the recording material with a specific hologram exposure method. For example, using a combination of angle and peristrophic multiplexing with a certain time delay between each exposure in a specific type of photopolymer. This is required because the exposure method and delay time affects the monomer diffusion mechanism of the hologram formation process (see Chapter 8 in Section 8.4). It may also be necessary to give the material a "pre-imaging" exposure to have the material exceed the threshold for polymerization.

The next step in developing an exposure schedule is to approximate the characteristic curve for the refractive index modulation (n_1) with a polynomial function. The refractive index modulation is determined indirectly using Equations 12.7 and 12.8 from the previous section that relates the diffraction efficiency, v_i, and n_1. Figure 12.9 shows an approximation to the refractive index modulation function with respect to the hologram exposure energy density similar to that used by Pu et al. [13] in which a sixth order polynomial approximates the experimental grating strength:

$$v_N = b_0 + b_1 E + b_2 E^2 + b_3 E^3 + b_4 E^4 + b_5 E^5 + b_6 E^6, \tag{12.12}$$

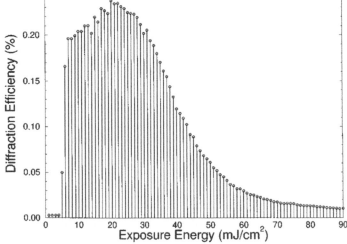

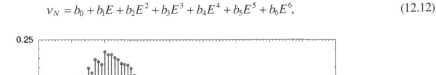

FIGURE 12.8 Diffraction efficiency for 90 holograms recorded in Dupont HRF-150 with equal 1 mJ/cm² exposure energy for each hologram. (From Pu, A. et al., *Opt. Eng.*, 35, 2824–2829, 1996. With permission from the SPIE.)

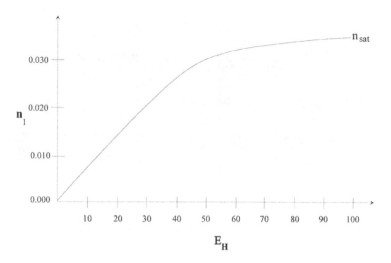

FIGURE 12.9 A representation of the refractive index modulation (n_1) for a holographic polymer as a function of the exposure energy per hologram (E_H). n_{sat} is the cumulative grating strength at the saturation point of the material. (From Pu, A. et al., *Opt. Eng.*, 35, 2824–2829, 1996. With permission from the SPIE.)

where v_N is the cumulative grating strength for N holograms, E is the cumulative exposure energy, and b_i are the polynomial coefficients determined from the curve fitting function. Note that N is less than M which is the maximum number of holograms that can be stored in the material at the point of saturation.

The derivative of Equation 12.12 with respect to the exposure energy gives the rate of change of the grating strength with exposure energy. If the entire dynamic range of the material is equally divided among all multiplexed holograms, the required exposure schedule can be written as:

$$\frac{v_{Sat}}{M} = \left[\frac{\partial v_N}{\partial E} \Big|_{E = \sum_{i=1}^{m-1} E_i} \right] \times E_m, \tag{12.13}$$

where v_{Sat} is the experimentally determined saturation grating strength value, M is the total number of holograms recorded to the point of saturation (therefore, v_{Sat}/M is the grating strength for recording M holograms with equal diffraction efficiency), E_m is the exposure energy density required to record the mth hologram in the sequence, and E_i is the amount of exposure energy required to record the previous ith holograms in the sequence. Re-arranging this expression and substituting in the polynomial function for v_N gives the exposure energy required for recording the mth hologram in the sequence with the same efficiency as the others:

$$E_m = \left(\frac{v_{N-Sat}}{M} \right) \cdot$$

$$\left[b_1 + 2b_2 \sum_{i=1}^{m-1} E_i + 3b_3 \left(\sum_{i=1}^{m-1} E_i \right)^2 + 4b_4 \left(\sum_{i=1}^{m-1} E_i \right)^3 + 5b_5 \left(\sum_{i=1}^{m-1} E_i \right)^4 + 6b_6 \left(\sum_{i=1}^{m-1} E_i \right)^5 \right]^{-1}. \tag{12.14}$$

Using this procedure, Pu et al. [12] were able to obtain near equal diffraction efficiency for 50 holograms colocated in Dupont HRF-150 photopolymer, and Waldman et al. recorded 500 equal efficiency holograms in the Aprilis ULSH-500 photopolymer (Figure 12.10) [13].

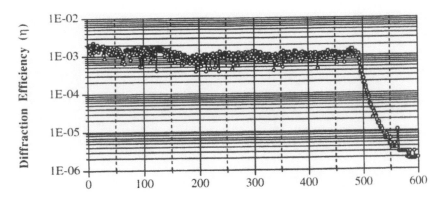

FIGURE 12.10   Diffraction efficiency of 500 holograms recorded in a 200 μm thick sample of Aprilis ULSH-500 photopolymer after scheduling the exposure dosage. (From Waldman, D.A. et al., *SPIE*, 3291, 89–103, 1998. With permission from the SPIE.)

12.5 Object Beam Conditioning

As shown earlier, the Fourier transform geometry is a very useful configuration for data storage as it confines most of the useful information of the object field in a compact region of the storage medium (Figure 12.11). However, since the data page consists of regularly spaced pixels across the aperture of an SLM or data mask, the Fourier transform of this pattern forms an array of high intensity peaks weighted by an envelope function (Figure 12.12a) [14]. For the image shown in this figure, the data mask has circular pixels of diameter d_p arranged on a grid with period P_x and P_y in the x- and y-directions, respectively. The functional form for the transform of this object field is:

$$E_o\left(x_o, y_o\right) = C \cdot \frac{J_1\left(kd_p r_o / 2f\right)}{\left(kd_p r_o / 2f\right)} \cdot \frac{\sin\left(kMP_x x_o / 2\right)}{\sin\left(kP_x x_o / 2\right)} \cdot \frac{\sin\left(kMP_x y_o / 2\right)}{\sin\left(kP_x y_o / 2\right)}, \tag{12.15}$$

where C is a constant, k is the propagation vector, f is the focal length of the transform lens, M is the number of pixels in the x- and y-directions, and $r_o = \sqrt{x_o^2 + y_o^2}$. A problem with this method is that the resulting intensity variations of the transformed object beam are difficult to record within the dynamic range of the holographic material and leads to artifacts in the reconstructed data (Figure 12.12b).

A simple approach to reduce the intensity variation of the transformed object beam is to shift the material away from the focal point of the transform lens as shown in Figure 12.13. Figure 12.11c shows the change in the intensity pattern (Figure 12.11a) when the evaluation plane is shifted by 1% of the focal length from the focal plane. However, although the intensity becomes more uniform, the area occupied by the beam increases and decreases the storage density.

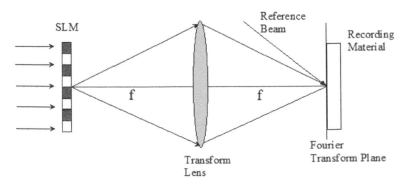

FIGURE 12.11   Fourier transform geometry for recording the object field from a SLM.

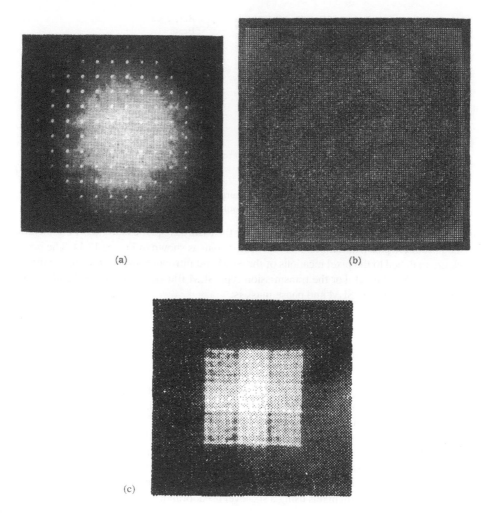

(a)

(b)

(c)

FIGURE 12.12 (a) Fourier transform of a data mask consisting of regularly spaced circular object data points, (b) Reconstructed image of the hologram showing unwanted intensity variations in the reconstructed data points, and (c) The object beam at the hologram plane after shifting the Fourier plane by 1% of the focal length of the transform lens. (From Takeda, Y., *Jap. J. Appl. Phys.*, 11, 656–665, 1972. With permission from IoP Science.)

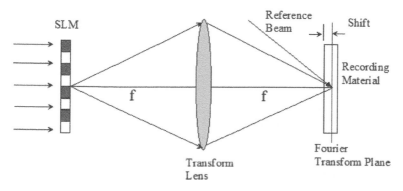

FIGURE 12.13 Method of decreasing the intensity variation of the transformed object field by shifting the recording material away from the focal plane of the transform lens.

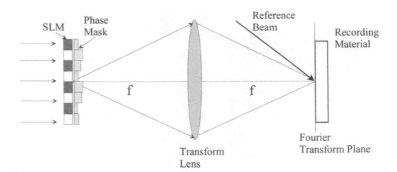

FIGURE 12.14 Fourier transform hologram recording configuration with a phase mask positioned next to the SLM to decreases the intensity peaks of the transformed data at the recording plane.

Another approach to reducing the intensity peaks at the recording plane is to incorporate either a random or pseudo random phase mask into the recording system as shown in Figure 12.14. The pixels of the phase mask are matched to the pixel locations of the SLM and introduce an arbitrary phase difference to the field from each SLM pixel. For the transmission type SLM illustrated in Figure 12.14, the resulting transmittance function for the SLM and phase mask is:

$$t'(x, y) = \sum_{p=1}^{N} \sum_{q=1}^{N} a_{p,q} \exp(j\varphi_{p,q}) rect[x - p \cdot d_{pix}] \cdot rect[y - q \cdot d_{pix}], \qquad (12.16)$$

where N is the number of pixels in the SLM in each orthogonal direction, $\varphi_{p,q}$ is the phase of the phase mask at the pixel (p,q), and d_{pix} is the width of the pixel. The Fourier transform of $t'(x, y)$ is the object field for the holographic recording. Randomizing the phase of the pixels modifies the diffraction pattern from the periodic SLM transmittance function. With proper design, this technique can reduce the intensity spikes at the recording material without significantly increasing the area of the transformed data. Phase masks with completely random phase variations and with patterns of $N \times N$ randomized bits that are repeated across SLM aperture have both been used with success [15–17].

Figure 12.15 illustrates the results of modifying the intensity distribution of the Fourier transform of a 256×256 input array with 50 μm pixels using a six-level pseudorandom phase mask [17]. The figure also

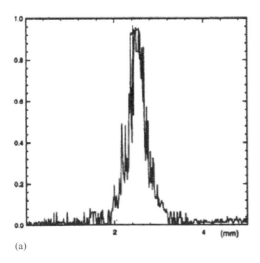

(a)

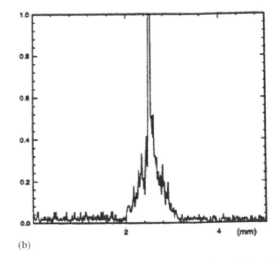

(b)

FIGURE 12.15 Intensity distribution of the Fourier transform of a 256×256 input array with 50 μm pixels using a six level pseudorandom phase mask (a). The figure on the right shows the intensity distribution of the Fourier transform of the same object using focal shifting to keep the width of the distribution the same as with the pseudorandom phase mask (b). (From Gao, Q. and Kostuk, R.K., *Appl. Opt.*, 36, 4853–4861, 1997. With permission from the Optical Society of America.)

shows the distribution obtained by shifting the focal plane by an amount to make the width of the distribution the same as that for the pseudo random phase mask. Note that the focal shifted distribution saturates the camera response while the pseudorandom phase mask method does not saturate the camera. This indicates that holograms recording an object field with a pseudorandom phase mask will require less dynamic range than focal shifting for the same storage density.

12.6 Representative Holographic Data Storage Systems

Several holographic data storage systems have been demonstrated that incorporate the advances discussed in this chapter [18,19]. In work reported by Orlov et al. [18], a disk system was demonstrated that was able to retrieve data stored in 8200 holograms (1 GByte) at a rate of 0.65 Gbits/sec. The holograms were recorded on a 16.5 cm diameter disk with a 200 µm thick polymer. Only 10% of the available media area was used to record the holograms, so significantly more storage density is possible. IBM researchers developed a holographic data storage test system for evaluating modulation and error correction codes and other performance metrics using a photorefractive crystal with a 1.6 mm² aperture as the recording material [19]. This system was capable of using data with $<10^{-12}$ bit error rate at 61% overall code rate and an area storage density of 250 Giga-pixels/in². However, perhaps the most successful holographic data storage system to date is the InPhase Tapestry 300r write once read many type system that is able to store 300 Gbyte on a 13 cm diameter disk with a transfer rate of 160 Mbit/sec [10]. It is also capable of an area storage density of 515 Gbit/in² and is a significant advance over serial optical disk storage densities (75 Gbit/in² theoretical). In spite of difficulties with commercializing HDS systems, significant research is still ongoing due to the increasing demands for very high capacity, low power usage, archival data storage.

PROBLEMS

1. A set of 145 holograms are multiplexed in a 200 µm thick photopolymer. There are 50 holograms with $\eta = 0.2\%$; 30 holograms with $\eta = 0.15\%$; and 65 holograms with $\eta = 0.17\%$. Determine the $M^{\#}$ of the photopolymer.

2. The index modulation as a function of exposure energy density for a holographic recording material is shown in Figure 12.16. 100 holograms are recorded in this material. (i) Plot the diffraction efficiency as a function of the hologram number from 1 to 100 when the exposures are equal and use the full index modulation range of the recording material. (ii) Design an exposure schedule to equalize the diffraction efficiency of the 100 holograms and plot the result.

3. A disk format holographic storage system is formed as shown in Figure 12.4 with laser light having a wavelength of 532 nm. For this problem, assume that the reference beam is collimated

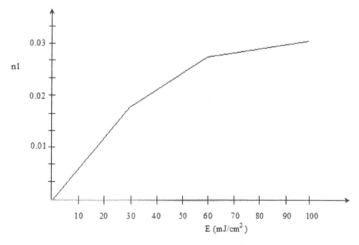

FIGURE 12.16 Refractive index modulation response with exposure energy density for Problem 2.

and at normal incidence to the disk surface. Compute the number of holograms that can be angle multiplexed in a 200 μm thick photopolymer with a refractive index of 1.52. Only consider one location on the disk—i.e., one rotation position. What factors will limit the range of object beam angles that can be used to form the holograms?

4. Assume that a peristrophic multiplexing scheme is now used with the holographic disk format storage system in problem 3. If the SLM has a width of 7.5 mm, the Fourier transform lens has a focal length of 15 cm, and the object beam angle relative to the disk is 30°, determine the number of holograms that can be multiplexed using the peristrophic method with these conditions. Assume that a full 360° of rotation is possible for the peristrophic multiplexing.

5. The holographic disk format storage system described in Problems 3 and 4 is now used with a shift multiplexing scheme. The reference beam in this case is focused with a 2.5 cm diameter, 10 cm focal length lens, and the object beam stays at an angle of 30° relative to a normal to the disk surface. (i) Determine the minimum shift distance and (ii) the number of shift multiplexed holograms that can be recorded along the circumference at a radius of 6 cm from the center of the disk. Assume that the diameter of the recording disk is 12 cm.

REFERENCES

1. K. Curtis, L. Dhar, A. Hill, W. Wilson, and M. Ayres, *Introduction to Holographic Data Recording*, Wiley Online Library, doi:10.1002/9780470666531, published June 30, 2010.
2. D. Gabor, "A new microscope principle," *Nature*, Vol. 161, 777–778 (1948).
3. P. J. van Heerden, "Theory and optical information storage in solids," *Appl. Opt.*, Vol. 2, 393–400 (1963).
4. K. Shimada et al., "High density recording using monocular architecture for 500 GB consumer system," *Opt. Data Storage Conf.*, Paper TuC2, Buena Vista, FL (2009).
5. H. J. Coufal, D. Psaltis, and G. T. Sincerbox, Eds., *Holographic Data Storage*, Springer Optical Sciences, Berlin, Germany (2000).
6. J. W. Goodman, *Introduction to Fourier Optics*, 4th Ed., Ch. 6, Freeman & Co., New York (2017).
7. F. H. Mok, "Angle multiplexed storage of 5000 holograms in lithium niobate," *Opt. Lett.*, Vol. 18, 915–917 (1993).
8. D. Psaltis, M. Levene, A. Pu, and G. Barbastathis, "Holographic storage using shift multiplexing," *Opt. Lett.*, 20, 782–784 (1995).
9. K. Curtis, A. Pu, and D. Psaltis, "Method of holographic storage using peristrophic multiplexing," *Opt. Lett.*, Vol. 19, 993–995 (1994).
10. L. Dhar, K. Curtis, and T. Facke, "Holographic data storage: Coming of age," *Nat. Photonics*, Vol. 2, pp. 403–406 (2008).
11. K. Anderson and K. Curtis, "Polytopic multiplexing," *Opt. Lett.*, Vol. 29, 1402–1404 (2004).
12. A. Pu, K. Curtis, and D. Psaltis, "Exposure schedule for multiplexing holograms in photopolymer films," *Opt. Eng.*, Vol. 35, 2824–2829 (1996).
13. D. A. Waldman, H.-Y. S. Li, and E. A. Cetin, "Holographic recording properties in thick films of ULSH-500 photopolymer," *SPIE*, Vol. 3291, 89–103 (1998).
14. Y. Takeda, "Hologram memory with high quality and high information storage density—hologram memory," *Jap. J. Appl. Phys.*, Vol. 11, 656–665 (1972).
15. C. B. Burkhardt, "Use of random phase mask for the recording of Fourier transform holograms of data masks," *Appl. Opt.*, Vol. 9, 695–700 (1969).
16. A. Emoto and T. Fukuda, "Randomly displaced phase distribution design and advantage in page-data recording of Fourier transform holograms," *Appl. Opt.*, Vol. 52, 1183–1191 (2013).
17. Q. Gao and R. K. Kostuk, "Improvement to holographic digital data-storage systems with random and pseudorandom phase masks," *Appl. Opt.*, Vol. 36, 4853–4861 (1997).
18. S. S. Orlov, W. Phillips, E. Bjornson, Y. Takashima, P. Sundaram, L. Hesselink, R. Okas, D. Kwan, and R. Snyder, "High-transfer rate high capacity holographic disk data storage system," *Appl. Opt.*, Vol. 43, 4902–4914 (2004).
19. G. W. Burr, C. M. Jefferson, H. Coufal, M. Jurich, J. A. Hoffnagle, R. M. Macfarlane, and R. M. Shelby, "Volume holographic data storage at an areal density of 250 gigapixels/in²," *Opt. Lett.*, Vol. 26, 444–446 (2001).

Appendix A: *Mathematical Relations*

A.1 Spatial Fourier Transform Operations

Many of the linear operations in optics such as diffraction propagation can be expressed as spatial Fourier transforms. One form of the spatial Fourier transform and inverse transform are [1]:

$$FT\{u(x,y)\} = \int\!\!\int_{-\infty}^{\infty} u(x,y)\exp\left[-j2\pi\left(v_x x + v_y y\right)\right]dxdy = U\left(v_x,v_y\right)$$

(A.1)

$$FT^{-1}\{U\left(v_x,v_y\right)\} = \int\!\!\int_{-\infty}^{\infty} U\left(v_x,v_y\right)\exp\left[j2\pi\left(v_x x + v_y y\right)\right]dv_x dv_y = u(x,y),$$

where v_x and v_y are the spatial frequencies in the transform plane.

Several theorems and operations can be used to simplify the analysis of problems involving Fourier transform integrals [2]. The following theorems are some of the more important ones for holography. (For this analysis, u and w are functions of (x,y), while a and b are constants.) In addition:

$$U\left(v_x,v_y\right) = FT\left[u(x,y)\right]; \quad W\left(v_x,v_y\right) = FT\left[w(x,y)\right].$$

(A.2)

1. *Linearity Theorem*:

$$FT\{au + bw\} = aFT\{u\} + bFT\{w\}$$

(A.3)

This theorem shows that the transform of a combination of linear scaled functions is separable.

2. *Similarity Theorem*:

$$FT\{u(ax,by)\} = \frac{1}{|ab|}U\left(\frac{v_x}{a},\frac{v_y}{b}\right)$$

(A.4)

This relation indicates that scaling the coordinates in one domain inversely scales both the amplitude and coordinates in the transform domain.

3. *Shift Theorem*:

$$FT\{u(x-a,y-b)\} = U\left(v_x,v_y\right)\exp\left[-j2\pi\left(av_x + bv_y\right)\right].$$

(A.5)

This important theorem shows that a translation in one domain results in a phase shift in the transform domain.

4. *Convolution Theorem:*

$$FT\left\{\int\int_{-\infty}^{\infty} u(x_1, y_1) w(x_2 - x_1, y_2 - y_1) dx_1 dy_1\right\} = U(v_x, v_y) W(v_x, v_y).$$ (A.6)

This theorem indicates that a convolution integral in one domain can be converted to a simple multiplication in the transform domain. In many cases it provides a great deal of simplification to problems involving the Fourier integral.

5. *Autocorrelation Theorem:*

$$FT\left\{\int\int_{-\infty}^{\infty} u(x_1, y_1) u^*(x_2 - x_1, y_2 - y_1) dx_1 dy_1\right\} = \left|U(v_x, v_y)\right|^2$$ (A.7)

This operation takes the Fourier transform of a function times its complex conjugate and gives the power spectrum of the function.

6. *Power Conservation (Parseval's Theorem):*

$$\int\int_{-\infty}^{\infty} |u(x, y)|^2 dx dy = \int\int_{-\infty}^{\infty} |U(v_x, v_y)|^2 dv_x dv_y$$ (A.8)

This theorem states that the power in one domain does not change after transformation into the other domain.

A.2 Some Common Fourier Transform Pairs

1. *rect and sinc* functions:
   The *rect* function is defined as:

$$rect(x) = \begin{array}{ll} 1 & |x| \le 1/2 \\ 0 & |x| > 1/2 \end{array},$$ (A.9)

and the sinc function as:

$$sinc(x) = \frac{\sin(\pi x)}{\pi x}.$$ (A.10)

These functions form a transform pair where:

$$rect(ax) \underset{\overleftarrow{}}{\overset{FT}{\longrightarrow}} \frac{1}{|a|} sinc(v_x/a)$$ (A.11)

2. *circ* and *Fourier Bessel* transform:
   The *circ* function is defined as:

$$circ(r) = \begin{array}{ll} 1 & |r| \le 1 \\ 0 & |r| > 1 \end{array},$$ (A.12)

where $r = \sqrt{x^2 + y^2}$ is a radial coordinate. The transform of this function is a Fourier Bessel function with:

$$B_F\left\{circ(r)\right\} = 2\pi \int\limits_0^1 rJ_o(2\pi r\rho)dr = \frac{J_1(2\pi\rho)}{\rho} \qquad\qquad (A.13)$$

And ρ is the radial coordinate in the transform plane.

REFERENCES

1. J. W. Goodman, *Introduction to Fourier Optics*, 4th ed., Freeman & Co., New York (2017).
2. R. Bracewell, *The Fourier Transform and Its Applications*, McGraw-Hill, New York (1965).

Appendix B: Practical Considerations for Hologram Construction

B.1 Holographic Material Considerations

For many holographic recording materials such as photopolymers, dichromated gelatin (DCG), and silver halide films the material properties before and after processing can change. Differences can occur in the lateral and axial dimensions of the film and the average refractive index. These differences can alter the diffraction efficiency and imaging properties of the hologram and must be characterized and incorporated into the design process.

B.1.1 Change in Film Dimensions (Film Shrinkage, Swelling, or Shearing)

The overall dimensional change in holographic emulsion and photopolymer film can be expressed as a relative volumetric shrinkage or swelling:

$$\gamma = \frac{\Delta V}{V}, \tag{B.1}$$

where V is the volume of the film, and ΔV is the volumetric material change. For materials such as silver halide that is processed using a fixation step, the material shrinks and results in a negative value for ΔV. This also occurs for photopolymers that undergo heat treatment. For DCG, the volume can either shrink or swell depending on the type of processing used. ΔV can be expressed as a combination of lateral and axial changes with:

$$\Delta V = \left(\Delta x \cdot \Delta y \right) \cdot \Delta z, \tag{B.2}$$

where $\left(\Delta x \cdot \Delta y \right)$ is the lateral, and Δz is the axial change in length. Since the volume of the recording material contains the grating period, the grating period will also change during processing:

$$\Delta V \propto \left(\Delta \Lambda_x \cdot \Delta \Lambda_y \right) \cdot \Delta \Lambda_z, \tag{B.3}$$

where $\Delta \Lambda_i$ represents the change in grating period in the x-, y-, or z-directions. As described earlier in Chapter 3, changes in the grating period also affect the grating (K) vector. For many important situations, the effects of lateral and axial change to the material can be treated separately. These conditions are examined in the following sections.

B.1.1.1 Effect of Axial Thickness Change in the Recording Material

For relatively thin holographic films (i.e., 2–25 μm) that are attached to a rigid substrate such as glass, the volumetric change, ΔV, predominately occurs in the z- or thickness direction. This situation is illustrated in Figure B.1 for a planar hologram where d is the original thickness and d' is the final thickness after processing. If $\Delta d = d' - d$ the resulting change in the z-component of the grating vector is:

$$K_z' = \frac{K_z}{1 + \dfrac{\Delta d}{d}}, \tag{B.4}$$

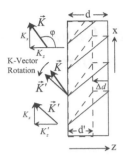

FIGURE B.1   Illustration of the change in grating vector for a planar grating resulting from a change in film thickness.

where K_z is the original z-component of the grating vector. If the original grating vector lies in the x–z plane, the magnitude after an axial thickness change becomes:

$$\left|\vec{K}'\right| = \sqrt{K_x^2 + K_z'^2}, \tag{B.5}$$

and the direction of the grating vector relative to the z-axis:

$$\varphi' = \tan^{-1}\left(K_x / K_z'\right). \tag{B.6}$$

To illustrate the consequence of a thickness change, consider a planar hologram formed with construction angles of 0° and 20° in a 16 µm thick recording material without change and with a shrinkage $\Delta d = -2$ µm. Figure B.2 shows a plot of the diffraction efficiency for these two cases. Note that the shrinkage results in a shift in the peak diffraction efficiency angle of ~1.5°. Also note that if the hologram with shrinkage is reconstructed at the original construction angle of 0°, the efficiency would drop to about 60% of the maximum efficiency value. For planar gratings, the effects of a thickness change can be compensated by changing the construction angles so that the maximum diffraction efficiency can be obtained at the desired reconstruction angles. The magnitude of the thickness change can be determined by experimentally comparing the measured shift in peak diffraction efficiency to a value computed using the approximate coupled wave model (Chapter 5). However, when the holographic grating is not planar, it is not possible to compensate this effect at all positions on the hologram aperture by a single change

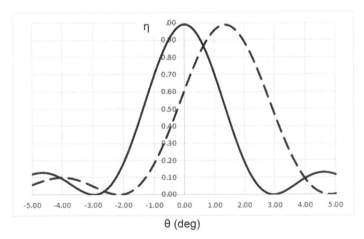

FIGURE B.2   Change in the diffraction efficiency as a function of reconstruction angle for a planar hologram formed with 0° and 20° construction beam angles in a 16 µm thick emulsion. Solid line is with no shrinkage, and dashed curves shows the angle shift that occurs with 2 µm of shrinkage.

in the reconstruction angle. It should also be noted that for unslanted transmission gratings, a thickness change does not shift the angle where maximum diffraction efficiency occurs. An unslanted transmission grating can therefore be used to determine the effects of lateral length changes of the emulsion with processing which are independent of the axial change.

B.1.1.2 Effect of Lateral Length Change in the Recording Material

The holographic film can also shrink or swell in the lateral directions $(\Delta x, \Delta y)$ after processing. Lateral changes in the surface grating period $\Delta\Lambda_x$ or $\Delta\Lambda_y$ will affect the imaging properties of the hologram. This can readily be seen by considering the grating equation:

$$\sin\theta_i - \sin\theta_p = \frac{\lambda}{\Lambda_x}, \tag{3.9}$$

where i refers to the image ray, p to the reconstruction ray, λ the reconstruction wavelength, and Λ_x the pre-processed grating period in one of the lateral directions (x). If the lateral period Λ_x changes to a new value Λ'_x, then the phase matched diffraction angle or wavelength must change if the same reconstruction angle, θ_p, is used. The change in lateral grating period can also be thought of as scaling the hologram and results in a magnification effect on the reconstructed image [1].

B.1.1.3 Grating Period and Slant Angle Chirp

In materials such as DCG, the development process can be used to *chirp* or vary the grating period and slant angle of the grating vector as a function of depth. In this case, the volume grating cannot be described with one grating vector, but requires a continuously varying one. At the glass emulsion interface $(z = 0)$, the pre- and post-processed grating vectors are the same. However, at increasing depths, the fringe curvature increases and the difference between the unprocessed grating vector, $\vec{K}$, and the grating vector after processing, $\vec{K}'$, increases as shown in Figure B.3. The chirp angle $\varphi(z)$ can be approximated by an exponential function of hologram depth (z) according to [2]:

$$\Delta x = \frac{z}{d}\exp\left(-a\cdot\frac{z}{d}\right)$$

$$\varphi(z) = \tan^{-1}\left(\frac{x - \Delta x}{z}\right), \tag{B.7}$$

where a is constant that must be determined by empirical comparison with experimental results, d is the pre-processed emulsion thickness, and $\varphi(z)$ is the angle that the grating fringe makes

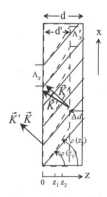

FIGURE B.3 Illustration of slant angle and grating period chirp in a volume hologram.

with the z-axis as a function of z. By using hot dehydration processing techniques, the spectral bandwidth of holograms formed in DCG can be extended to more than 300 nm in ~30 μm thick emulsions [2,3].

B.2 Effect of Average Refractive Index and Modulated Refractive Index Change

For many types of holographic materials, processing after the exposure step changes the average refractive index of the material. This will affect the Bragg condition of the reconstructed field and impacts the performance of the hologram. For example, Figure B.4 shows the shift in the Bragg angle when the average refractive index changes from 1.50 during exposure to 1.54 after processing for a transmission hologram in a 16 μm thick film with construction beams at 0° and 20°. The result is a shift in the Bragg angle of ~0.20°. However, if the hologram is recorded as an unslanted transmission grating, a change in the average refractive index does not produce a shift in the Bragg angle of the reconstruction beam outside the grating medium. This is due to Snell's law changing the angle inside the grating medium by the exact amount needed to maintain the Bragg match condition.

Several methods have been evaluated for measuring the average refractive index of holographic materials after exposure and processing [4,5]. The most effective methods found were those that compare the diffraction efficiency of experimental test holograms to results computed with rigorous coupled wave analysis with different average refractive index values.

During hologram exposure, the irradiance of the object and reference beams decrease as a function of depth due to absorption by the photosensitizing compound in the film or polymer. This has the effect of producing a gradient in both the average refractive index and the amplitude of the refractive index modulation with depth as shown in Figure B.5. The variation in the amplitude of the refractive index modulation can be approximated as an exponential decay with depth according to:

$$n_1(z) = n_1(0)\exp(-\alpha z),\tag{B.8}$$

where $n_1(0)$ is the amplitude of the refractive index modulation at the surface of the material, and α is the absorption coefficient for the material. Both α and $n_1(0)$ are experimentally determined for a specific material. One of the main results of n_1 not being a constant is that the effective grating thickness can be significantly different from the geometrical film thickness. In some cases, $t_{eff} \approx 0.6 \cdot t_{geom}$ and has a large impact on the selectivity of volume holograms [6].

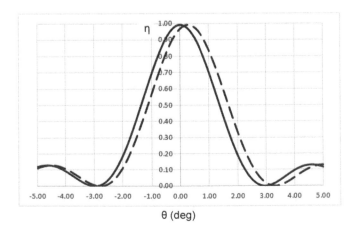

FIGURE B.4 Diffraction efficiency for a hologram formed in a 16 μm thick film with construction beam angles of 0° and 20° and 532 nm light. The solid curve is the diffraction efficiency when the average refractive index does not change after processing (1.50), and the dashed curve when the index changes to 1.54.

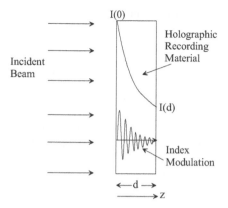

FIGURE B.5 Illustration of the effect of absorption on the irradiance and the resulting amplitude of the refractive index modulation in a holographic recording material.

B.3 Relation of Hologram Exposure and Visibility to Diffraction Efficiency

Figure B.6 shows a typical response of the refractive index modulation (Δn) to the exposure energy density of a holographic phase material such as DCG, photopolymer, or a bleached silver halide emulsion. Several distinct regions exist for the formation of the index modulation. For instance, for photopolymer materials, there is usually a threshold energy density that must be overcome before a measurable index modulation occurs. In addition, materials such as the Bayer/Covestro Bayfol photopolymers must be exposed with a certain irradiance (mW/cm²) level in order for the material to respond. Other materials such as DCG and bleached silver halide have a gradual and non-linear increase in Δn before reaching the linear response range where:

$$\Delta n \propto E. \tag{B.9}$$

Finally, as the exposure energy density continues to increase, the material response becomes saturated and does not increase significantly with additional exposure.

In the early 1970s, C.H. Lin outlined an approach that combined the index modulation response to exposure energy density and the visibility of the interference fringes used to form the hologram [8].

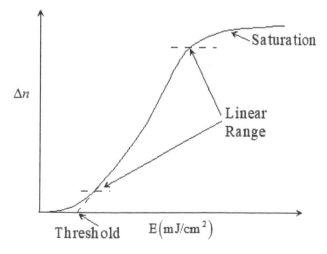

FIGURE B.6 Illustration of the change in the amplitude of the refractive index modulation as a function of exposure energy density for a typical phase hologram.

The relation provides a useful first-order estimate for comparing different recording materials and is summarized in the following paragraphs.

For this analysis, assume that the complex object and reference fields are:

$$\tilde{o}(x) = o\exp(jk_o x); \quad k_o = 2\pi \sin\alpha_o / \lambda$$
$$\tilde{r}(x) = r\exp(jk_r x); \quad k_r = 2\pi \sin\alpha_r / \lambda, \tag{B.10}$$

where α_o and α_r are the angles that the object and reference beams make with the normal to the recording film plane. The resulting combined irradiance at the film plane becomes:

$$I(x) = (\tilde{o} + \tilde{r}) \cdot (\tilde{o} + \tilde{r})^*. \tag{B.11}$$

The resulting hologram transmittance function after processing the film is:

$$t(x) = t_{avg} + \beta\tau\left[\tilde{o}\tilde{o}^* + \tilde{r}\tilde{r}^* + \tilde{o}^*\tilde{r} + \tilde{o}\tilde{r}^*\right], \tag{B.12}$$

where t_{avg} is the average transmittance of the film and is not a function of exposure and processing, β is the film response constant with exposure, and τ is the exposure time. If the resulting hologram transmittance function is illuminated with a wave $\tilde{r}_p$, the reconstructed object wave $\tilde{o}$ is given by:

$$\tilde{u}_o(x) = 2(\beta\tau)\tilde{r}_p\tilde{r}^*\tilde{o}, \tag{B.13}$$

where an additional constant 2 has been added to simplify the final result (Note that since β is a constant, the factor 2 can also be included in that term). Examination of Equation B.12 also shows that the average exposure energy density is:

$$E_o = \tau\left(\left|\tilde{o}\right|^2 + \left|\tilde{r}\right|^2\right), \tag{B.14}$$

Previously it was shown that Equation B.11 for the intensity can be expressed as:

$$I(x) = \left|\tilde{o}\right|^2 + \left|\tilde{r}\right|^2 + 2\left|\tilde{o}\right| \cdot \left|\tilde{r}\right|\cos\left[(k_r - k_o)x\right]. \tag{B.15}$$

Letting $o = \left|\tilde{o}\right|$; $r = \left|\tilde{r}\right|$, the corresponding fringe visibility becomes:

$$V = \frac{2or}{(o^2 + r^2)} = \frac{2\sqrt{R}}{1 + R}, \tag{B.16}$$

with the beam ratio $R = r^2/o^2$. Combining these results with the average energy exposure density gives:

$$2or = (o^2 + r^2) \cdot V$$
$$(\beta\tau)2or = (\beta\tau)(o^2 + r^2) \cdot V = \beta E_o V, \tag{B.17}$$

where the second equation was obtained by multiplying by the factor $(\beta\tau)$.

If the reconstruction beam $\tilde{r}_p = \tilde{r}$, the reconstructed field is:

$$\tilde{u}_o(x) = 2(\beta\tau)\tilde{r}_p\tilde{r}^*\tilde{o} = \left|\tilde{r}\right|\beta E_o V e^{jk_o x}, \tag{B.18}$$

with a magnitude:

$$\left|\tilde{u}_o\right| = \left|\tilde{r}\beta(E_o V)\right|. \tag{B.19}$$

Since the diffraction efficiency is the ratio of the diffracted beam intensity relative to the incident reconstruction intensity, it can be written as:

$$\eta = \frac{|\tilde{u}_o|^2}{|\tilde{r}|^2},$$

(B.20)

or

$$\sqrt{\eta} = \beta E_o V.$$

(B.21)

This result shows that the square root of the diffraction efficiency is proportional to the exposure energy density and is also linearly proportional to the visibility of the fringes used to form the hologram. It should also be recalled that for an unslanted transmission type grating, the diffraction efficiency at the Bragg condition is:

$$\eta = \sin^2\left(\frac{\pi \Delta n d}{\lambda \cos\theta}\right),$$

(B.22)

and for small modulation values:

$$\eta = \sin^2\left(\frac{\pi \Delta n d}{\lambda \cos\theta}\right) \approx \left(\frac{\pi \Delta n d}{\lambda \cos\theta}\right)^2$$

(B.23)

$$\therefore \ \sqrt{\eta} \propto \Delta n \propto \beta E_o V,$$

and shows the direct dependence of Δn on the average energy exposure density and visibility of the interference pattern.

B.4 Effects of Polarization and Beam Ratio on Hologram Recording

As shown in the previous section, hologram diffraction efficiency is proportional to the square root of the visibility of the interference pattern used to expose the recording material. Therefore, different factors that affect the fringe visibility are important for insuring optimal hologram performance. To obtain a more general expression for the visibility, consider the interference of complex reference $(\tilde{r})$ and object $(\tilde{o})$ beam vectors [9] as shown in Figure B.7:

$$I = \left\langle |\tilde{o} + \tilde{r}|^2 \right\rangle = \left\langle (\tilde{o} + \tilde{r}) \cdot (\tilde{o} + \tilde{r})^* \right\rangle$$

$$= \left\langle \tilde{o} \cdot \tilde{o}^* \right\rangle + \left\langle \tilde{r} \cdot \tilde{r} \right\rangle + \left\langle \tilde{o} \cdot \tilde{r}^* \right\rangle + \left\langle \tilde{o}^* \cdot \tilde{r} \right\rangle$$

$$= I_o + I_r + \left\langle \tilde{o}\tilde{r}^* + \tilde{o}^*\tilde{r} \right\rangle \cos\Omega_{or}$$

$$= I_o + I_r + 2\,\mathrm{Re}\left\{ \left\langle \tilde{o}\tilde{r}^* \right\rangle \right\} \cos\Omega_{or}$$

(B.24)

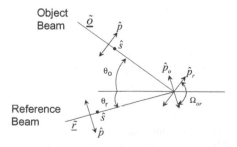

FIGURE B.7 General recording of a reference and object beam with arbitrary linear polarization vectors.

where Ω_{or} is the angle between the p-type polarization vectors for the reference and object beams. When the polarization state of the reference and object beams are perpendicular to the plane of incidence (s-state), the field vectors are aligned, $\Omega_{or} = 0°$, and the fringe visibility is optimum. However, when the polarization for the construction beams are in the plane of incidence (p-state), the angle between the field vectors is $\Omega_{or} = \theta_o + \theta_r$, and the interference term is reduced.

Using the general expression for the fringe visibility:

$$V = \frac{I_{max} - I_{min}}{I_{max} + I_{min}}, \tag{B.25}$$

and using Equation B.24, it is possible to express the visibility as:

$$V = \frac{2(I_o I_r)^{1/2} \cos \Omega_{or}}{I_o + I_r}. \tag{B.26}$$

Furthermore, if the beam ratio is defined as $R = I_r / I_o$, then the generalized visibility can be written as:

$$V = \frac{2R^{1/2} \cos \Omega_{or}}{R + 1}. \tag{B.27}$$

Note that if the beam ratio, R, is large, then $V \propto 1/\sqrt{R}$.

B.5 Noise Gratings

Many holographic recording materials such as silver halide, photorefractive crystals, and photo polymers consist of photosensitive compounds suspended in a transparent host medium. When illuminated, the difference in polarizability between the suspended photosensitive compounds and the host material causes light scattering. Optical inhomogeneity in the host material also produces light scattering [10]. During hologram formation, the scattered illumination interferes with the desired object and reference beams producing scatter or noise gratings as shown in Figure B.8 [11–14]. Since noise gratings are essentially formed with a single beam within the recording material, a very stable environment exists during the exposure, and the resulting gratings can be very efficient (Figure B.9). This occurs even though the beam ratio and visibility of the fringes are less than 1. When reconstructing a hologram formed between a reference and object beam, an efficient noise grating will cause a large drop in the diffraction efficiency of the desired hologram at the Bragg condition (Figure B.10). This can be detrimental to the performance of the desired hologram and steps should be taken to reduce the scatter that gives rise to the formation of noise gratings. This can be accomplished with careful control of the development process or by changing the reconstruction conditions so that the noise grating is no longer Bragg matched with the desired hologram.

Another characteristic of noise gratings results from spatial distribution of the scattered light [15,16]. For many recording materials, the scattered light has a dipole distribution pattern with the axis of the dipole aligned with the direction of polarization of the incident electric field (Figure B.11). As a result, if the noise grating is recorded with s- (perpendicular to the plane of incidence) or p- (parallel to the plane of incidence) polarized light, a different scatter pattern (object beam) is encoded. If the noise grating is then reconstructed with one polarization state of light, but was recorded with the orthogonal polarization, the noise grating will be much weaker as shown in Figure B.12. Therefore, choosing different polarization states for recording and reconstruction can be used as a method to reduce the effect of noise gratings with respect to a desired holographic reconstruction. This also suggests that it might be possible to use noise gratings as polarization sensing elements.

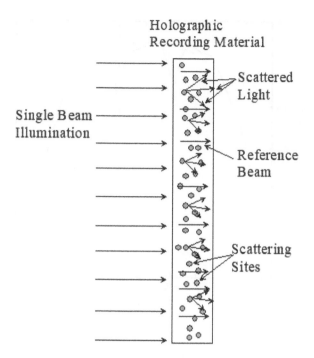

FIGURE B.8 Light illuminating a holographic recording material produces a scatter field that interferes with the remaining illumination.

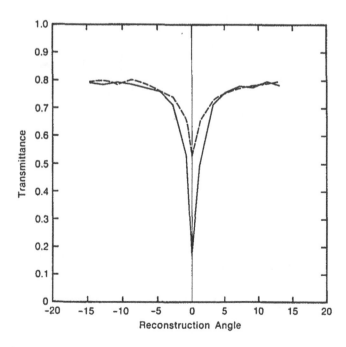

FIGURE B.9 Transmittance through a noise grating formed in a silver halide recording material. The solid line is the transmittance when the polarization of the construction and reconstruction illumination is the same. The dashed curve is when the reconstruction illumination is orthogonal to the construction illumination. (From Kostuk, R.K. and Sincerbox, G.T., *App. Opt.*, 27, 2993, 1988. With permission from the Optical Society of America.)

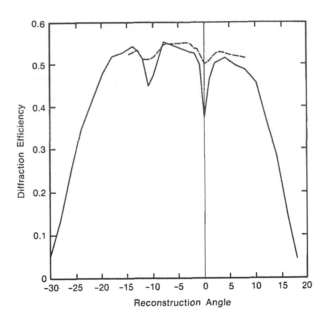

FIGURE B.10   The diffraction efficiency of a reflection hologram formed in a silver halide material. The dip in diffraction efficiency at 0° and −10° result from the formation of noise gratings. The solid curve is the efficiency when the polarization during formation and reconstruction of the noise grating are the same. The dashed curve is the efficiency when the construction and reconstruction polarizations are orthogonal. (From Kostuk, R.K. and Sincerbox, G.T., *App. Opt.*, 27, 2993, 1988. With permission from the Optical Society of America.)

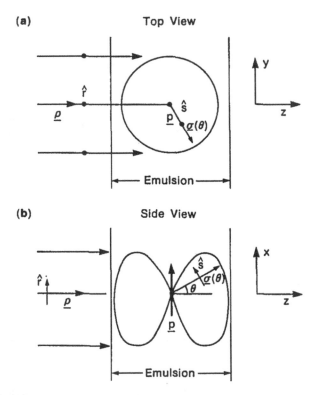

FIGURE B.11   Scatter profiles from silver halide grains in an emulsion for light polarized: (a) perpendicular to the plane of incidence and (b) polarized in the plane of incidence. (From Kostuk, R. K. and Sincerbox, G. T., *App. Opt.*, 27, 2993, 1988. With permission from the Optical Society of America.)

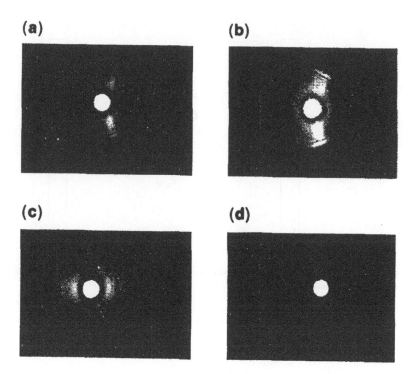

FIGURE B.12 Scatter profiles from noise gratings: (a) formed with p-polarized light and reconstructed with s-polarized light, (b) formed with p- and reconstructed with p-polarized light, (c) formed with s-polarized light and reconstructed with s-polarized light, and (d) formed with s-polarized and reconstructed with p-polarized light. (From Kostuk, R.K. and Sincerbox, G.T., *App. Opt.*, 27, 2993, 1988. With permission from the Optical Society of America.)

B.6 Multiplexed Holograms Recorded in Materials with Finite Dynamic Range

The ability to multiplex several holograms in the same recording material is an important attribute of holography. However, the way in which multiplexing is performed can have a significant impact on the efficiency of the resulting holograms. This is caused by the finite dynamic range of holographic recording materials [17].

In order to quantify this difference, consider two ways of recording N object waves in a multiplexed hologram as shown in Figure B.13 [17]:

1. All object beams illuminate the film simultaneously with one reference beam (a single exposure), and
2. Each object beam is recorded with a separate reference beam (N sequential exposures).

When N object beams (E_i) are simultaneously recorded with a single reference beam, $E_{R,S}$, the resulting intensity illuminating a linear recording material can be written as:

$$
\begin{aligned}
I_s &= \left| \left(E_{R,S} + \sum_{i=1}^{N} E_i \right) \right|^2 \\
&= \left| E_{R,S} \right|^2 + \sum_{i=1}^{N} \left| E_i \right|^2 + \sum_{i=1}^{N} \left(E_{R,S} E_i^* + E_{R,S}^* E_i \right) \\
&\quad + \sum_{j=1}^{N} \sum_{k=1}^{N} \left(E_j^* E_k + E_j E_k^* \right)
\end{aligned}
\tag{B.28}
$$

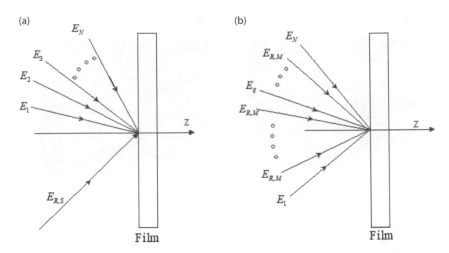

FIGURE B.13 (a) Simultaneous recording of N object beams and one reference beam and (b) Sequential recording of N object beams and N reference beams.

If $E_{R,S} \gg E_i$ and $|E_{R,S}|^2 \gg \sum_{i=1}^{N}|E_i|^2$, the cross terms can be neglected and the resulting exposing intensity can be simplified to:

$$I_s = |E_{R,S}|^2 + \sum_{i=1}^{N}\left(E_{R,S}E_i^* + E_{R,S}^*E_i\right). \tag{B.29}$$

Now consider the case where each object beam is sequentially recorded with a reference beam $E_{R,M}$. The intensity exposing the film now has the form:

$$\begin{aligned} I_M &= \sum_{i=1}^{N}\left(E_{R,M}+E_i\right)\left(E_{R,M}^*+E_i^*\right) \\ &= N|E_{R,M}|^2 + \sum_{i=1}^{N}|E_i|^2 + \sum_{i=1}^{N}\left(E_{R,M}E_i^* + E_{R,M}^*E_i\right). \end{aligned} \tag{B.30}$$

In order to make the two situations comparable, it is assumed that: (1) the exposure time for each sequential hologram is equal to the exposure time for the simultaneously exposed gratings, and (2) the average exposure irradiance for the two cases are equal:

$$|E_{R,S}|^2 = N|E_{R,M}|^2 = I_R. \tag{B.31}$$

As discussed earlier in this Appendix, the diffraction efficiency of the ith reconstructed object beam is proportional to the visibility of the interference pattern used to expose the recording material, i.e., $\eta_i \propto V_i^2$. For the single exposure hologram with two recording beams, the visibility is proportional to the modulation term of the basic interference expression:

$$I = I_1 + I_2 + 2\sqrt{I_1 I_2}\cos\varphi_{12}. \tag{B.32}$$

The visibility of a fringe for this case is:

$$V = \frac{2\sqrt{I_1 I_2}}{I_1 + I_2}. \tag{B.33}$$

For the simultaneously exposed multiplexed hologram with N object beams, the visibility of the ith fringe is:

$$V_i = \frac{2\sqrt{I_i I_R}}{\displaystyle\sum_{i=1}^{N} I_i + I_R}. \tag{B.34}$$

Considering the case where $I_i = I_j$, i.e., all of the object beam intensities are equal, $I_R = RI_i$, and the visibility becomes:

$$V_i = \frac{2\sqrt{I_i RI_i}}{RI_i + \displaystyle\sum_{i=1}^{N} I_i} = \frac{2\sqrt{R}}{R + N}. \tag{B.35}$$

The optimum beam ratio for maximum diffraction efficiency can be found by taking the derivative of the visibility with respect to R and setting it equal to zero:

$$\frac{\partial V_i}{\partial R} = 2\frac{1}{2}R^{-1/2}(N+R)^{-1} - (N+R)^{-2} 2\sqrt{R} = 0. \tag{B.36}$$

This gives $N = R$ for the optimum beam ratio. The corresponding visibility for an individual hologram in the simultaneously exposed multiplexed hologram case now becomes:

$$V_i^S = \frac{2N^{1/2}|E_{R,M}||E_i|}{N|E_{R,M}|^2 + \displaystyle\sum_{i=1}^{N}|E_i|^2}, \tag{B.37}$$

where $E_{R,S} = N^{1/2}E_{R,M}$. In contrast, for the sequentially exposed multiplexed hologram, the visibility of an individual hologram is:

$$V_i^S = \frac{2|E_{R,M}||E_i|}{N|E_{R,M}|^2 + \displaystyle\sum_{i=1}^{N}|E_i|^2} = \frac{1}{N^{1/2}}V_i^M. \tag{B.38}$$

Therefore, the ratio of the resultant diffraction efficiencies for the simultaneous and sequential exposure cases are related by:

$$\frac{\eta_{S,i}}{\eta_{M,i}} = \frac{V_{S,i}^2}{V_{M,i}^2} = N. \tag{B.39}$$

This result implies that the efficiency of the simultaneously exposed multiplexed hologram is $N \times$ larger than the sequentially exposed multiplexed holograms. However, this result is achieved at the expense of forming inter-modulation gratings that diffract light into unwanted directions. As a result many multiplexed holograms are recorded sequentially.

REFERENCES

1. J. N. Latta, "Computer-based analysis of holography using ray tracing," *Appl. Opt.*, Vol. 10, 2698–2710 (1971).
2. H. Schuette and C. G. Stojanoff, "Effects of process control and exposure energy upon the inner structure and optical properties of volume holograms in dichromated gelatin films," *SPIE Proc.*, Vol. 3011, 255–266 (1997).
3. J. M. Russo Miranda, "Holographic grating-over-lens dispersive spectrum splitting for photovoltaic applications," PhD Dissertation, University of Arizona, Tucson, AZ, pp. 68–90 (2014).
4. R. D. Rallison and S. R. Schicker, "Polarization properties of gelatin holograms," *SPIE Proc.*, Vol. 1667, 266–275 (1992).
5. G. Campbell, T. J. Kim and R. K. Kostuk, "Comparison of methods for determining the bias index of a dichromated gelatin hologram," *Appl. Opt.*, Vol. 34, 2548–2555 (1995).
6. J. M. Castro, J. Brownlee, Y. Luo, E. de Leon, J. K. Barton, G. Barbastathis and R. K. Kostuk, "Spatial-spectral volume holographic systems: Resolution dependence on effective thickness," *Appl. Opt.*, Vol. 50, 1038–1046 (2011).
7. R. K. Kostuk, "Dynamic hologram recording characteristics in DuPont photopolymers," *Appl. Opt.*, Vol. 38, 1357–1363 (1999).
8. L. H. Lin, "Method of characterizing hologram-recording materials," *J. Opt. Soc. Am.*, Vol. 61, 203–208 (1971).
9. R. J. Collier, C. B. Burckhardt and L. H. Lin, *Optical Holography*, Academic Press, New York (1971).
10. E. Hecht and A. Zajac, *Optics*, Chapter 8, Addison-Wesley, Reading, MA (1974).
11. M. R. B. Forshaw, "Explanation of the diffraction fine-structure in overexposed thick holograms," *Opt. Commun.*, Vol. 15, 218 (1975).
12. M. R. B. Forshaw, "Explanation of the two-ring diffraction phenomenon observed by Moran and Kaminow," *App. Opt.*, Vol. 13, 2 (1974).
13. R. Magnusson and T. K. Gaylord, "Laser scattering induced holograms in lithium niobate," *App. Opt.*, Vol. 13, 1545 (1974).
14. R. R. A. Syms and L. Solymar, "Noise gratings in silver halide volume holograms," *App. Phys. B*, Vol. 30, 177 (1983).
15. R. K. Kostuk and G. T. Sincerbox, "Polarization sensitivity of noise gratings recorded in silver halide volume holograms," *App. Opt.*, Vol. 27, 2993 (1988).
16. J. Frantz, R. K. Kostuk, D. A. Waldman, "Model of noise grating selectivity in volume holographic recording materials be use of Monte Carlo simulations," *J. Opt. Soc. Am. A*, Vol. 21, 378–387 (2004).
17. J. T. LaMacchia and C. J. Vincelette, "Comparison of the diffraction efficiency of multiple exposure and single exposure holograms," *Appl. Opt.*, Vol. 7, 1857–1858 (1968).

Appendix C: Laser Operation and Properties Useful for Holography

C.1 Basic Laser Operation

While a full discussion of lasers and laser operation is beyond the scope of this book, they are such important optical sources for holography that an overview of some of their basic characteristics is warranted. For a more comprehensive treatment please see references 1–4 below.

A laser consists of three components as shown in Figure C.1: an *optical gain medium* that provides a mechanism for *light-matter interactions*; a *pump* that supplies power to the gain medium; and a *resonator* to provide optical feedback and sets the *threshold conditions* for lasing. The light-matter interactions in the optical gain medium were introduced in Chapter 2. They are: spontaneous emission, absorption, and stimulated emission and are illustrated in Figure 2.7 which is repeated below as Figure C.2. In absorption, the energy of an incident photon is transferred to an electron causing it to transition from a low to high energy state (E_{1i}–E_{2i}). The energy difference $E_{2i} - E_{1i} = h\nu_{12}$ is equal to the energy of the incident photon. For stimulated emission, an incident photon illuminates a material with electrons in the upper or excited state and causes a downward transition of an electron. The released energy is in the form of a photon with energy and phase equal to that of the incident photon. Absorption and stimulated emission are complementary processes in the sense that they are optically induced transitions between two energy levels. The third light-matter interaction, spontaneous emission, requires that the upper energy level is populated, and then after a certain lifetime photons are randomly emitted again with energy $E_{2i} - E_{1i} = h\nu_{12}$.

In order to extract optical power through the stimulated emission process, a non-equilibrium condition must be established called a population inversion. In this situation, the population of the upper energy level (N_2) must be greater than that of the lower energy level N_1, (i.e., $N_2 > N_1$). Specific characteristics of the energy levels of the gain material and pumping between energy levels are required to achieve population inversion. Pumping between energy levels can be accomplished by electrical, chemical, or optical methods.

Two material structures that allow for an inversion are those with 3- and 4-energy level configurations as shown in Figure C.3. In the 3-level system, electrons are pumped from the ground state (1) to the upper energy level (2). Electrons in level 2 undergo a fast decay to level 3 which has a longer spontaneous emission lifetime than the 2- to 3-level decay time. An incident photon with energy $h\nu_{31} = E_3 - E_1$ induces stimulated emission between these levels and lasing. Notice that more than half of the atoms in the ground state must be pumped to the upper energy level in order have a population inversion ($N_3 > N_1$). This requires significant pumping power to achieve lasing in a 3-level system.

For a 4-level laser, the lower energy level for the stimulated emission transition (E_4) is elevated above the ground state. Therefore, a population inversion exists between levels E_3 and E_4 as soon as electrons are collected in E_3. This makes 4-level laser systems much more efficient than 3-level systems.

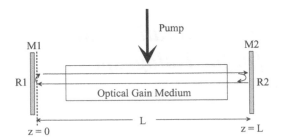

FIGURE C.1   Basic components of a laser: pump; optical gain medium; and the optical resonator.

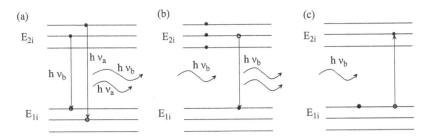

FIGURE C.2   (Figure 2.7) Illustration of the three light-matter interactions: spontaneous emission (a), stimulated emission (b), and absorption (c).

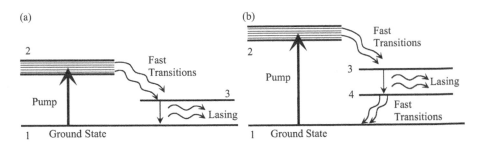

FIGURE C.3   (a) Three-level energy system for laser operation and (b) Four-level energy system for laser operation.

C.2 Conditions for Lasing

In order for lasing to occur with the basic system shown in Figure C.1, two conditions must be met: (a) a threshold power must be achieved within the cavity; and (b) the field within the resonator must be phase matched.

The power threshold condition can be determined by considering a round trip path of the beam within the laser system shown in Figure C.1. For this analysis, it is assumed that the gain medium occupies the full space between the two mirrors (i.e., over the full distance L), and that the power reflectivity of the two mirrors are R_1 and R_2. The power density of the beam within the laser can be expressed as:

$$I(z) = I(0)\exp\left\{\left[\gamma(v) - \bar{\alpha}_s\right] \cdot z\right\}, \tag{C.1}$$

where I_o is the intensity at $z = 0$, $\gamma(v)$ is the spectral gain per unit length acquired by propagating through the gain medium, and $\bar{\alpha}_s$ is the optical scatter loss per unit length within the gain medium. After the beam propagates through one round trip of the cavity the power is:

$$I(2L) = I(0)R_1R_2\exp\left[2L \cdot \left(\gamma(v) - \bar{\alpha}_s\right)\right]. \tag{C.2}$$

The *threshold condition* for the power density within the laser cavity is $I(2L) = I(0)$, which is a simple statement that the *gain = loss*. Combining the threshold condition with Equation C.2 provides the threshold value for the gain to overcome the cavity loss:

$$\gamma_{th} = \bar{\alpha}_s + \frac{1}{2L}\ln\left(\frac{1}{R_1R_2}\right). \tag{C.3}$$

Any gain provided above this value will result in useful power output. It should be noted that the reflectivity of one of the mirrors is typically less than the other and serves as the output mirror.

The *phase matching condition* for lasing can be found by considering the time independent form for the electric field that propagates within the cavity:

$$E(z) = E_o\exp(-j\beta z), \tag{C.4}$$

where $\beta = 2\pi n/\lambda$, λ is the free space wavelength, and n is the refractive index within the gain medium. After the field propagates over a round trip within the laser cavity, the phase matching condition can be written as:

$$e^{-j2\beta L} = 1$$
$$2\beta L = 2\pi q, \tag{C.5}$$

where q is an integer. Rewriting this condition as:

$$q = \frac{1}{2}\frac{L}{(\lambda/n)}, \tag{C.6}$$

it can be seen that q is the number of half wavelengths within the cavity. The corresponding stable optical frequencies within the cavity are:

$$v_q = q\frac{c}{2nL}, \tag{C.7}$$

and the separation between the allowed frequencies is:

$$\Delta v = q\frac{c}{2nL} - (q-1)\frac{c}{2nL} = \frac{c}{2nL}. \tag{C.8}$$

An illustration of the oscillating modes for a laser with an optical gain medium with a gain spectrum of $\gamma(v)$ and a threshold loss γ_{th} is shown in Figure C.4. Only the modes with $\gamma(v) > \gamma_{th}$ will oscillate and be emitted from the laser. These modes are often referred to as the temporal longitudinal modes of the cavity.

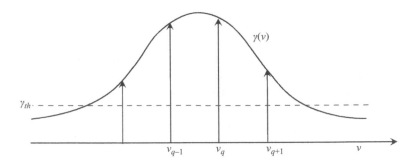

FIGURE C.4   Illustration of an optical gain medium $\gamma(v)$ and temporal longitudinal modes of a laser.

C.3 Temporal Coherence of a Multi-mode Laser

In order to form an interference pattern to expose a hologram, the optical source must have sufficient coherence length to accommodate the path difference in the optical system and a coherence time that exceeds the film exposure time. As discussed in Chapter 2, the coherence time for an optical source is related to the full width half maximum (FWHM) of the spectral bandwidth of the source by:

$$\Delta v \, \Delta t = 1. \tag{C.9}$$

For a laser source with multiple oscillating modes at different frequencies, the coherence time can be found using the complex degree of coherence defined as [5]:

$$\gamma_{12}(\tau) = \frac{\lim T \to \infty \dfrac{1}{2T} \displaystyle\int_{-T}^{T} E(t+\tau)E^*(t)\,dt}{\lim T \to \infty \dfrac{1}{2T} \displaystyle\int_{-T}^{T} E(t)E^*(t)\,dt}, \tag{C.10}$$

where τ is the coherence time and is related to the coherence length through $l_{coh} = c\tau$. Using the Fourier shift relation from Appendix A, the complex degree of coherence can be rewritten in terms of the power spectral density $\Phi(v)$ as:

$$\gamma_{12}(\tau) = \frac{\displaystyle\int_{-\infty}^{\infty} \Phi(v)\exp(-j2\pi v\tau)\,dv}{\displaystyle\int_{-\infty}^{\infty} \Phi(v)\,dv}. \tag{C.11}$$

As an example of analyzing the temporal coherence properties of a laser with multiple temporal modes, consider the case where only two frequency modes oscillate within the laser cavity. In this case, the power spectral density can be written as [6]:

$$\Phi(v) = I_m \delta(v - v_m) + I_{m+1}\delta(v - v_{m+1}), \tag{C.12}$$

where I_m and I_{m+1} are the power densities in the mth and $m+1$ modes, and v_m and v_{m+1} are the frequencies associated with these modes. Substituting this power spectral density relation into the complex degree of coherence relation results in:

$$\gamma_{12}(\tau) = \frac{1}{I_m + I_{m+1}}\left\{I_m \exp(j2\pi v_m \tau) + I_{m+1} \exp(j2\pi v_{m+1}\tau)\right\}. \tag{C.13}$$

Noting that the frequency modes are related through:

$$v_{m+1} = v_m + \frac{c}{2L}, \tag{C.14}$$

and using normalized intensity coefficients:

$$c_m = \frac{I_m}{I_m + I_{m+1}}; \; c_{m+1} = \frac{I_{m+1}}{I_m + I_{m+1}}, \tag{C.15}$$

the complex degree of coherence function becomes:

$$\gamma_{12}(\tau) = \exp(j2\pi v_m \tau)\left\{c_m + c_{m+1}\exp\left[j2\pi\tau\left(c/2L\right)\right]\right\}. \tag{C.16}$$

Taking the magnitude of $\gamma_{12}(\tau)$, squaring it, and then taking the square root results in:

$$\begin{aligned}|\gamma_{12}(\tau)| &= \left|\left\{c_m + c_{m+1}\exp\left[j\pi\tau\left(c/L\right)\right]\right\}\right| \\ &= \left|\left\{c_m^2 + c_{m+1}^2 + 2c_m c_{m+1}\cos\left[\pi\tau\left(c/L\right)\right]\right\}^{1/2}\right|.\end{aligned} \tag{C.17}$$

If $c_m = c_{m+1} = 1/2$, the complex degree of coherence becomes:

$$|\gamma_{12}(\tau)| = \left|\left[\frac{1}{2} + \frac{1}{2}\cos(\pi c\tau/L)\right]^{1/2}\right| = |\cos(\pi c\tau/2L)|. \tag{C.18}$$

This implies that the complex degree of coherence is an oscillatory function that is zero when τ is an odd multiple of L/c and a maximum at even multiples. The result can be also be applied to the case when M modes oscillate in the laser leading to:

$$|\gamma_{12}(\tau)| = \left|\left(\frac{1}{M}\right)\frac{\sin(M\pi c\tau/2L)}{\sin(\pi c\tau/2L)}\right|. \tag{C.19}$$

For high power argon and krypton ion lasers, the mirror separation can be close to a meter in length and gives rise to 10–20 oscillating longitudinal modes. This can reduce the coherence length to only a centimeter or so and may not be sufficient for some holographic setups. The coherence length can be increased by placing a secondary resonator or etalon within the laser cavity as shown in Figure C.5.

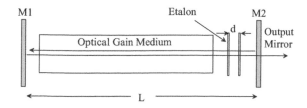

FIGURE C.5 Illustration of a laser cavity with an etalon to increase the coherence length.

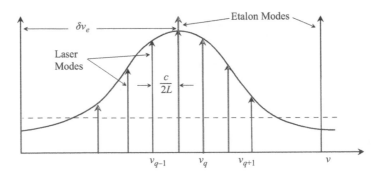

FIGURE C.6   Illustration of the laser and etalon modes.

The mode spacing for the etalon is given by:

$$\delta v_e = \frac{c}{2d},$$ (C.20)

and is considerably larger than the laser resonator (Figure C.6). With proper design, an etalon allows one mode to propagate in the laser cavity and greatly extends the coherence length of the laser.

C.4 Spatial Modes and Gaussian Beam Properties

Laser resonators can be formed from a variety of mirror configurations with different radii of curvature and separation distances. A full treatment and analysis of different configurations are provided in reference [3]. Different resonator configurations give rise to different stable spatial field distributions or laser modes. Perhaps the most useful mode for holography is the lowest order which has a Gaussian intensity profile given by:

$$I(r) = I_o e^{\left(\frac{-2r^2}{w^2}\right)},$$ (C.21)

where I_o is the irradiance at $r = 0$, and w is the radius of the profile when the irradiance is $1/e^2$ of the maximum value. "w" is also referred to as the *spot size* of the Gaussian beam and is a function of the distance z from the minimum beam radius or beam waist (w_o). An expression for $w(z)$ is given by:

$$w(z) = w_o \left[1 + \left(\frac{\lambda z}{\pi w_o^2} \right)^2 \right]^{1/2}.$$ (C.22)

At large distances from the beam waist, the spot size can be approximated with the expression:

$$w(z) \approx \left(\frac{\lambda}{\pi w_o} \right) z,$$ (C.23)

and the corresponding diffraction angle is:

$$\theta = \frac{\lambda}{\pi w_o}.$$ (C.24)

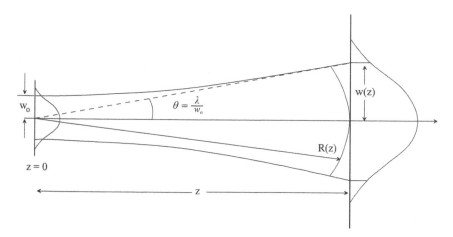

FIGURE C.7 Gaussian beam parameters corresponding to the lowest spatial mode emitted from a laser cavity.

Combining these approximate relations, the size of the Gaussian beam at a distance z from the beam waist can be estimated with:

$$w(z)^2 = w_o^2 + \theta^2 z^2. \tag{C.25}$$

The different Gaussian beam parameters are illustrated in Figure C.7. For many laser cavities, the beam waist is located at the output mirror and is used as the location for $z = 0$.

C.5 Lasers Commonly Used for Holography

When selecting a laser for a holography application, it is important to check the following:

1. Match the laser wavelength with the recording material spectral response sensitivity—ensure that the laser wavelength falls well within the spectral sensitivity range of the material. Otherwise the material will either require very long exposure times or may not respond at all.
2. Make sure that the temporal coherence length is sufficient for the anticipated optical path length difference in the optical setup.
3. Ensure that the laser provides adequate power for the setup and material exposure requirements. Often times a large area beam is required and necessitates greater rated laser power to meet the minimum power density requirement of a recording material.
4. Review the laser specifications for the TEM_{00} spatial mode. This mode corresponds to the Gaussian beam described in the previous section and is the most useful for recording holograms.
5. Note the type of polarization and orientation of the laser output beam. A linear polarization that can be set perpendicular to the optical table provides the highest contrast interference fringes. Some lasers come with "random" polarization and will require an external polarizer that reduces the overall output power. If the polarization is horizontal to the plane of the table, and the laser cannot be physically rotated, a half-wave plate will be required to rotate the polarization direction. For gas lasers, the polarization is typically set with Brewster windows on both ends of the gas tube as shown in Figure C.8.

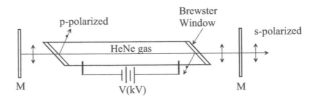

FIGURE C.8 Schematic diagram of the components in a HeNe gas discharge laser showing placement of the Brewster windows to force emission to linear s-polarized light.

Several types of lasers meet these requirements and are briefly outlined below.

C.5.1 Gas Discharge Lasers

In this type of laser, a mixture of gases such as helium and neon (HeNe) are contained in a sealed tube. This combination forms a 4-level energy system for lasing. Pumping is accomplished by applying a high voltage (~1.5 kV) across the gas through electrodes that are inserted into the tube. The primary stimulated emission wavelength is 632.8 nm and typical laser power ranges from 1 to 20 mW. Most HeNe lasers have 2–3 longitudinal modes with resonator mirror separations of 20–30 cm. The gas discharge bandwidth is approximately 1.5 GHz and a coherence length of approximately 20 cm which is suitable for many holography applications. The beam divergence is about 1 mrad and the $1/e^2$ points for the Gaussian beam is <1 mm. A schematic for a typical HeNe laser is shown in Figure C.8.

The helium cadmium (HeCd) laser is another type of gas discharge laser that is useful for making holograms in photoresist materials. Holograms in photoresist film are typically used for mastering embossed holograms for security and identification applications. The primary wavelengths of HeCd lasers are at 325 and 441 nm.

C.5.2 Ionized Gas Lasers

An important class of lasers for holography is the ionized gas laser. These lasers can provide several watts of power at different emission lines that are within the spectral sensitivity range of a variety of holographic recording materials such as DCG, photopolymers, and photoresist. The argon ion laser is perhaps the most widely used of this type with the primary emission wavelengths at 488 and 514.5 nm, and secondary emission at 496.5, 476.5, and 457.9 nm. An argon ion laser with a nominal output power of 5 W for all emission lines has approximately 2.0 W at 514.5 nm, 1.5 W at 488 nm, 0.6 W at 457.9 nm, and the remaining power distributed over the other emission lines. The krypton laser is the second most frequently used ionized gas laser with emission at 476.2, 520.8, 568.2, and 647.1 nm.

The ionized gas laser is a complex system that requires high discharge voltage and magnetic confinement of the ionized gas plasma. A diagram of an ionized gas laser is shown in Figure C.9. Since the ions

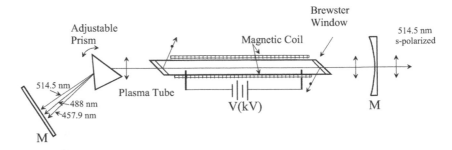

FIGURE C.9 Schematic diagram for the components in an argon ion laser. The prism is used to separate the laser emission wavelengths and allow single line operation.

are charged particles, a magnetic coil can be used to confine the ions to a small region near the axis of the plasma tube. This helps to reduce the power required to form a population inversion. The emission from the 4-level argon/krypton medium occurs at all lasing lines at the same time. Therefore, a tunable glass prism is typically used to disperse the lines and allow only one wavelength to be amplified in the laser cavity at a time. The emission at the single "line" will typically have multiple longitudinal modes since the cavities are on the order of 1 meter in length. Therefore, as described earlier, an intra-cavity etalon is used to filter and allow only one temporal mode to oscillate with high gain. A typical 5 W argon ion laser will require approximately 20 kW of power at 220 V and water cooling. The laser tubes last between 3 and 5 years depending on use and care and can be replaced many times to prolong the system lifetime.

C.5.3 Diode Pumped Solid State (DPSS) Lasers

These lasers use semiconductor diode lasers to optically pump Nd^{3++}-doped vanadate crystals and produce stimulated emission at 1064 and 914 nm. The primary output wavelengths can be frequency doubled using a non-linear conversion process in a crystal such as KD*P [3]. The resulting output at 532 and 457.5 nm falls within the spectral sensitivity of DCG, silver halide emulsions, photoresist (with 457.5 nm), and a variety of photopolymers. The output powers range from a few 10's of milli-watts to several watts, the coherence length is typically on the order of 5 m, and the divergence is <5 mrad making them useful for a variety of holography applications. The output polarization, while linear, can be either vertical or horizontal to the required mounting position. Therefore, attention should be paid to this factor when purchasing.

REFERENCES

1. A. E. Siegman, *Lasers*, University Science Books, Palo Alto, CA (1986).
2. J. T. Verdeyen, *Laser Electronics*, 3rd ed., Prentice Hall, Upper Saddle River, NJ (1995).
3. A. Yariv, *Optical Electronics in Modern Communications*, 5th ed., Oxford University Press, New York (1997).
4. C. Pollock, *Fundamentals of Optoelectronics*, Irwin, Chicago, IL (1995).
5. M. Born and E. Wolf, *Principles of Optics*, p. 501, 5th ed., Pergamon Press, Oxford, UK (1975).
6. R. J. Collier, C. B. Burkhardt, and L. H. Lin, *Optical Holography*, Academic Press, New York (1971).

Appendix D: Holographic Material Processing Techniques

Chapter 8 provided a review of many types of holographic recording materials. This Appendix gives detailed information on several commercially available silver halide emulsions as well as developing, fixation, and bleaching methods that have proven useful for different applications. Overviews are also given on the preparation, exposure, and development methods for dichromated gelatin film, and for phenanthrenequinone-doped poly(methyl methacrylate) photopolymer (pq-MMA photopolymer). Both of these materials can be prepared in a modest chemistry lab with inexpensive chemicals and equipment making them ideal for experimentation and use in different applications.

D.1 Silver Halide Emulsions

Table D.1 is a list of common silver halide emulsions used for holography as well as a summary of their physical properties. Table D.2 lists the main silver halide film processing methods and the properties of the resultant holograms. The excellent book by Bjelkhagen [1] provides an extensive review of silver halide materials and their processing chemistries.

The following processing chemistries work well with Kodak 649F, Agfa-Gevaert 8E75HD 10E75HD, and Ilford 737T and SP673 silver halide emulsions. The JD2 and four chemistries were designed for use with the PFG emulsions distributed by Integraf [2].

D.1.1 Silver Halide Hologram Developers

D-19 Develop 4–5 minutes at 20°C [1]

Metol 2 g
Sodium sulfite (anhydrous) 90 g
Hydroquinone 8 g
Sodium carbonate (monohydrated) 52.5 g
Potassium bromide 5 g
Distilled water 1 L

CPA-1 (Solution physical developer) [3]

Phenidone 0.02 g
Hydroquinine 0.65 g
Sodium sulfite 13 g
Potassium hydroxide 1.4 g
Ammonium thiocyanate 3.1 g
Distilled water 1 L

TABLE D.1

Material	Emulsion Thickness (µm)	Spectral Sensitivity (µJ/cm²)	Resolution (lp/mm)	Grain Size (nm)	Comments
Agfa 8E75HD	7	10@633	~5000	35/44	&
Agfa 8E56HD	7	25@442	~5000	35/44	&
Ilford 695T	6	100@515	7000	30	&
Kodak 649F pl.	17	90@633	2000	58	Plates; Panchromatic
PFG-01	7–8	100@694	3000	40	&
PFG-03M	6–7	500@633	5000	8–12	Optimized for reflection
PFG-03C	9–10	500@633	5000	8	Panchromatic
VRP-M	6–7	100@515	3000	35–40	Refl/Trans

TABLE D.2

Process	Hologram Type	Thickness Change	Average Index Change	Scatter	DE
Develop-Fix	Absorption	Yes	Yes	Low	Low
Develop-Fix-Bleach	Phase	Yes	Yes	High	High
Reversal Bleach	Phase	Yes	Yes	Low	High
Develop-Bleach	Phase	No	No	Low	High

CW-C1 (Modified tanning developer without fixing) [4]

Part A:
    Catechol 10 g
    l-ascorbic acid 9 g
    Distilled water 500 mL
Part B:
    Sodium carbonate 30 g
    Distilled water 500 mL
    Mix equal parts of A and B.

CW-C2 [4]

Catechol 10 g
Sodium sulfite (anhydrous) 5 g
L-ascorbic acid 5 g
Urea 50 g
Sodium carbonate (anhydrous) 30 g
Distilled water 1 L

Pyrochrome Developer: (for use with dichromate bleach) [5]

Part A:
    Pyrogallol 10 g
    Distilled water 1 L
Part B:
    Sodium carbonate 60 g
    Distilled water 1 L

JD-4 (Developer component) [2]
  Part A:

    Metol 4 g
    Ascorbic acid (powder) 25 g
    Distilled water 1 L
  Part B:
    Sodium carbonate 70 g
    Sodium hydroxide 15 g
    Distilled water 1 L

JD-2 (Developer component) [2]
  Part A:

    Catechol 20 g
    Ascorbic acid (powder) 10 g
    Urea 75 g
    Sodium sulfite 10 g
    Distilled water 1 L
  Part B:
    Sodium carbonate 60 g
    Distilled water 1 L

Phenidone, Ascorbic Acid, Phosphate Developer (PAAP) (Developer for silver halide sensitized gelatin [SHSG]) [6]

  Phenidone 0.5 g
  l-ascorbic acid 18 g
  Sodium hydroxide 12 g
  Sodium phosphate diboric 28.4 g
  Distilled water 1 L

D.1.2 Silver Halide Hologram Fixer Solution

Generic Hologram Fixer: Fixing time twice the visible clearing time at 20°C [1].

  Sodium thiosulfate (Hypo) 240 g
  Sodium sulfite (anhydrous) 10 g
  Sodium bisulfite 25 g
  Distilled water 1 L

NOTE: Fixers such as Kodak Rapid Fixer are not recommended for fine grain holographic emulsions.

D.1.3 Silver Halide Hologram Bleaches

Direct Bleach: [7]

  Potassium alum 20 g
  Sodium sulfate 25 g
  Copper sulfate 40 g
  Potassium bromide 20 g
  Sulfuric acid (concentrated [conc.]) 5 mL
  Distilled water 1 L

R10 Bleach: [8,9]

Potassium dichromate (oxidizer) 2 g
Potassium bromide 36 g
Sulfuric acid 10 mL
Distilled water 1 L

Fixation Free Bleach PBQ-2: (Protects oxidation products and promotes tanning) [4]

p-Benzoquinone 2 g
Citric acid 15 g
Potassium bromide 50 g
Distilled water 1 L

Fixation Free Bleach Ferric-ethylene diamine tetra acetic acid (EDTA) Bleach: (Protects oxidation products but less toxic than PBQ) [10]

Ferric sulfate 30 g
Di-sodium EDTA 30 g
Potassium bromide 30 g
Sulfuric acid (conc.) 10 mL
Distilled water 1 L

Reversal Bleach: (Pyrochrome bleach) [5]

Potassium dichromate 4 g
Sulfuric acid (conc.) 4 mL
Distilled water 1 L

JD-4 (Bleach component) [2]

Copper sulfate pentahydrate 35 g
Potassium bromide 100 g
Sodium hydroxide 15 g
Distilled water 1 L

JD-2 (Bleach component) [2]

Potassium dichromate 4 g
Sodium bisulfate 80 g
Distilled water 1 L

R-10 Bleach (Modified for SHSG) [11]

Part A:
Distilled water 500 mL
Ammonium dichromate 20 g

Sulfuric acid 14 mL

Distilled water to 1000 mL

Part B:

Potassium bromide 92 g

Distilled water 1 L

GENERAL NOTES: Most development times are approximately ~2 minutes—vary the time to optimize; bleach the hologram until the exposed area clears plus~30 seconds; running water rinse solutions should be kept at a constant temperature of ~20°C

D.1.4 Silver Halide Sensitized Gelatin Hologram Processing

This method is a "transition" type process between that of a bleached silver halide and a dichromated gelatin hologram formed in fine-grained silver halide emulsions [12–15]. The sensitized gelatin hologram is exposed and then processed with a tanning developer and bleach. Tanning is a cross-linking of bonds between gelatin molecules. Higher levels of tanning occur near the sites of the exposed silver. The corresponding chemical process is:

$$6Ag^\circ + Cr_2O_7^{2-} + 14H^+ \rightarrow 6Ag^+ + 2Cr^{3+} + 7H_2O. \tag{D.1}$$

After developing, the emulsion is fixed to remove both the exposed and unexposed silver halide grains. The emulsion is then dehydrated using isopropyl alcohol and sealed to prevent moisture from penetrating the film. The finished hologram results in a low scatter, high efficiency grating. A typical set of processing steps for SHSG holograms are listed below [16]:

1. Develop with PAAP developer (given above): 4 minutes
2. Running water rinse: 1 minute
3. Bleach with R-10 mod (given above) until cleared plus 30 seconds
4. Running water rinse: 30 seconds
5. Soak in fixer: 2 minutes
6. Running water rinse: 20 minute
7. Dehydrate in 50% isopropanol: 3 minutes
8. Dehydrate in 90% isopropanol: 3 minutes
9. Dehydrate in 100% isopropanol: 3 minutes
10. Dry in a clean environment.

All solutions should be at 20°C except bleach.

D.2 Dichromated Gelatin Emulsion Preparation, Exposure, and Processing

Dichromated gelatin has been used since the 1960s to realize high efficiency low scatter holograms. Although it provides a near ideal recording material, it is somewhat difficult to work with and obtain repeatable results. In order to achieve good results, the relative humidity and temperature must be carefully controlled to the same value within a few percent. A basic preparation method and development process are given below and give good results provided some degree of temperature and relative humidity control is maintained.

D.2.1 DCG Film Preparation and Exposure Guidelines

A recipe for a DCG emulsion formed using a molding technique consists of [17]:

1. Bovine gelatin: 13 g
2. Ammonium dichromate: 3 g
3. De-ionized water (DI) water: 100 mL.

The first step is to dissolve bovine gelatin in de-ionized water at 40°C. Next, the ammonium dichromate is dissolved into the solution. Molds are prepared by cleaning plates of glass (4" × 5") and applying strips of polyvinyl tape around the perimeter of the glass. A cover glass is also prepared with one side coated with a mold release agent such as Rain-X windshield cleaner. The DCG film thickness can be controlled by changing the thickness of the tape. For a typical 7 mil thick tape, the resulting film thickness is ~36 μm.

A small amount of DCG solution is poured onto the mold, the cover glass is applied (Rain-X side down), and then the glass sandwich is placed in a freezer at 10°C for 15 minutes to induce gelling of the solution. After gelling, the cover glass is removed and the plate is placed in a dark fume hood for 24 hours to allow for drying. A typical exposure energy density for this film is 250 mJ/cm² at 532 nm for transmission holograms and 150 mJ/cm² at 514.5 nm for reflection holograms.

D.2.2 Basic DCG Development Process

After exposure, the holograms are processed using the following steps:

1. Rapid fixer (part-A)
2. Rapid fixer A + B
3. Rinse in room temperature tap water (swelling)
4. Room temperature 50/50% iso/water bath
5. Dehydration 75/25% iso/water at 35°C
6. Dehydration 99%–100% isopropanol at 45°C
7. Dry in an oven at 80°C for 5 minutes
8. Seal the hologram with a glass cover plate using an optical cement.
   - Part A (fixer) reduces Cr^{+6} to Cr^{+5} without binding it to the gelatin matrix
   - Part B (hardener) increases the hardness of gelatin in the unexposed areas—reduces the amount of swelling
   - Part A fixer (200 g ammonium thiosulfate in 1 L of water) and Part B hardener (formalin, alum, chrome alum or potassium alum [1]).

D.3 Phenanthrene Quinone-Doped Poly Methyl Methacrylate

This material can be formed from commercially available chemicals to fabricate a very thick (1–3 mm) volume hologram recording material [18,19]. The system consists of a: host material, MMA; thermal initiator, Azobisisobutyronitrile (AIBN); and a photo initiator, pq. A good starting point solution consists of:

(MMA): 100 g
(AIBN): 0.5 g
(pq): 0.7 g.

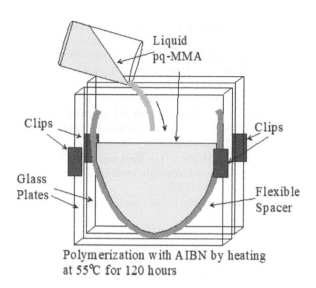

Polymerization with AIBN by heating at 55°C for 120 hours

FIGURE D.1   Schematic of a simple mold for fabricating pq-MMA photopolymer holographic recording material.

The compounds are mixed until dissolved, and then placed in a mold as shown in Figure D.1 below. The glass plates should be treated with a mold release agent (Rain-X windshield treatment), and flexible tubing serves as the spacer that sets the thickness of the resulting hologram. The spacer must be flexible to allow for shrinkage during the heat treating process. Heat curing consists of placing the prepared mold into an oven set at a temperature of 50°C–55°C for approximately 120 hours. This allows the AIBN to solidify the solution into a plastic material that can then be cut into a convenient size for recording. The material is exposed with blue-green wavelengths (488–532 nm) to record holograms with approximately 1J/cm² exposures. After exposure, the polymer is placed in a dark environment for several hours to allow monomer diffusion into the high intensity regions of the interference pattern. The remaining monomer is then fixed either with laser light and a rotating diffuser or with strong sunlight (~100 mW/cm²). The material can also be doped with nanoparticles to increase the diffraction efficiency and modify the spectral and angular bandwidth of the holograms [20,21].

REFERENCES

1. H. I. Bjelkhagen, *Silver-Halide Recording Materials*, Springer-Verlag, Berlin, Germany (1993).
2. www.integraf.com. (accessed on 12 December 2018)
3. R. Aliaga, H. Chuaqui and P. Pedraza, "Solution physical development of Agfa-Gevaert emulsions for holography," *Opt. Acta*, Vol. 30, 1743–1748 (1983).
4. D. J. Cooke and A. A. Ward, "Reflection hologram processing for high efficiency in silver halide emulsions," *Appl. Opt.*, Vol. 23, 934–941 (1984).
5. W. Spierings, "Pyrochrome processing yields color-controlled results with silver halide materials," *Holosphere*, Vol. 10, 9 (1983).
6. H. J. Caufield, Ed., *Handbook of Optical Holography*, Academic Press, New York (1979).
7. M. Lehmann, J. P. Lauer and J. W. Goodman, "High efficiencies, low noise, and suppression of photochrome effects in bleached silver halide holography," *Appl. Opt.*, Vol. 9, 1948–1949 (1970).
8. D. H. McMahon and A. R. Franklin, "Efficient, high-quality, R-10 Bleached holographic diffraction gratings," *Appl. Opt.*, Vol. 8, 1927–1929 (1969).
9. J. N. Latta, "The bleaching of holographic diffraction gratings for maximum efficiency," *Appl. Opt.*, Vol. 7, 2409–2416 (1968).
10. N. Philips, "Benign bleaching for healthy holography," *Holosphere*, Vol. 14, 21–22 (1986).

11. D. K. Angell, "Improved diffraction efficiency of silver halide (sensitized) gelatin," *Appl. Opt.*, Vol. 26, 4692–4702 (1987).

12. K. S. Pennington, J. S. Harper, and F. P. Laming, "New phototechnology suitable for recording phase holograms and similar information in hardened gelatin," *Appl. Phys. Lett.*, Vol. 18, 80–84 (1971).

13. W. R. Graver, J. W. Gladden and J. W. Estes, "Phase holograms formed by silver halide sensitized gelatin processing," *Appl. Opt.*, Vol. 19, 1529–1536 (1980).

14. J. M. Kim, B. S. Choi, Sun Il Kim, J. M. Kim, H. I. Bjelkhagen and N. J. Phillips, "Holographic optical elements recorded in silver halide sensitized gelatin emulsions. Part I. Transmission holographic optical elements," *Appl. Opt.*, Vol. 40, 622–632 (2001).

15. J. M. Kim, B. S. Choi, Sun Il Kim, J. M. Kim, H. I. Bjelkhagen and N. J. Phillips, "Holographic optical elements recorded in silver halide sensitized gelatin emulsions. Part 2. Reflection holographic optical elements," *Appl. Opt.*, Vol. 41, 1522–1533 (2002).

16. A. Fimia, I. Pascual and A. Belendez, "Silver halide sensitized gelatin as a holographic storage medium," *Proc. SPIE*, Vol. 952, 288–291 (1988).

17. J. M. Russo Miranda, "Holographic grating-over-lens dispersive spectrum splitting for photovoltaic applications," PhD. Dissertation, Chapter 4, University of Arizona, Tucson, AZ (2014).

18. O. Beyer, I. Nee, F. Havermeyer, and K. Buse, "Holographic recording of Bragg gratings for wavelength division multiplexing in doped partially polymerized poly(methyl methacrylate)," *Appl. Opt.*, Vol. 42, 30–37 (2003).

19. A. Sato and R. K. Kostuk, "Holographic grating for dense wavelength division optical filters at 1550 nm using phenanthrenequinone-doped poly(methyl methacrylate)," *SPIE Conference*, Vol. 5216 (2003).

20. J. M. Russo, J. E. Castillo and R. K. Kostuk, "Effect of silicon dioxide nanoparticles on the characteristics of PQ/PMMA holographic filters," *Proc. SPIE*, Vol. 6653 (2007).

21. J. M. Russo and R. K. Kostuk, "Temperature dependence properties of holographic gratings in phenanthrenequinone doped poly(methyl methacrylate) photopolymers," *Appl. Opt.*, Vol. 46, 7494–7499 (2007).

Appendix E: Holography Lab Experiments

This Appendix contains a collection of laboratory experiments that can be used in an introductory course on holography or for individual instruction. Any recording material can be used, but the new Bayer/Covestro photopolymer gives excellent results with minimal processing complication.

E.1. Lab 1: Transmission Holograms

The purpose of this lab experiment is to become familiar with basic holographic recording procedures and to evaluate the properties of transmission holograms.

E.1.1 Handling and Processing Holographic Materials

For this experiment, a photopolymer material is used that does not require wet processing. This feature greatly simplifies the hologram fabrication process and gives very good and repeatable results. Nonetheless, variations from ideal material properties still occur, and we will investigate the effects of these changes on the resulting hologram performance characteristics in this lab.

The photopolymer used in this experiment is the Bayfol 102 HX photopolymer from Covestro. The material is panchromatic (i.e., sensitive over a broad spectral range) with manufacturer specified sensitivities of 18 mJ/cm² at 632.8 nm; 25 mJ/cm² at 532 nm; and 30 mJ/cm² at 473 nm exposing wavelengths. Optimum exposure of the photopolymer can lead to refractive index modulations of $\Delta n \sim 0.03$. The photopolymer thickness is approximately 16 μm and is deposited on a flexible polycarbonate film substrate (Makrofol) that is approximately 175 μm thick. Typically, the flexible substrate is laminated onto a rigid substrate such as glass to make a stable platform for exposure. The photopolymer requires a minimum power density to activate polymerization of 0.1 mW/cm² at 632.8 nm and 0.5 mW/cm² for green and blue exposures.

The panchromatic feature of the photopolymer is advantageous for fabricating multicolor holograms, however, it also makes handling and processing the film more difficult since a safelight cannot be used for prolonged periods of time. Prolonged exposure to dim light will eventually degrade the performance of the photopolymer. However, low level lighting at shorter wavelengths can be used for limited periods of time to help in handling.

After the polymer is exposed to the laser illumination to form the hologram, the film is illuminated with incoherent illumination (sunlight or laser light with a rotating diffuser) to polymerize the remaining monomer and bleach the remaining dye. This helps clear the film, making it more transparent and reduces residual absorption.

E.1.2 Transmission Type Holograms

For the actual experiment, transmission holograms will be formed with a 30° inter-beam angle between the two construction beams. The optimum diffraction efficiency is found by making a set of holograms with different exposure energy densities near the nominal value of 18 mJ/cm² at 632.8 nm.

E.1.3 Experimental Procedures

Set up the optical system shown in Figure E.1. It is usually easier to align the system without the spatial filter. Retro-reflect the beams from each mirror and the film plate to make sure that the beams are parallel to the table. Also check the polarization of the laser and make sure that is perpendicular to the table,

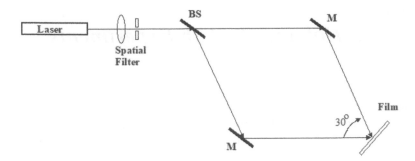

FIGURE E.1   Transmission hologram setup for Experiment 1.

and the optical path lengths are approximately equal. This last point ensures that the system is operating within the coherence length of the laser. After you are reasonably assured that the beams are aligned, center the beam on the collimating lens located at a distance approximately one focal length (of the collimating lens) from the pinhole of the spatial filter. Next, place the microscope objective into the spatial filter mount, and adjust the mount so that the beam is symmetrically located at the film plane. The beam should be Gaussian in profile across the aperture. Finally, insert the pinhole and center it with the micrometers. It is usually easiest to back the microscope objective away from the pinhole along the z-axis to get some light through the pinhole, and then slowly move it forward, adjusting the pinhole micrometers, until a relatively large and smooth Gaussian beam can be observed. Next, check the beam collimation by placing a mirror in the beam and reflecting it back to the pinhole. A small spot should be observed at the pinhole plane. Once collimated, set up an interferometer as shown below to record an unslanted grating with ~30° inter-beam angle.

Place an adjustable iris in the beam after the collimating lens to restrict the size of the beam on the film plate. A beam diameter of about 1–1.5 cm will allow several exposures to be recorded on the same film plate. In order to reduce spurious reflections and gratings from being formed, use some index matching oil to match a piece of blackened glass to the glass side (back) of the film plate (the photopolymer side should face the beams).

The exposure range for this film at 632.8 nm is near 18 mJ/cm^2. In order to find the optimal exposure, start with an exposure of about 15 mJ/cm^2 and increment by 2 mJ/cm^2. After processing, make a plot of the diffraction efficiency versus exposure. Using the grating that gave max diffraction efficiency (DE), make a plot of DE versus angle covering an angle range of 15° across the peak angle with 1.0° intervals.

Make another hologram with one beam normal to the hologram plane and one beam at 30° to the normal with the optimum exposure energy density. Use this hologram to determine if there are any changes to the photopolymer after processing.

E.1.4 Questions

1. What grating period does the 30° inter-beam angle correspond to inside the photopolymer? Assume a refractive index of 1.52 for the film.
2. What will happen if the polarization is not perpendicular to the table? If it is perpendicular to the table does this correspond to the transverse electric (TE) or transverse magnetic (TM) polarization relative to the film plane?
3. Is there any shrinkage or swelling of the photopolymer? How do you know if the thickness changes? Determine the amount of thickness change.
4. Do you observe any other diffraction orders from the grating? What orders are they? Does the efficiency add to 1.0? Determine the absorption/photopolymer thickness by measuring the total power versus incident power at an angle where the diffraction efficiency is low. Why should it be done at this condition?

E.2 Lab 2: Transmission Type Holographic Lenses

This lab demonstrates the basic properties of a transmission type holographic lens. The focal point of the lens is on the optical axis and reduces the magnitude of off-axis aberrations (see Chapter 4). It also shows the variation in diffraction efficiency that occurs across the aperture of the lens when compensation methods are not applied.

E.2.1 Experimental Procedures

1. Set up an optical system to record a transmission type holographic lens as illustrated in Figure E.2. Use an available refractive plano convex or convex-convex lens with a focal length of 15–20 cm. Record the position of the point focus relative to the hologram plane and the diameter of the beam at the hologram plane. Set the collimated reference beam at an angle roughly 20° to the normal of the film plane. Adjust the beams so that they overlap on the film plane. Keep the emulsion side toward the beams, and the path length difference within 3 cm

2. Measure the intensities of the recording beams at the hologram plane. Adjust the beams for a 1:1 intensity ratio at the hologram center

3. Make three exposures approximately equal to the optimum exposure value found in Lab 1. Vary the three exposures by ±10% about the optimum exposure value and at the optimum exposure value found in Lab 1. Use the highest efficiency holographic optical element (HOE) to answer the following questions. Be sure to index match the glass backside of the plate during your exposures with a piece of glass that coated with absorbing black paint.

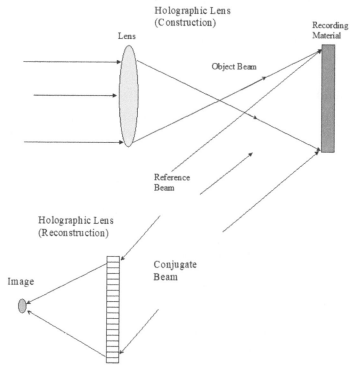

FIGURE E.2   Holographic lens setup for Experiment 2.

E.2.2 Questions

1. What is the variation in the beam intensity ratio across the surface of the hologram? Use a small diameter detector (1–2 mm) or a small aperture over a detector to sample the beam

2. What is the range in inter-beam angles across the aperture in the plane of the table? Calculate five inter-beam angle values across the aperture (one at center, two at the edge, and two at half the maximum aperture value). Use these positions to determine the grating vectors at the corresponding locations

3. Using a small diameter collimated beam, measure the peak DE at the five locations used in question 2 when reconstructed at the recording angle. Plot your results and indicate the period corresponding to each efficiency value. (Remember that you reverse the hologram to reconstruct a real image see Figure below.)

4. Set the system so that the reconstruction angle is the same as the reference beam angle used during construction and fill the aperture of the hologram as much as possible. Observe the beam pattern formed at the focal point on a piece of white paper or an index card. Expand the image with a lens if necessary. Also measure the DE at this reconstruction angle. Change the angle slightly from the construction angle on either side of the construction angle. What reconstruction angle gives the best point image? Measure the DE at this angle. How does it compare to the DE when reconstructed at the recording angle? Identify the dominant aberration as the angle deviates from the construction angle

5. Compare the DE when reconstructed at the recording angle for the five evaluation points across the aperture found in part 3 to the total DE integrated over the aperture as found in part 4. How well do they compare? Which locations are most efficient and why?

E.3 Lab 3: Reflection Type Display Holograms

The purpose of this lab is to record a single beam reflection type display hologram. You can use your own object, but it should reflect red light and be about 2" in maximum dimension for best results. This hologram can be taken home and viewed in white light.

E.3.1 Experimental Procedures

Set up the recording system as shown in Figure E.3. Holograms of this type are also referred to as Denisyuk holograms after Yuri Denisyuk who popularized their use in the early 1960s. Keep the object close to the film plane. Place the pinhole for the spatial filter at least 25 cm from the film plane—this helps when reconstructing the hologram and viewing the object. Use a low power objective in the spatial filter to maximize the power on the object and maximize the distance of the pinhole from the film plane.

Use a half wave plate or rotate the laser in its mount so that the polarization is in the plane of the table. Adjust the tilt angle of the film plane so that it is at the Brewster angle ($\tan\theta_B = n_2/n_1 \sim 57°$) for the center ray through the film. This will help reduce secondary reflections during recording since it's not so easy to index match a reflection hologram. The higher angle of incidence will also help to view the hologram with a white light point source located behind an observer.

The single beam provides both object and reference beams. Take a small piece of unexposed film and measure the absorption. Illuminate your object and get a rough value for the object beam intensity that will arrive at the film plane. Combine this value with the measured film absorption to improve your estimate. The best exposure level for reflection holograms will be 1.5 to 2 times greater than the value used for transmission holograms. The spatial filter will approximately be located at the point for a reconstruction source for the best reconstruction fidelity.

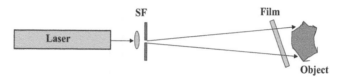

FIGURE E.3   Reflection hologram setup for Experiment 3.

E.3.2 Questions

1. Assume that the film has a refractive index of ~1.54 and compute the exact value for the Brewster angle

2. Measure the maximum efficiency of your holograms with different exposure values. Can you increase the efficiency by rotating the film plane away from the construction angle?

3. Examine your hologram in white light. What color is it? Did the hologram shrink or swell during processing? The resulting color should give an indication of how the film was changed during processing

4. Calculate the fringe spacing at the center of the hologram

5. Now assume that the incident beam is a plane wave along the optical axis and that the object is a mirror placed perpendicular to the optical axis. The hologram should remain tilted at the Brewster angle. Draw a diagram of the fringe planes for the resulting hologram. Assuming the plane wave geometry and no lateral shear, what angle of incidence at the same reconstruction wavelength gives the maximum reconstruction efficiency if the 16 μm thick emulsion shrinks by 2% (use the Bragg circle diagram to determine the shrinkage).

E.4 Lab 4: Hologram Multiplexing

One of the advantages of holographic recording is the ability to multiplex gratings in the same region of the recording material. Depending on the material and multiplexing technique, it is possible to record thousands of holograms in the same region of the holographic material. In this experiment, you will examine some of the issues related to angle multiplexing. For this experiment, we will use a relatively thin recording material (Bayer polymer ~16 μm thick) and is not ideal for this application. However, it can still be used to illustrate the basic concepts of hologram multiplexing.

Set up a transmission hologram recording system with a 20° inter-beam angle as shown in Figure E.4. In previous labs, we found that the maximum exposure for the Bayer photopolymer is about 18 mJ/cm². This exposure is optimum for maximum Δn and DE of a single hologram. When multiplexing, we have to divide the single grating Δn into equal values for each multiplexed grating. This is a bit more difficult to do than simply dividing the maximum exposure by N (number of gratings) due to the reciprocity failure effect of a latent image material. If all the holograms were made with the same exposure, the first holograms recorded will be more efficient than the later ones. This results from subsequent exposures enhancing the first holograms made in the sequence of multiplexed gratings. To offset this affect, the first exposure energy density values should be less than the later ones.

E.4.1 Experimental Procedures

- Make the first exposure with the beam angles at 20° and 40° to the normal of the film plane
- Make a second exposure with the incident beams at −20° and −40° to the normal of the film plane

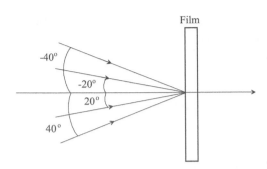

FIGURE E.4   Transmission hologram multiplexing configuration for Experiment E.4.

- Keep the diameters of the beams about 1 cm and make sure that the beams overlap for the two gratings. Also account for the cosine function reduction to the normally incident power when computing the exposure times
- For the first multiplexed grating keep the exposure energy density equal for the two sequential exposures
- Determine the angular bandwidth for a hologram with a 20° inter-beam angle using the Kogelnik model and a film thickness of 16 µm. This is the angle between the first two nulls on either side of the maximum DE
- Were the angles used adequate to separate the beams diffracted from the two gratings without interaction?
- Make a second multiplexed grating in a different area of the film, however, this time make the first exposure less than the second by about 25% to see if this offsets the reciprocity affect
- Make plots of DE versus angle for both multiplexed holograms.

E.4.2 Questions

1. Compute the minimum angular spacing between adjacent holograms formed in a material that is 200 µm thick with 632.8 nm light. Assume that the criterion for minimum spacing is that the peak efficiency of one hologram coincides with the minimum efficiency for an adjacent hologram. How many holograms can be multiplexed in this case?
2. For the material described in Question 1 and assuming a lossless phase grating, what is the index modulation that gives 100% diffraction efficiency for a single hologram?
3. Was the exposure compensation for the second multiplexed grating adequate? How much should the exposures be changed to equalize the efficiency of the individual gratings?
4. How much interference (diffraction efficiency overlap) exists between the multiplexed gratings? This indicates the inter-symbol interference for data stored in the holograms.

E.5 Lab 5: Digital Holography

Digital holography was first demonstrated in the late 1960s and has become an important imaging technique especially with the advent of inexpensive, large format, high resolution digital cameras. In this technique, an interference pattern is formed on the camera aperture and is then digitally reconstructed using a diffraction propagation algorithm. For this method, there is no resulting physical hologram as obtained with analog holography recorded on film. However, the properties of the camera are very important for determining the hologram geometries that can be recorded. In this lab, an in-line and off-axis hologram geometry will be evaluated.

E.5.1 Experimental Procedures

The Figure E.5 below shows the in-line and off-axis geometry that will be used in the experiments. The geometries are similar to those used for analog recording in film type materials. However, in digital holography the optically formed interference pattern is recorded on a digital camera and then numerically reconstructed. This gives a lot of flexibility in manipulating the fields and the reconstruction process. For the basic digital hologram recordings for this lab, a fast fourier transform (FFT) algorithm encoded in Matlab is used for the reconstruction. The algorithm is attached at the end of this handout.

The digital camera used for this experiment has a pixel size of 3.5 μm and an aperture width of 25 mm. The hologram is recorded at a wavelength of 632.8 nm and an Air Force resolution chart (AFRC) is used as the object. This object consists of a series of horizontal and vertical bars with decreasing line width. The minimum width in the 7–6 group has 228 lines/mm or 4.4 μm minimum feature width. It is important to match the object and reference beam powers in order to obtain high contrast interference fringes. This can accomplished by using neutral density filters in the higher power beam (usually the reference beam).

The process in both digital holography (DH) cases is similar and consists of:

*   Setting up the interference pattern
*   Capturing the pattern on the camera aperture

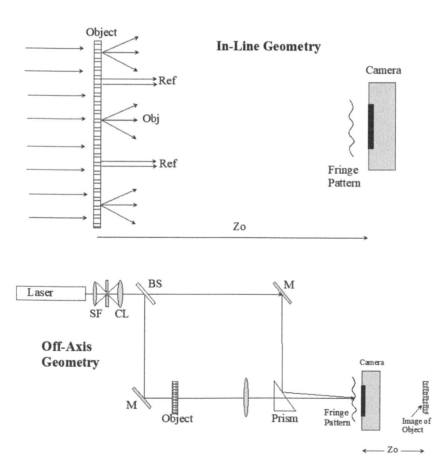

FIGURE E.5 In-line and off-axis digital hologram geometries for Experiment 5.

- Reconstructing the image using a beam propagation method. This is accomplished by entering recording parameters into a FFT Matlab algorithm for Fresnel diffraction
- Evaluating the resulting images.

E.5.2 Questions

1. What is the maximum inter-beam angle that can be recorded on the camera?—assume that the object and reference waves are plane waves
2. What is the minimum feature size that can be expected to be recorded for the in-line DH geometry? What feature size is resolved experimentally? Assume that "resolved" means that the contrast difference between the feature and the background is at least 20%. This can be quantified by measuring the intensity across the resolution bars by taking a line scan across the pattern and using the visibility relation:

$$V = \frac{I_{max} - I_{min}}{I_{max} + I_{min}}.$$

Explain any difference between measured and expected resolution
3. Change the object distance in the algorithm slightly. What reconstruction distance (from the camera aperture) gives the highest reconstruction resolution?
4. What is the minimum feature size that can be expected to be recorded for the off-axis DH geometry? What feature size is resolved experimentally? Explain any difference.

E.5.3 DH Algorithm

```
%---------------- ECE/OPTI 527 ------------------------------
% Function: the reconstruction of a hologram using the Fresnel
% propagation algorithm.
% Written by: Yuechen Wu (UA PhD Graduate Student)
%
%Procedure:
% (1) Read an image file
% (2) Get the parameters asked for on the screen
% (3) Reconstruct the hologram by applying Fresnel propagation with FFT
% (4) Adjust the saturation of the image
%
% Variables:
% Ih: hologram;
% h: wavelength (um);
% L: width of the hologram (mm);
% L0: width of the diffracted object field (mm);
% z0: reconstruction distance (mm);
% U0: complex amplitude in the reconstruction plane;
%------------------------------------------------------------
clear;clc;close all;
disp('Hologram reconstruction')
% Input the hologram file
file_name = input('Please enter the hologram file name: ','s');
I1=imread(file_name);
% Display the hologram
figure(1);
imagesc(I1);colormap(gray);axis equal;axis tight;title('Digital hologram');
```

```
% Change the format, and adjust the photo into square (N*N)
Ih1=double(I1);
% m = input('The number of division on the hologram(>=0):')
% Ih1= interp2(Ih1,m); % "increase" the hologram resolution, increase the
image size.
[N1,N2]=size(Ih1);
N=min(N1,N2);
Ih=Ih1(1:N,1:N);


% Input parameters: pix, h, z0
% For our lab: pixel pitch:0.0035mm, wl:0.6328
pix=input('Pixel pitch (mm): ');
% pix=pix/(2^m); % when "increase" the resolution of the hologram
h=input('Wavelength (um): ');
h=h/1000;
% pix=0.0035; % For Lumenera Lw625M
% h=0.6328; % For HeNe laser


z0=input('Reconstruction distance z0 (+ for a real image, - for a virtual
image)(mm): ');
L=pix*N; % Size of the hologram


%-----------------------Reconstruction by FFT
% create the axes for the hologram
n=-N/2:N/2-1;
x=n*pix;y=x;
[xx, yy]=meshgrid(x, y);
k=2*pi/h; % value of k vector


% The Fresnel transform using FFT
Fresnel=exp(i*k/2/z0*(xx.^2+yy.^2)); % quadratic phase term
f2=Ih.*Fresnel; % Adding quadratic phase term to the hologram
Uf=fft2(f2,N, N); % FFT
Uf=fftshift(Uf);


% calculate and create the new axes at the reconstruction plane
ipix=h*abs(z0)/N/pix;
x=n*ipix;
y=x;
[xx, yy]=meshgrid(x, y);


% Additional phase term for the Fresnel transform
phase=exp(i*k*z0)/(i*h*z0)*exp(i*k/2/z0*(xx.^2+yy.^2));
U0=Uf.*phase;
%-----------------------End of FFT


% Calculate the intensity pattern on the image plane
If=abs(U0).^0.75;
Gmax=max(max(If));
Gmin=min(min(If));


L0=abs(h*z0*N/L); % image size
disp(['Width of the reconstruction plane =',num2str(L0),' mm']);


% Display the reconstruction image
figure(2);
```

```
imagesc(If, [Gmin, Gmax]),colormap(gray);axis equal;axis
tight;ylabel('pixels');
xlabel(['Width of the reconstruction plane =',num2str(L),' mm']);
title('Image reconstructed by S-FFT');

% Asking for higher sateration?
p=input('Display parameter (>1): ');
while isempty(p) == 0
figure(2);
imagesc(If, [Gmin Gmax/p]),colormap(gray);axis equal;axis
tight;ylabel('pixels');
xlabel(['Width of the reconstruction plane =',num2str(L),' mm']);
title('Image reconstructed by S-FFT ');
p=input('Display parameter (>1) (0=end): ');
if p==0,
break
end
end
```

Index

Note: Page numbers in italic and bold refer to figures and tables, respectively.

Printed and bound by CPI Group (UK) Ltd, Croydon, CR0 4YY

24/10/2024

01778296-0001